Valorisation of Food Waste

This book explores innovative strategies for repurposing food waste, with a strong focus on process and product design. It highlights diverse food waste sources, including expiring materials and often overlooked resources, offering a comprehensive evaluation of technologies, methods, and sustainable solutions to address the global challenge of food waste.

Key Features:

- Offers guidelines and a food waste valorisation framework.
- Presents unique pathways for valorising food waste and expired food.
- Delves into principles and strategies for food waste valorisation, transforming food waste into building and construction materials, upcycling food waste, and extracting bioactive compounds from expired sources.
- Explores converting food waste into biochar, biofuel, and maximising biogas production, as well as utilising bioconversion technology with Black Soldier Fly larvae and integrating the water–energy–food nexus.
- Each chapter includes exercises and case studies to enhance understanding and practical application.

This text is aimed at academics, engineers, technologists, and researchers in the food, pharmaceutical, and chemical industries.

Chien Hwa Chong is Associate Professor in the Department of Chemical and Environmental Engineering at the University of Nottingham Malaysia. He is a Professional Engineer registered with the Board of Engineers Malaysia, a Chartered Member of IChemE, and a Chartered Engineer registered with the Engineering Council (UK). Additionally, he holds the titles of International Professional Engineer, IntPE (Malaysia), APEC Engineer, Fellow of the Institution of Engineers Malaysia (FIEM), and Fellow of the Higher Education Academy (FHEA). He is expert in the field of drying of biomaterials, 3D printing technology, wastewater treatment, and engineering education. Dr. Chong has made significant contributions to the field, including the publication of his book, *Drying and Valorisation of Food Processing Waste* (CRC Press, 2024).

Green Chemistry and Chemical Engineering

Series Editor
Dominic C.Y. Foo
University of Nottingham, Malaysia

The Water-Food-Energy Nexus: Processes, Technologies, and Challenges
I. M. Mujtaba, R. Srinivasan, and N. O. Elbashir

Hemicelluloses and Lignin in Biorefineries
Jean-Luc Wertz, Magali Deleu, Séverine Coppée, and Aurore Richel

Materials in Biology and Medicine
Sunggyu Lee and David Henthorn

Resource Recovery to Approach Zero Municipal Waste
Mohammad J. Taherzadeh and Tobias Richards

Hydrogen Safety
Fotis Rigas and Paul Amyotte

Nuclear Hydrogen Production Handbook
Xing L. Yan and Ryutaro Hino

Water Management: Social and Technological Perspectives
Iqbal Mohammed Mujtaba, Thokozani Majozi and Mutiu Kolade Amosa

Efficiency and Sustainability in the Energy and Chemical Industries: Scientific Principles and Case Studies, Third Edition
Krishnan Sankaranarayanan

Process Integration for Resource Conservation, Second Edition
Dominic C.Y. Foo

Industrial Decarbonization: Materials, Methods, and Developments
Qingxu "Bill" Jin, Ida Chen, Fan Shi and Dushyant Shekhawat

Valorisation of Food Waste: Process and Product Design
Edited by Chien Hwa Chong

For more information about this series, please visit: https://www.routledge.com/Green-Chemistry-and-Chemical-Engineering/book-series/CRCGRECHECHE

Valorisation of Food Waste

Process and Product Design

Edited by Chien Hwa Chong

CRC Press
Taylor & Francis Group
Boca Raton London New York

CRC Press is an imprint of the
Taylor & Francis Group, an **informa** business

Designed cover image: © 2026 UPCYCLES Technologies Co. Ltd.

First edition published 2026
by CRC Press
2385 NW Executive Center Drive, Suite 320, Boca Raton FL 33431

and by CRC Press
4 Park Square, Milton Park, Abingdon, Oxon, OX14 4RN

CRC Press is an imprint of Taylor & Francis Group, LLC

ISBN: 978-1-032-71651-0 (hbk)
ISBN: 978-1-032-71847-7 (pbk)
ISBN: 978-1-032-71850-7 (ebk)

DOI: 10.1201/9781032718507

Typeset in Times
by codeMantra

Contents

Series preface

Towards the late 20th century, the development of various environmentally friendly processes, techniques, and methodologies saw significant growth in the scientific community. The main driving force for such growth was due to the rising awareness of sustainable development, more stringent environmental regulation, and increasing costs of raw materials and waste treatment. After several decades of development, we now broadly term these environmentally friendly processes, techniques, and methodologies as *green/clean technologies.*

In the 21st century, the global community has experienced many extreme weather events such as prolonged draught, extreme heat, tornadoes, and wildfires. The scientific community believes that these extreme weather events are closely linked to climate change, and they are expected to increase in frequency and intensity in the future. Following the Paris Agreement and Glasgow Climate Pact, there is now an international commitment to limit the rise of global average temperature to well below 2°C, and pursuing efforts to limit the temperature increase to 1.5°C above pre-industrial levels. Hence, it is believed that green/clean technologies have a much bolder role to play in combating the global climate change in the coming years.

It is also worth mentioning that the United Nation Sustainable Development Goals (UNSDG) that were launched in 2015 define 17 important goals to transform the world by 2030. It is believed that some of these goals may be addressed with the development of green/clean technologies. These include:

- Goal 6—Clean Water and Sanitation: Ensure access to water and sanitation for all
- Goal 7—Affordable and Clean Energy: Ensure access to affordable, reliable, sustainable, and modern energy
- Goal 12—Responsible Consumption and Production: Ensure sustainable consumption and production patterns
- Goal 13—Climate Action: Take urgent action to combat climate change and its impacts

The Green Chemistry and Chemical Engineering book series by CRC Press/Taylor & Francis focuses on the subset of green technologies dedicated to address the "2E" agenda, i.e., *environment* and *energy.* It involves the development of materials (e.g., catalysts, nanomaterials), methodologies (e.g., process optimisation, footprint reduction, artificial intelligence), and processes (e.g., waste treatment) that will bring forth solutions to address pressing problems such as:

- Greenhouse gas management and reduction
- Sustainable water production
- Wastewater treatment and recycling
- Circular economy and waste reduction

- Renewable energy
- Sustainable use of energy resources

I am hopeful that this Green Chemistry and Chemical Engineering series will serve as a de facto source of reference materials and practical guides for academics and industrial practitioners looking to advance the discipline and aims of green chemistry and chemical engineering.

Dominic C. Y. Foo
Series editor, Green Chemistry and Chemical Engineering

Preface

This book explores innovative strategies for repurposing food waste, with a strong focus on process and product design. It highlights the diverse sources of food waste, including expiring materials and often overlooked resources, offering a comprehensive evaluation of various technologies, methods, and sustainable solutions to address the global challenge of food waste.

This book covers a wide range of topics, presenting unique pathways for valorising food waste and expired food. It delves into principles and strategies for food waste valorisation, transforming food waste into building and construction materials, upcycling food waste, and extracting bioactive compounds from expired sources. Additionally, it explores converting food waste into biochar, biofuel, and maximising biogas production, as well as utilising bioconversion technology with Black Soldier Fly larvae (BSFL) and integrating the water–energy–food nexus.

Each chapter includes exercises and case studies to enhance understanding and practical application.

Edited by Dr. Chien Hwa Chong and authored by experts from Malaysia, the Philippines, China, India, Mexico, the United Kingdom, Poland, Indonesia, Saudi Arabia and Iran. This book serves as an essential guide for researchers, engineers, practitioners, entrepreneurs, and policymakers, offering innovative solutions to tackle food waste through process and product design, ultimately generating valuable products and materials.

Acknowledgements

The insights and practical applications presented in this book are significantly enriched by the generous contribution of UPCYCLE Technologies, Japan, Renikola Sdn. Bhd, Malaysia, Henan Mingjie Environmental Equipment Co., Ltd, China, Edafos Agro PLT, Malaysia, AgSTAR (collaborative program sponsored by EPA and USDA), Zorg Biogas GmbH, Germany, STRABAG Umwelttechnik GmbH, Germany, Agrivert, Hong Kong and North London. Their provision of detailed case studies, photos, and information provided real-world examples that illuminate the core concepts of this book.

I also wish to express my deepest gratitude to the individuals who generously lent their time and expertise to review this book. Their insights and feedback were invaluable in shaping the final form of this book. Their thoughtful and constructive suggestions have been used to improve the narrative, clarity, and its impact.

Special thanks to the reviewers: Professor Dorota Konopacka, Professor C.A. Kalpana, Associate Professor Dr. Rafeah binti Wahi, Associate Professor Dr. Chua @ Yeo Gek Kee, Dr. Ng Yong Sing, Dr. Cheok Choon Yoong, Dr. Safa Senan Mahmod, Dr. Kuan Shiong Khoo, Dr. Ivy Chai Ching Hsia, Dr. Lim Jun Wei, Dr. Abby K Jones, and Dr. Nor Faizah Binti Razali.

Contributors

Yuana Elly Agustin
Xiamen University Malaysia
Malaysia

Aan Mohammad Nusrat Aman
University of Nottingham Malaysia
Malaysia

Gelyn L. Bongabong
University of the Philippines Diliman
Philippines

Yi Jing Chan
University of Nottingham Malaysia
Malaysia

Chien Hwa Chong
University of Nottingham Malaysia
Malaysia

Yin Hui Chow
Taylor's University
Malaysia

Isaac Jerome C. Dela Cruz
University of the Philippines Diliman
Philippines

Nesa Dibagar
Wroclaw University of Environmental and Life Sciences
Poland

Adam Figiel
Wroclaw University of Environmental and Life Sciences
Poland

Soorya Haridas
Amrita University
India

Haripriya Ravikumar
Amrita University
India

Wan Thing Hong
Heriot-Watt University Malaysia
Malaysia

Yew Woh Hui
Xiamen University Malaysia
Malaysia

Olufemi Adebayo Johnson
East Dunbartonshire Council
United Kingdom

Jocelyn Jean Yi Lim
Xiamen University Malaysia
Malaysia

Brenda Cansino-Loeza
Universidad Michoacana de San Nicolás de Hidalgo
Mexico

Malvin Ma
Xiamen University Malaysia
Malaysia

Athira Mundassery
Amrita University
India

Moslem Namjoo
University of Jiroft
Iran

Chui Kim Ng
Tunku Abdul Rahman University of Management and Technology
Malaysia

José María Ponce-Ortega
Universidad Michoacana de San Nicolás de Hidalgo
Mexico

Don Nelson C. Potato
University of the Philippines Diliman
Philippines

Norazira Abdu Rahman
Xiamen University Malaysia
Malaysia

Janci Rani Ramaswamy
Amrita University
India

César Ramírez-Márquez
Universidad Michoacana de San Nicolás de Hidalgo
Mexico

Alma Yunuen Raya-Tapia
Universidad Michoacana de San Nicolás de Hidalgo
Mexico

Xate Geraldine Sánchez-Zarco
Universidad Michoacana de San Nicolás de Hidalgo
Mexico

Anurita Selvarajoo
University of Nottingham Malaysia
Malaysia

Ryan Zhe Hse Soh
University of Nottingham Malaysia
Malaysia

Jian Ping Tan
Xiamen University Malaysia & Xiamen University China
Malaysia and China

Swee Tiam Tan
Xiamen University Malaysia
Malaysia

Lee Hong Tee
Xiamen University Malaysia
Malaysia

Man-Kei Wong
Xiamen University Malaysia
Malaysia

Collen Yang
University of Nottingham Malaysia
Malaysia

Said Ismail Amer Aref Yaser
Xiamen University Malaysia & King Fahd University of Petroleum and Minerals
Malaysia & Saudi Arabia

Teck Ann Yeow
Xiamen University Malaysia
Malaysia

1 Principles and guidance for food waste valorisation

Challenges, limitations, risk, and barriers to adoption

Chien Hwa Chong

1.1 INTRODUCTION

According to the Cambridge dictionary, valorisation means the act of thinking or stating that something has value or is valuable, making something valuable or useful from an existing substance and deciding or increasing the price or value of goods, services, etc. (Valorisation, 2025). It is usually used for energy recovery, biofertilisers and biomaterials. Valorising the food waste (FW) into biofertilisers is easiest to implement by shopping outlets. The Mid Valley Megamall, Kuala Lumpur, Malaysia, turned FW into a compost product (Figure 1.1). A brief process of FW processing was printed on the back of the paper bag. Southeast Asia's largest on-site FW management system can be found in Malaysia at 1 Utama Shopping Centre (W.I.S.E with Food, 2025). The current under the W.I.S.E programme recycled FW daily as high as 4,000 kg from 160 F&B tenants to produce 100% organic fertiliser. Figure 1.2 shows a typical FW valorisation process for organic FW which all shopping outlets can implement. All organic FW will be separated from inorganic FW manually by consumers into respective bins or collectors. The organic FW will be collected and lifted to the machine to grind. After grinding, the organic FW will be placed in inoculated microbes to break down all the fats, oils and grease. Then, the organic FW will be heated at 80°C to kill the pathogens or bacteria. In the end, processed organic FW will be dewatered or dehydrated using different technologies to obtain the final products which can be combined with other materials to valorise it. Organic FW needs to be processed as soon as possible due to the growth of pathogens or bacteria if it is not kept in the freezer as putting it in a freezer will increase the cost of the entire process.

Governments and mall operators must set regulations and rules to valorise FW. Businessmen or shopping malls with many restaurant outlets will not take the initiative to separate FW and convert FW into valuable products without stating the roles in the tenancy agreement. The government agency shall collaborate with shopping

DOI: 10.1201/9781032718507-1

FIGURE 1.1 Compost from kitchen waste of shopping mall F&B outlets for MVClub of Mid Valley Shopping mall (IGB Berhad, Malaysia).

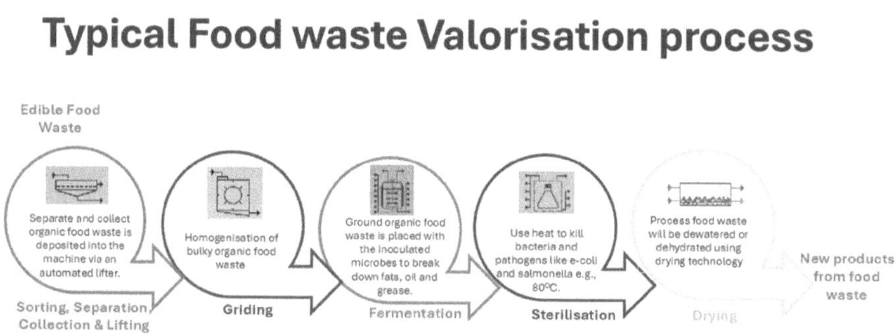

FIGURE 1.2 Typical FW valorisation process.

malls' top management to address the FW issue by repurposing it into valuable products, aligning with the region's sustainability goals. Meeting stringent regulations and compliance standards for FW recycling and product development can be challenging without rules and regulations. Companies like Veolia Environment, SUEZ, Renewi, Biogen, ReFood, Andisgestion, TerraCycle, ReFood, FoodCycle, Enval, Tönsmeier, HZI, Leafield Environmental and Organic Waste Systems and UPCYCLES Technologies Co. Ltd. are leaders in the valorisation of FW. The typical types of FW recycled or repurposed are coffee waste, tea waste, mangoes, apples, grapes, citrus fruits, carrots, beetroot and berries wastes. Valorised products are cosmetics, soil amendment, bio brick, biomedical products, electronics, aviation, automotive, bio composites, biogas, energy and household products. Figure 1.3 shows the products made from coffee waste by UPCYCLES Technologies Co. Ltd. In the other chapters, different valorisation processes involve chipping, drying/soaking, grinding, thermal conditioning, catalyst, solvents, enzymatic treatment, composting, hydrolysis, fermentation, liquefaction, incineration, anaerobic digestion (AD), gasification and hydrothermal carbonisation will be discussed in detail.

1.2 BARRIERS AND OPPORTUNITIES

According to the United Nations Environment Programme (2024), more than USD 1 trillion worth of food was thrown away and the top ten countries are China, India, Pakistan, Nigeria, the United States, Brazil, Egypt, Indonesia, Bangladesh and Mexico from 13.4 million tons to 108.7 million tons annual FW 2024 (United Nations Environment Programme, 2024). For instance, using AD to produce renewable energy is one of the promising methods to valorise FW, but the main challenges are the cost of developing AD facilities, the cost of storage of perishable FW, acquiring approval from local authorities, process instability, liquid effluent, FW separation, etc.

Multiple policies, laws and acts were set by different countries to reduce or valorise FW. In Malaysia, the National Solid Waste Management Policy and the National Food Waste Reduction Roadmap are initiatives to reduce the FW. However, the FEED ACT (Act 698) Law of Malaysia 2009 limited the FW that can be used as animal feed. Not all FW can be used for animal feed for livestock except for harvesting carcass bones and eggshells. Additional support from the country is the 10th Prime Minister of Malaysia; Datuk Seri Anwar Ibrahim announced RM 2.5k relief for purchasing domestic FW composting machines from the Year 2025 to 2027. In Malaysia, the famous equipment available for FW processing is waste sorting, processing and AD. Mentari Alam Eko (M) Sdn Bhd or MAEKO is one of the companies that leases and sells composting equipment for hard FW with big bones, typical

FIGURE 1.3 Chairs, cutleries, comb, cups and plates made from 51% coffee waste (Provided by UPCYCLES Technologies Co. Ltd, Japan).

kitchen waste and turns all organic waste into liquids. Some multinational companies like Panasonic Malaysia and Green Steps organised separate waste contests to initiate the FW valorisation process.

In Asia, various policies can be found in Japan, Taiwan, Thailand, South Korea, Indonesia, The Philippines and Vietnam. The National Development Plan-Green Industry-Resource Recycling and Reuse Programme (2003–2008) Thailand, the National 3R strategy (2011), BCG Economy model, Plastic Waste Management and the National Waste Management Master Plan 2016–2021. In South Korea, the Daejeon Bioenergy Center can convert 15,000 tons of daily FW into fertiliser, fed to livestock or turned into biogas. There is a weight-based waste fee (WBWF) system that comes with a scale and RFID and implemented volume-based FW fee system at the household level. Garbage cans that can automatically weigh food get disposed of as trash. In Japan, the Food Waste Recycling Law (2001) states that FW should be recycled into composts, animal feedstock and biogas and the Food Loss Reduction Promotion Act (2019). Japan has advanced technology to convert FW into various products such as biofuels to fertilisers, contributing to resource efficiency. Indonesia has a National Strategy on Circular Economy and a National Food Loss and Waste Reduction. The Philippines has an Ecological Solid Waste Management Act of 2000 and various digital platforms and technologies to facilitate food donation programs. Vietnam has a National Action Plan on Sustainable Food Systems 2021–2030 and a national strategy for food loss and waste reduction. The latest technology found in Vietnam is bioconversion technologies that convert FW into biogas or other bioproducts.

Among the countries in Southeast Asia, Singapore has a recommended FW management system set by the National Environment Agency (NEA), Singapore. The FW recycling rate ranged from 17 to 19% in Singapore from 2018 to 2022. They have a framework and identify companies to do the collection and processing of FW (3R Programmes and Resources, 2025). The National Environment Agency (NEA), Singapore, has initiated the Resource Sustainability Act (RSA), 2019, to address the FW for commercial and industrial buildings. Different requirements for occupants and owners of new and existing buildings were set. Fraser's property in Singapore valorised more than 27 tons of FW collected during the pilot run in 2023. The companies involved are Dian Xiao Er, Kuriya Japanese Market, PastaMania, Shabu Sai, Tokyo Shoduko and Ichiban Boshi https://www.frasersproperty.com/sg/press-releases/2024/01-jan/fps-food-waste-valorisation

For Europe, FW-generated at any stage of the food supply chain is to be collected and treated separately. European Financial has subsidy policies for renewable energy products. For example, the Environmental Regulations on Development of Anaerobic Digestor Installation Emission Reduction Fund (ERF) and Renewable Energy Target (RET). Germany has National Strategy for Food Waste Reduction, "Too Good for the Bin!" and Circular Economy Act; France has Anti-Food Waste Law (2016), National Pact against Food Waste and Anti-Waste and Circular Economy Law (AGEC Law); Belgium has Flanders Food Supply Chain Roadmap for Food Loss and National Pact against Food Waste; Denmark has a "Stop Wasting Food" movement food preservation, upcycling technologies and waste-to-energy solutions; Ireland has Waste Action Plan for a Circular Economy and National Food Waste Prevention Roadmap.

In Ireland, the technology is focused on FW in bioenergy and fertiliser. Among the top shopping mall owners, Simon Property Group has initiated a 100% recycling initiative since 2012. General Growth Properties unveiled its first food composting program in 2013.

In Mexico, they have FW to Carbon Credit unit. The Mexican Food Banking Network and Core Zero developed a carbon credit to reduce carbon dioxide and methane emission from FW.

Integrate ------ Quantity -------Monetise -Voluntary Carbon Market (VCM)

- Small-scale Technology Certificates (STC) <100 kW
- Large-scale Generation Certificates (LGC) >100 kW

1.3 RESPONSIBILITY OF DIFFERENT STAKEHOLDERS

All the stakeholders that generate FW has a responsibility. Table 1.1 shows the proposed responsibility of stakeholders in contributing to the valorisation of FW. The stakeholders are food service provider's tenants, frozen and fresh food and drink processing manufacturers, shopping centre owners, restaurant owners and local authorities. The food service tenant, frozen or fresh food and drinking processing manufacturers shall identify and record edible, non-edible FW and expired food to the management or building managers. The FW-generated or expired food should be sent to the general facilities for further processing. In addition, they should own a storage facility for edible and non-edible FW. As for the owner of a shopping mall or restaurant, they should provide designated facilities for tenants or staff to store and process FW or storage for external parties to collect the separate FW. Local authorities should categorise the FW and offer companies to provide valorisation services and publish on the official platform to inform the stakeholders. Not all FW can be processed on-site.

A FW valorisation framework was proposed for buildings in this chapter. The proposed template for FW valorisation and framework for shopping malls, hotels and F&B-related factories is shown in Table 1.2. The local authorities should set rules for valorisation of FW based on Table 1.2 threshold. Any unit smaller than 20 m^2 might be considered a centralised system in particular area. For example,

a. Food with cracked shells can be turned into ready-to-use raw materials by removing the eggs that are cracked and selling them to bakers or restaurants. So, the cracked eggs can be used and processed into food. It helps the F&B industry to save man hours to remove the eggshells.
b. A small-scale AD unit is used for cooking purposes.
c. A large-scale AD unit used to produce power or compressed biomethane. The cost of a single digester is USD 953 per m^3 (Chen et al., 2023).

It is proposed to set it into at least four different categories: fruit processing industry, dairy industry, meat and poultry industry and grain industry. Other categories are the coffee industry, tea industry and seafood industry.

TABLE 1.1
Proposed responsibilities and requirements for different stakeholders in the valorisation of FW or expired FW

Stakeholders	Responsibilities and requirements
Foodservice tenant, frozen or fresh food and drink processing manufacturer	Identify edible and non-edible FW or expired FW. Assess the shelf life of edible and non-edible FW during storage. Submit to the owner/building managers. Segregate FW/expired FW and send it to general facilities provided by the owner or building manager. Provide storage for edible and non-edible FW by the owner or building manager.
Shopping centre owners, restaurant owners	Provide designated facilities within the building to allow tenants to dispose of segregated FW or expired FW by occupier or tenant. Provide on-site facilities to convert organic FW into fertiliser or building materials like pellets/fillers into products. Provide FW on-site or engage a licensed waste collector to send the FW for treatment at an off-site facility. Schedule collecting and transporting segregated FW. Examples of off-site treatment facilities are converting it into biogas and performing extraction to recover valuable compounds and natural dyes. Provide an on-site FW treatment system that removes moisture from FW through dewatering or drying. Detailed and recommended drying of food processing waste can be referred to Chong et al. 2024. Optional to provide sanitation to the FW.
Local authorities	Provide an approved FW processing company name, person in charge, address, processing capacity status and types of FW or expired food. Types of accepted FW or expired FW is spent grains, soya bean waste, used coffee grounds, used tea leaves, rejected milk powder, bread waste, food manufacturer waste, fluor waste, dried FW, canned FW, rice, drink, premix food flavouring, fruits, vegetables, expired/spoilt food, off-specs food, pre-mixes, food by-products, excess food and expired can fruits/food. Provide approved on-site and off-site companies on official authority websites.

1.4 GUIDING PRINCIPLES FOR FOOD WASTE CONVERSION AND VALORISATION:

1.4.1 Technological innovation

a) Provide a list of combinations or hybrid technologies needed to improve the sustainability of FW.
b) Promote AD to produce biogas from low-quality or non-edible FW.

TABLE 1.2
FW valorisation and framework for buildings

Types of buildings	Location	Threshold (m²)
Neighbourhood centre	Supermarket with food court and restaurants	3,000 to 12,000
Community centre	Department stores or supermarkets with food courts and restaurants	4,000 to 12,000 or up to 40,000
Regional centre	Full-line department stores with food courts and restaurants	8,000 to 40,000
Hotels (National Environment Agency. Food Waste Management, 2025).	Food and beverage area and function area	More than 3,000
Single-user factor (National Environment Agency. Food Waste Management, 2025).	Large food manufacturers	Operation area >750
Multi-user factory (National Environment Agency. Food Waste Management, 2025).	Gross floor area (GFA) and food tenants	GFA > 20,000 and with > 20 food tenants
	At least one large food manufacturer	Operation area >750
Food additive, dried food or spices	Food and beverage area and function area	More than 20
Bottled drinking water	Beverage area and function area	More than 20
High-pressure processing of food	Food and beverage area and function area	More than 20
A la carte, self-service and fast service restaurants	Food and beverage area and function area	More than 20
Restaurants	Food and beverage area and function area	More than 20

c) Provide a list of emerging biorefinery technology for FW conversion processes and equipment to produce biofuels, green energy, green chemicals, chemicals, materials, animal feed, extracts of protein, carbohydrates and other nutrients.
d) Educate farmers on the new technology to valorise FW.
e) Renting equipment/facilities to farmers at agricultural processing sites.
f) Artificial intelligence (AI) and machine learning (ML) to monitor, manage and centralise FW collection and processes. The ML techniques are the least squares-support vector machine, response surface methodology, time series and Pearson's correlation analysis, artificial neural network (ANN), genetic algorithm and ANN, recurrent neural network ANN, Random Forest, long short-term memory networks and support vector machine (SVM) (Said et al., 2023).
g) Use ML to select high-value FW via quality identification, odour-based categorisation and supply chain management.

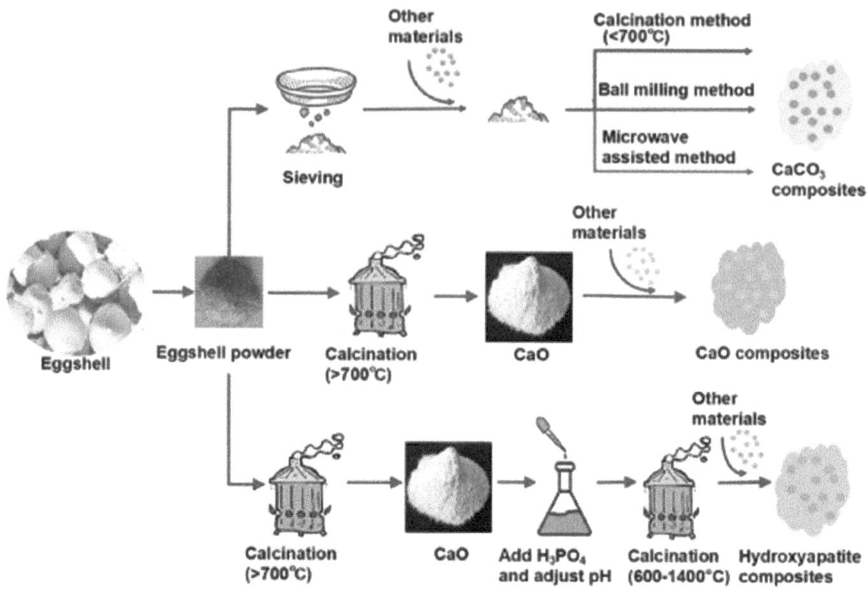

FIGURE 1.4 Eggshells valorised into hyaluronic acid, lyophilised collagen and peptide (Adapted from Shang et al., 2022).

1.4.2 Examples of innovations

(a) Natural dye from FW

SAGES turn FW into natural dyes which is 100% natural made from FW. The colour pigments in a vibrant range of 12 colours. The products have lower CO_2 emissions, net neutrality, high in colour strength and colour fastness (https://www.sageslondon.com/)

(b) Eggshells

Egg shells are extracted into peptides with antioxidants and antimicrobial properties, dietary supplement, fertiliser, cosmetic product, biomedical medication product and collagen drink. Figure 1.4 shows the FW converted to green technology-based extraction methods and used for industrial applications.

(c) Chicken feathers

Chicken feathers collected from the poultry industry are used to produce protein via degradation of bacterium. Total soluble protein and bioactive peptides produced are 260 mg/g and 86.16%, respectively (Figure 1.5).

1.4.3 Food and beverage recovery and redistribution

Provide an established system to collect and distribute surplus food, FW, expired food and beverages to the factory for recovery and further processing for producers, manufacturers, distributors and retailers.

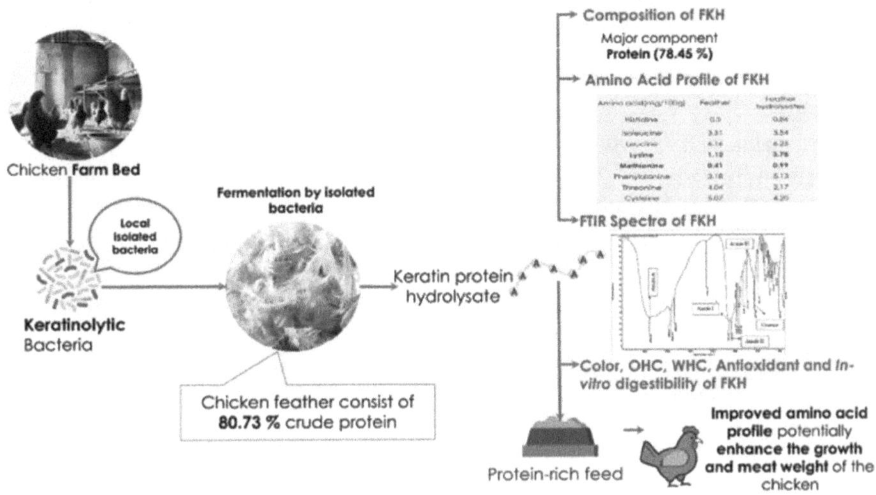

FIGURE 1.5 Chicken feathers valorised into soluble protein and bioactive peptides (Adapted from Prajapati et al., 2021).

1.4.4 Mapping business and supply chain

a) Collaboration between companies, government, non-profit organisations and research institution.
b) Sharing of resources and best practices in FW conversion and valorisation between producers, manufacturers, distributors and retailers.
c) AI application in optimising FW conversion and valorisation supply chain.

1.4.5 Legislative and policy

a) Provide incentives to companies to valorise FW.
b) FW conversion and valorisation must be standardised and regulated to avoid economic, environmental and social risks on investment and rebound effects.
c) Generic and specific food legislation regulates the suitability and safety of products valorised from FW.
d) Guarantee consumer's safety.
e) Authorisation from state entity.
f) National target for biogas production.
g) Power exports renewable energy policy.

1.5 CHALLENGES AND PROSPECTIVE: LEGISLATIVE, LOGISTICAL AND ECONOMIC

The FW comes from semi-finished food, extra hot and cold served dishes, expired or semi-finished food and beverages, wilted fruits and vegetables, expired products/exceed the best-before date, food is not fresh, food and beverages with signs of spoilage, not tasty, take up too much space for storage, forgot to consume, allergic to food, nobody want to eat and bought too much. Therefore, FW consists of many categories and different levels of spoilage.

The main challenges are FW collecting cost, selection of tools, availability of technologies, a skilled hand, FW characterisation, chemical contaminants, potential pathogens and process compatibility. At this moment, there are no specific work safety evaluation studies, physicochemical and microbiological assessment policies and properties to be determined and evaluated from different FW. For example, moisture content (dry basis), water activity, pH values, ash content, colour and nutrients. The nutrients are fiber, carbohydrates, lipids and proteins content. Additionally, the mechanical properties like texture profile, water-holding capacity (WHC) and oil-holding capacity (OHC) should be considered (Socas-Rodríguez et al., 2021).

Environmental factors where the location of FW stored and duration to transport at different temperature, oxygen level and humidity can create a conducive environment to spoil microbial growth. The development of legislative procedures and physicochemical and microbiological assessment policies supports the adoption of sustainable technology to convert and valorise good waste into useful goods and energy. In addition, modern computing tools like Internet of Things (IoT), sensors, ML and AI enhanced the process and product of FW conversion and valorisation.

1.6 RISK AND BARRIER ADOPTION

Financial risk, market uncertainty, technological uncertainty, public perception, source consistency, consumer scepticism, market education, regulatory hurdles, institutional resistance and emerging risks that are unknown are the risks of the valorisation of FW. To overcome these issues, a good financial viability study and return of investment plan must be prepared. In this book, the latest and feasible technologies will be discussed in detail. The concerns about the safety and quality of valorised products can be addressed by getting certified laboratories to perform the analysis. In the end, a joint venture or collaboration between government, non-government organisations and businesses can address institutional resistance. Risk treatment plans to mitigate, eliminate, transfer or accept the identified risks are required. The control measures include foodborne illnesses on animal or aquaculture feeds, cleanliness and hygiene, operational and financial risk. Comprehensive monitoring and review mechanism are required for risk treatment. The risk that needs to overcome are vetting of third party, compliance risks, reputational risks and operational risks for valorisation of FW.

1.6.1 Certifications

Valorisation of FW companies can set their standards based on ISO 31000 for risk management. The principles and guidelines can be used as they identify and evaluate potential risks, including likelihood and potential impact on hygiene, food safety, hazards, supply chain disruptions, financial stability and reputation. The techniques used are hazard analysis, HACCP (Hazard Analysis and Critical Control Points) and environmental scanning.

1.7 INDUSTRY CASE STUDIES

Case study: UPCYCLE Technologies Co., Ltd., Japan

AirX Coffee is one of the brands that produce cutlery from coffee waste. In the product design, coffee waste can be replaced with up to 51% plastic materials. Figures 1.6 shows coffee grounds, waste and other organic matter before putting them in the carboniser.

Currently, the mainstream carbonisation method is to make charcoal in a kiln heated to over 800°C in an oxygen-free environment, but this requires large equipment and is expensive, and since high temperatures and large amounts of smoke are generated, installation locations are limited. As a result, a primitive method is used in which large amounts of wood, which is raw material for charcoal, are transported to the kiln, and the charcoal after carbonisation is transported to the desired location. At this moment, the company developed a prototype to process coffee waste. The other FW will be considered in the future, including general FW and flower waste. Figure 1.7 shows the photos and specifications of the low temperature catalytic carboniser to valorise FW like coffee and tea waste. This carbonisation method uses a catalyst, making it possible to carbonise at low temperatures at low cost. This catalytic carboniser is able to be compact at low temperatures, and the volume of waste is reduced to about 1/10 by carbonising it at this prototyping stage. It provides great

FIGURE 1.6 Coffee waste collected for processing. (Provided by UPCYCLE Technologies Co., Ltd., Japan)

FIGURE 1.7 Low temperature catalytic carboniser (120–140°C). (Provided by UPCYCLE Technologies Co., Ltd., Japan)

FIGURE 1.8 A uniform biochar produced from organic waste. (Provided by UPCYCLE Technologies Co., Ltd., Japan)

economic and environmental benefits from reducing the cost of transporting waste as it can be installed in shops or factories that generate large amounts of FW.

The current test machine has a processing capacity of carbonising 30 kg of waste in 2 h, but the second version, scheduled for completion in March 2025 is planned to be upgraded to a device that processes 100 kg in 2 h. The current unit can be further designed to improve its functions and reduces smoke, water vapour and odours. This low temperature catalytic carbonisation equipment is a very groundbreaking technology. Until now, there has been catalytic carbonisation technology that uses concentrated

FIGURE 1.9 Pellets from semi-carbonised FW. (Provided by UPCYCLE Technologies Co., Ltd., Japan)

sulfuric acid, but it has rarely been put to practical use because charcoal becomes highly acidic and is expensive. The major advantages of this technology are the catalyst itself is very inexpensive, and the charcoal produced remains alkaline, making it usable in a variety of products as is.

A uniform biochar produced from organic waste using the low-temperature catalytic carboniser is shown in Figure 1.8. The semi-carbonised waste can be converted into Polyethylene Terephthalate (PET) and used as fuel, and if the carbonisation degree is increased, it can be used as a bioplastic, and if the unit price is further increased, it can be used for various purposes such as a substitute for graphite.

The company is working on manufacturing bioplastic products, upcycling them into various products, and selling them. This initiative will contribute to reducing petroleum-based plastics and reducing waste, reducing greenhouse gases generated in the process and reducing carbon dioxide and economic costs during transportation. The company is also working on a project to convert semi-carbonised FW into pellets and sell them as charcoal fuel (Figure 1.9).

1.9 CONCLUSIONS

The important aspect of FW or expired FW valorisation for different stakeholders engaged in it was present in this chapter as a guideline and opportunity. The stakeholders, including government agency, mall operators/owners, F&B owners and FW waste specialist rules were proposed and discussed in detail. Successful case studies related to managing FW and valorise FW into household products were presented. From the case study, the increasing percentage of FW used in the products design are the opportunities to valorise FW. In the following chapters, detailed process and design including case studies to extracts and make more values from FW into building materials, fortification of meat-based products with up-cycled FW, extraction of polyphenols, fibres, protein and aromas from FW, converting FW into biochar, biofuel; bioconversion of FW into value-added products via black soldier fly larvae, maximising biogas production from FW and synergies in the water–energy–food nexus that presents transforming challenges into opportunities for FW valorisation.

EXERCISES

1. Discuss FW valorisation and understand the scale of problem and potential solutions.
2. Conceive and design processes to valorise FW and expired food.

REFERENCES

3R Programmes and Resources. (2025). https://www.nea.gov.sg/our-services/waste-management/3r-programmes-and-resources/food-waste-management

Chong, C. H., Wahi, R., Choo, C. M., Chew, S. J., & Dassanayake, M. K. (2024). *Drying and Valorisation of Food Processing Waste.* Taylor and Francis Group. https://www.routledge.com/Drying-and-Valorisation-of-Food-Processing-Waste/Chong-Wahi-Choo-Chew-Dassanayake/p/book/9781032320885#:~:text=Drying%20and%20Valorisation%20of%20Food%20Processing%20Waste%20is%20a%20comprehensive,the%20issue%20of%20food%20waste.

National Environment Agency. Food Waste Management. (2025). Buildings covered under the food waste segregation and reporting framework, Singapore.https://www.nea.gov.sg/our-services/waste-management/3r-programmes-and-resources/food-waste-management.

Prajapati, S., Koirala, S., & Anal, A. K. (2021). Bioutilization of chicken feather waste by newly isolated keratinolytic bacteria and conversion into protein hydrolysates with improved functionalities. *Applied Biochemistry and Biotechnology*, *193*(8), 2497–2515. https://doi.org/10.1007/s12010-021-03554-4

Said, Z., Sharma, P., Thi Bich Nhuong, Q., Bora, B. J., Lichtfouse, E., Khalid, H. M., Luque, R., Nguyen, X. P., & Hoang, A. T. (2023). Intelligent approaches for sustainable management and valorisation of food waste. *Bioresource Technology*, *377*(February), 128952. https://doi.org/10.1016/j.biortech.2023.128952

Shang, B., Wang, S., Lu, L., Ma, H., Liu, A., Zupanic, A., Jiang, L., Elnawawy, A. S., & Yu, Y. (2022). Poultry eggshell-derived antimicrobial materials: current status and future perspectives. *Journal of Environmental Management*, 314. https://doi.org/10.1016/j.jenvman.2022.115096

Socas-Rodríguez, B., Álvarez-Rivera, G., Valdés, A., Ibáñez, E., & Cifuentes, A. (2021). Food by-products and food wastes: are they safe enough for their valorization? *Trends in Food Science and Technology, 114*, 133–147. https://doi.org/10.1016/j.tifs.2021.05.002

United Nations Environment Programme. (2024). *Food Waste Index Report 2024. Think Eat Save_ Tracking Progress to Halve Global Food Waste*. https://wedocs.unep.org/20.500.11822/45230

Valorisation. (2025). https://dictionary.cambridge.org/dictionary/english/valorization

W.I.S.E with Food. (2025). https://www.1utama.com.my/wise/

2 Valorisation of food waste into building and construction materials

Wan Thing Hong, Chui Kim Ng, and Olufemi Adebayo Johnson

This chapter examines the potential of repurposing food waste into value-added construction materials, covering the analysis of its diverse impacts across multiple industries, the potential food-waste-derived building materials, the proposed production process, and the challenges faced for its practical application.

2.1 THE MULTIFACETED IMPACTS OF FOOD WASTE VALORISATION ON SUSTAINABLE BUILDING MATERIALS

Valorisation of food waste into building or construction materials has gained attention for its potential to enhance sustainability in construction. The practice emerges as an innovative approach offering a range of benefits to the environmental, social, and economic aspects. A conceptual framework is developed in Figure 2.1 illustrating the interdependent relationships of the multifaceted impacts brought by food waste valorisation into building materials. As presented in the framework, the key positive impacts are not limited to decreasing food waste, conserving natural resources for construction, and evolving new industries for food waste-derived materials but also extended to involve political decisions, legal development, and technological advancement.

The environmental impact due to food waste generation is the most concerning issue at present due to the stagnant pollution streams which is against the direction of the UN Sustainable Development Goals. This impact also carries significant weight in the societal and economic repercussions, including resource depletion, greenhouse gas emissions, economic losses across supply chains, food insecurity, and public health challenges. Repurposing food waste into building materials will undeniably decrease food waste in landfills, reducing CO_2 and methane emissions associated with organic waste decomposition. For example, composite building materials incorporating agricultural waste reduce the need for virgin timber, thereby decreasing deforestation and associated carbon emissions. The biochar materials derived from food waste can act as carbon sinks, helping to sequester carbon and reduce the overall carbon footprint of buildings. These actions therefore contribute to the mitigation of climate change. Moreover, many building materials derived from agricultural by-products are free from harmful chemicals throughout the production and

 DOI: 10.1201/9781032718507-2

installation process. This makes them safer materials for both construction workers and building occupants. The concept is aligned with a growing trend towards non-toxic, biodegradable, environmentally friendly materials that promote better indoor air quality and occupant well-being. As building materials can be set from a minimum of 5% to 100% from food waste in terms of volume, large-scale activity is possible to conserve natural resources by reducing the need for virgin materials in construction, e.g. cement, bricks, insulation materials, paints, and coatings, thereby decreasing the carbon footprint of both food waste disposal and building material production within the construction industry.

On the economic aspect, investment in food waste-derived building materials transforms food waste streams into valuable resources, fostering resource efficiency and promoting a circular economy across different industries. It saves a lot on waste disposal costs and raw material procurement for traditional building materials in the long run. Using food waste processing by-products as raw materials for construction can reduce the demand for virgin materials like timber and cement which require significant extraction, processing, and energy consumption. The virgin materials often come with costs like procurement, transportation, processing, and energy usage. This shift promotes not only a more resource-efficient model in maximising the utility of what would otherwise be discarded and lowers the construction costs of raw materials over time, providing affordable housing solutions for the nations. Further to this, valorising food waste into building materials allows for more localised sourcing of resources as the material production must bind with the local food waste channel. Communities can use locally sourced, food waste-derived materials to construct or renovate buildings, thereby boosting the local economies and fostering a sense of pride and ownership in community projects. Adopting food waste-derived materials in construction can also reduce costs related to carbon taxes and environmental impact mitigation. For example, bio-bricks and bio-based thermal insulation incorporating agricultural by-products such as straw, husks, and vegetable pith have good insulation properties due to their porous structures. Such materials exhibit notable thermal resistance and acoustic properties for building energy savings, leading to lower operational costs over the building's lifespan. The initiative of food waste valorisation into building materials supports the circular economy model where the food waste is designed to be reused and optimised for construction use, connecting diverse industries such as agriculture, construction, manufacturing, and beyond, demonstrating how a holistic approach can drive financial sustainability across the entire industrial landscape. This idea also creates a financial incentive for food waste valorisation. When all these materials become more widely adopted, economies of scale will likely reduce the cost of production, further increasing cost efficiency. Nevertheless, the cost-saving margins have also relied on the composition of food waste, selected valorisation processes, and the types of end-used products produced from food waste. A broader sustainability check shall be associated with food waste valorisation to avoid any risk on investment and rebound effects.

The initiative of food waste valorisation into building materials not only addresses environmental concerns but also yields significant social benefits, including enhanced food security, improved community well-being and resilience, new industry and job creation, as well as educational and research advancement growth. When food waste is reused, the risks of pollution and pest infestations decline, leading to a healthier living environment. Additionally, food waste-derived products are eco-friendly, thus promises a cleaner environment for community well-being and enhances the green credential of the construction industries. This action additionally fosters community involvement and awareness about sustainable practices among the public and construction stakeholders and enhances their quality of life by providing affordable housing solutions made of green and sustainable materials. The concept also promotes market growth with the development of new industries for recycled building materials. This is not only bound to resource recovery and circular economy initiative but also leads to industrial symbiosis that creates more green job opportunities. The emerged new industries such as the construction waste processing sector and manufacturing of recycled building materials will foster more community engagement and stakeholder participation and provide opportunities for workforce development and training. With a competitive environment evolving among the new industries, more research activities will be initiated to explore innovative solutions for food waste management and food waste upcycling processes and technologies. This fosters a culture of learning and innovation, exploring advanced knowledge and skills for future challenges. Effective conversion and innovative technologies applied to food waste valorisation will then increase the chances of producing high-value building materials. Successful food waste-derived building materials will finally promote the adopted technologies and attract international cooperation opportunities, which further bolster the national economy.

The shift towards utilising food waste-derived building materials also encourages the development of new policies and regulations for supporting food waste reduction, green material production, and sustainable construction practices. However, regulatory challenges may arise in establishing and enforcing standards for the safety and effectiveness of these new green materials. The National Recycling Strategy and Circular Economy frameworks published under the U.S. Environmental Protection Agency (EPA) emphasise strategies for recycling and using organic waste, including its potential for applications like building materials. This action is also supported by the European Union Circular Economy Action Plan (CEAP, 2020) developed by the European Commission on waste management and recycling, for implementing the use of bio-based materials in construction. Despite limited related policies specifying the importance of food waste-derived materials, more developing countries are demonstrating proactive strategies in the relevant sectors, highlighting the potential for cross-industry applications of food waste valorisation. The legal actions will become more evident when there is a necessity to follow the steps of the U.S. and European countries.

The idea to promote food waste valorisation may also result in increased funding and subsidies from the government, for further incentivising the adoption of these innovative solutions through various research and industrial projects. The notable initiatives are from the U.S. (EPA, 2024), i.e. (1) The EPA's Consumer Recycling Education and Outreach Grant ($39 million) to support consumer food waste recycling education, (2) The EPA's Recycling Infrastructure Grants ($117 million) to advance recycling infrastructure and boost food waste management, and (3) The U.S. Department of Agriculture (USDA) Composting and Food Waste Reduction

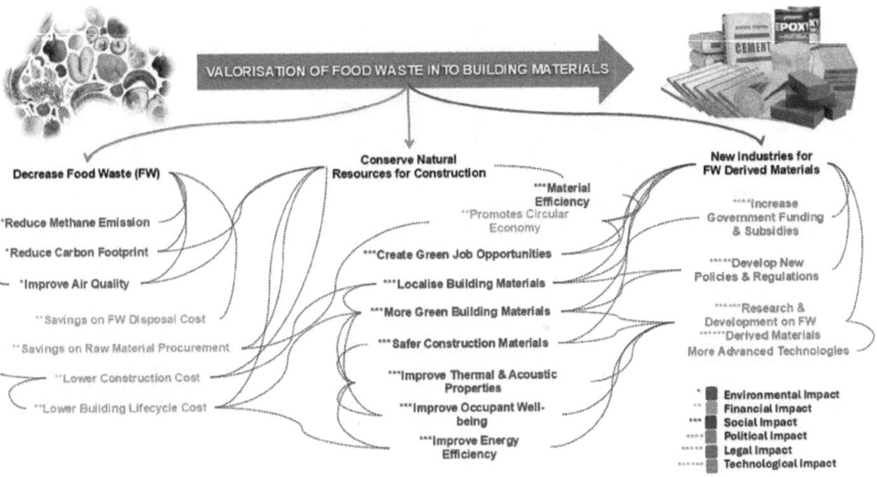

FIGURE 2.1 The interdependent relationships of the multifaceted impacts on food waste valorisation into building materials.

Cooperative Agreements ($10.2 million) for food waste reduction and composting projects. These findings supported the development of consumer knowledge, technologies, and processes that involved food waste conversion into value-added materials in general, contributing to a circular economy and reducing environmental impact. The specific funding focusing on recycle building materials from food waste is not yet available worldwide.

In summary, converting food waste into valuable building materials presents multifaceted impacts in addressing the environmental, social, and economic challenges. Embracing this practice collectively support the concept of reducing, reusing, and recycling for achieving the sustainable closed-loop system and a more resilient future.

2.2 FOOD WASTE-DERIVED BUILDING MATERIALS

Food loss and food waste occur at various phases along the food supply chain, from production to consumption. During the production and process stage, inefficient harvesting can generate crop residues such as risk husk and wheat straw, while improper transportation and storage will cause damage and spoilage of the harvested/ manufactured products, including fruits and vegetables. At the commercial stage, food may be discarded from restaurants, supermarkets, and hotels, due to approaching expiry dates, not meeting quality standards, aesthetic imperfections, or overstocking that may go to waste. These include rotten fruits and vegetables, expired eggs, and dairy products. At the consumer stage, food discarded from the household is peels, bones, or shells during meal preparation, plate waste, or due to overbuying, expiry dates, and improper storage that leads to spoilage of perishable items. Figure 2.2 illustrates the potential food waste categories from different supply chains which can be repurposed into useful building materials.

FIGURE 2.2 Food waste categories that can be repurposed into useful building materials.

Innovative building products integrating food waste have emerged as sustainable alternatives within the construction industry. Examples include bio-bricks made of dry agro-waste, concrete additive derived from crushed eggshells, bioplastics synthesised from fruit and vegetable (green) waste, finishing tiles formulated with coffee grounds mix, and crushed seashells used as a source of lime in mortar production. Table 2.1 presents a variety of food waste that can be repurposed and integrated into the production of innovative building materials to promote sustainability.

Based on the FAO (2022) Food Loss and Waste Database (2022), fruits and vegetables ranked the highest, i.e. 45% of waste generated from the total supply. Fruit and vegetable waste holds significant potential for valorisation into building materials through several processes, e.g. recycling, upcycling, or conversion into biodegradable materials. The building materials from green waste are bio-based insulation, bio-cement, fibreboards and particleboards, bio-based adhesives, biodegradable packaging materials, and bio-composite materials used for finishes or structural components. Besides, composting fruit and vegetable waste can produce nutrient-rich soil amendments that can be used for landscaping or urban agriculture projects associated

TABLE 2.1
Examples of food waste that can be incorporated into building material production

Types of food waste	Potential functions in building materials production
Fruit and vegetable waste/ peels	Extracted oils and fibres from citrus peels can be used as additives in construction materials. Banana peels can be used in the production of biodegradable plastics and as reinforcement in composites. Potato peel fibres can be processed into bio-based composites to create building materials, e.g. insulation and panelling components and used as an additive in cement and concrete. It can also be mixed with clay or other binding materials to produce biodegradable/eco-friendly bricks. Coconut husks or corn stover can be processed into insulation materials to replace traditional insulation materials like fiberglass.
Agriculture waste (crop residues such as rice husks, wheat straw, corn stalks, and sugarcane bagasse)	Rice husks are commonly used to reinforce composites, such as particleboard, ceramic glazes, and cement, improving their strength and thermal insulation. Wheat straw can be used in the production of straw bale construction, as building insulation or as reinforcement in composite materials.
Eggshell waste	Crushed eggshells can be used as a filler in concrete, pavement bricks and tiles, providing a sustainable alternative that reduces the need for traditional aggregates.
Nutshells	Peanut shells can be used as a filler in composites, providing a lightweight and low-cost alternative for certain construction applications. Coconut shells can be utilised as reinforcement in concrete and as a substrate in green roofs.
Seafood waste	Chitosan extracted from crab shells can be used as a biodegradable additive in coatings and films. It has potential applications in waterproofing and fire resistance.
Coffee grounds	Coffee grounds can be incorporated into materials like bricks and composites, offering insulation properties and reducing the overall weight of the material.
Dairy waste	Casein protein derived from milk can be used in the production of adhesives and coatings for building materials.
Waste cooking oil	Waste cooking oil can be used as an eco-friendly binder in the formulation of bonding agents, sealants, or binders for bricks, roof tiles, and compressed wood products. It can also be used as an additive in cement production to improve the workability and strength of concrete. Waste cooking oil can be a component in the production of bio-based insulation to improve thermal properties. When treated with appropriate additives, waste cooking oil can be used for wood preservation to enhance the durability and resistance of wood against decay and pests.

(Continued)

TABLE 2.1 (Continued)

Types of food waste	Potential functions in building materials production
Carcass bone waste	Bone waste, when converted into bone char, can be integrated into building materials like concrete to produce bio-cement.
	Bone ash derived from bone-burned bones is used in the production of ceramic tiles or other ceramic-based building materials.
	Bone waste can be used in the production of biodegradable plastics.
Brewery waste	The byproduct of beer brewing, spent grain, can be used in the production of biodegradable packaging materials and as a component in building materials like particleboard.

with buildings. This involves a series of biological processes where organic material is broken down by microorganisms under controlled conditions. The process starts with organic waste segregation, shredding and preparation, followed by layering and aeration, moisture and temperature control, and then allowing some decomposition period before the harvesting and application for urban agriculture projects. The duration for composting depends on factors like the method used (e.g. hot composting: 2–6 months, cold composting: 6 –12 months), the size of the pile, temperature, moisture, and the type of waste (Wang et al., 2021). After the active decomposition, an additional curing period of 1–2 months is typically needed to stabilise the compost.

Among so many construction materials, cement used in concrete and bricks are the most utilised products in construction projects. The Global Cement and Concrete Association (GCCA) reported that concrete is responsible for a large portion of global CO_2 emissions (Malsang, 2021). Cement is a binding agent mainly used for concrete structures in construction, such as foundations, slabs, columns and beams, walls, bridges, roads, and pavements. Therefore, the massive production of bio-cement incorporating food waste can effectively address the environmental issues in both food waste management and the construction industry. Recent research from the University of Tokyo in Japan developed a technique to produce bio-cement from food and green waste which includes cabbage leaves, orange peels, onions, bananas, and a few slices of pumpkin. As no adhesives are used in the process, the product is edible and has four times better-bending resistance (Ghisleni, 2023). While this technology holds promise for sustainable construction in reducing carbon emissions, edible cement remains in the research and development phase and is not yet commercially available in the construction industry. Further studies and industrial trials are necessary to assess its feasibility, scalability, and economic viability before it can be adopted in mainstream construction practices.

Besides the green waste, rice husks, a byproduct of agricultural production makes up about 20–25% of the total weight of harvested rice (Fathurahman & Surjosatyo, 2022). The seasoned risk husk ash (RHA) can be used as a supplementary cementitious material in concrete to improve its strength, durability, and resistance to chemicals. It can also be used as filler materials in ceramics, refractory materials, and lightweight concrete blocks. Mounika, Baskar and Sri Kalyana Rama (2022) utilised RHA in manufacturing concrete and found that the product has low workability and less density compared to conventional concrete. Concrete with RHA also shows better bond strength when compared to conventional mixes. It reduces chloride diffusion and efflorescence and increases resistance to sulphate and chemical attacks. Wheat straw is also a widely available agricultural waste which typically constitutes 1.3–1.5 times the grain yield. Ahmadi, Souri and Ebrahimi (2020) confirmed the effectiveness of wheat straw as building insulation through its application in the fired clay hollow bricks for wall construction. Besides, straw cement-based building materials have excellent thermal insulation, impact resistance, and other social and economic benefits (Chen et al., 2018).

Vasudevan and Wei (2020) showed that the use of dried eggshell powder (optimum volume at 15%) as a partial replacement for ordinary Portland cement has greatly improved the compressive and flexural strength of concrete. It was decided that the eggshell powder could be used in interlock brick production without changing the normal industrial process.

On top of the above major food waste contributors such as fruits and vegetables, wheat straw, and risk husks, other food waste components such as seafood shells, meat bones, waste cooking oil, dairy waste, and nutshells also have good potential to be valorised into green building materials. However, there are impediments to large-scale availability due to several factors, e.g. region, food consumption patterns, and supply chain inefficiencies.

Not all food waste is suitable to be repurposed due to its lifespan and storage issues. For example, restaurant food waste like leaves, vegetables, and bamboo sticks are highly perishable. Worms and bugs grow easily within 2–3 days of storage. Likewise, not all types of food waste meet the required engineering properties suitable for construction works. Consequently, the food waste below the minimum bending strength, compressive strength, and tensile strength requirements must be modified with other types of waste, e.g. pumpkin waste is inadequate to produce bio-cement, hence blending with Chinese cabbage leaves is necessary (Davey, 2023). It is also essential to consider the safety and regulatory standards of the final products before widely adopted in construction projects. Continued research and innovation are desired to overcome challenges such as cost-effectiveness, regulatory barriers, and scalability in the industry.

2.3 VALORISATION PROCESS, APPROACHES, AND CHALLENGES TO CONVERT FOOD WASTE INTO BUILDING MATERIALS

Food waste presents a significant global challenge, and valorisation, the process of transforming food waste into valuable products, offers innovative and promising solutions. This section explains the required process and approaches of food waste valorisation for construction materials.

Food waste undergoes various pre-treatment steps depending on the chosen valorisation approach. These may involve drying, grinding, sieving, or even composting to convert the waste into a usable form for construction materials. A few examples of the valorisation process are outlined below:

Bio-aggregates: Food waste can be incorporated as a binder or filler in bio-aggregates, which are sustainable alternatives to conventional construction materials. For example, fruit peels and rice husks can be combined with binders like clay or lime to create lightweight and insulating bio-composite bricks.

Bio-based polymers: Food waste can be used to produce bio-based polymers through fermentation or chemical processes. These polymers can then be used in various construction applications, such as adhesives, coatings, or even 3D printing materials.

Geopolymers: Food waste with high silica content, like rice husk ash or sugarcane bagasse ash, can be used as a source of reactive precursors for geopolymer production. Geopolymers are inorganic polymers with superior strength and fire resistance compared to traditional concrete (Garcia-Garcia et al., 2017).

2.3.1 The Process of Converting Food Waste into Value Building Products

There are different methods employed to convert food waste into useful construction materials. The common food waste conversion methods adopted in the industry involves biological conversion, chemical conversion, and direct valorisation.

Biological conversion is a process where food waste is converted into building materials through biological means, predominantly using microorganisms. The main biological conversion processes comprise anaerobic digestion, composting, and fermentation as discussed below: -

a) Anaerobic digestion: This process mimics natural decomposition in a controlled environment, producing biogas (a source of renewable energy) and digestate (a nutrient-rich fertiliser). It is suitable for various food waste types, including fruits, vegetables, and even meat scraps (Xu et al., 2018).
b) Fermentation: This process utilises microorganisms like bacteria or fungi to convert food waste into valuable products like bioplastics, biofuels, or organic acids used in various applications.

Chemical conversion is the process of transforming food waste into valuable products through chemical reactions. These processes usually involve the application of catalysts, heat, and/or chemical reagents to produce biofuels, bioplastics, and other useful substances. The key chemical conversion methods include the following processes:

a) Hydrothermal conversion: This method uses high pressure and temperature to convert food waste into bio-oil, a liquid fuel like crude oil. It is suitable for various organic materials, including food waste and woody biomass.
b) Chemical extraction: Valuable chemicals like sugars, organic acids, or antioxidants can be extracted from specific food waste streams through various chemical processes. These extracted components can be used in various industries, including food, pharmaceuticals, and cosmetics.

Direct valorisation of food waste refers to the immediate conversion of food waste into valuable products that can be used directly in construction without extensive preprocessing. Here are some common methods and examples of direct valorisation:

a) Bio-composite materials: Food waste like fruit peels, eggshells, and coffee grounds can be processed and combined with binders to create biocomposites which can be used for making panels, tiles, and other construction materials. Sivakumar et al. (2022) investigated the use of eggshell powder as a filler in bio-composites, highlighting its potential to improve the mechanical properties of the material while providing an environmentally friendly disposal route for eggshell waste.

b) Mycelium-based materials: Mycelium composites can grow on food waste substrates such as agricultural waste, transforming them into solid, lightweight materials that can be moulded into various shapes. Mycelium-based materials can be used for insulation, acoustic panels, and bricks. They are biodegradable, sustainable, and have good thermal and acoustic insulation properties (Jones et al., 2018).

2.3.2 Technical challenges and considerations

While the concept of transforming food waste into construction materials offers exciting possibilities, several technical challenges need to be overcome for widespread adoption. These cover challenges on pre-treatment, processing techniques, material performance, and regulatory and environmental concerns.

2.3.2.1 Pre-treatment challenges

a) Heterogeneity: Food waste is highly diverse in terms of composition, moisture content, and presence of contaminants. This heterogeneity makes it difficult to establish standardised pre-treatment processes, impacting consistency in the final product.
b) Moisture content: High moisture content in food waste can hinder its processing and make it susceptible to microbial growth and degradation. Drying the waste requires energy, potentially increasing the environmental footprint and production cost.
c) Contaminants: Food waste can contain non-organic materials like packaging remnants, which can negatively affect the performance and quality of construction materials. Efficient and cost-effective methods for removing these contaminants are needed.

2.3.2.2 Processing challenges

a) Standardisation and consistency: Developing standardised processing methods for different types of food waste is crucial for ensuring consistent quality and performance of the final construction materials.
b) Scalability: Scaling up current processing methods to handle large volumes of food waste needed for widespread application in construction requires robust logistics and infrastructure (Papargyropoulou et al., 2014).
c) Energy consumption: Some processing methods, like hydrothermal conversion, can be energy-intensive, raising concerns about the overall environmental impact.

2.3.2.3 Regulatory and environmental considerations

a) Building code compliance: Establishing relevant building codes and regulations for the use of food waste-based construction materials is necessary for their wider acceptance in the construction sector.

b) Environmental impact: The overall environmental impact of valorising food waste into construction materials must consider the life cycle of the materials, energy consumption in processing, and potential emissions.

2.3.2.4 Material performance challenges

a) Durability and longevity: As the building codes and regulations specifically for food waste-derived building materials is not yet available, long-term durability and performance of such materials under various environmental conditions should be considered for practice for wider acceptance in the construction industry.
b) Fire resistance: Meeting the standards of fire resistance can be challenging for some food waste-based materials. Research and development of fire-retardant additives or modifications are needed.
c) Structural integrity: Depending on the specific application, achieving the desired structural strength and other mechanical properties in food waste-based materials can be challenging, requiring optimisation of processing techniques and formulations.

Despite these challenges, ongoing research and development efforts hold promise for overcoming these obstacles.

2.4 CASE STUDIES

This section includes case studies for construction products where organic waste has been used as a resource. These are mainly obtained from agricultural waste; however, some waste streams can be identified at the city level from leaves, grass, and stalks collected from green areas. The products are generally produced with local reach, but it is important to note that the availability of resources applies at a global scale. Therefore, reusing organic waste for construction purposes is a global opportunity that can be exploited locally depending on country-specific and regional circumstances both from an environmental standpoint – such as climate – and a socio-economic standpoint – when considering the supply chain. It shall be noted that data collected in this report refer to the European context rather than the global one. Thus, the global perspective is addressed from a qualitative standpoint rather than quantitative.

The example case studies below are assessed depending on several factors such as their likelihood for application in the construction sector, the availability of resources at the global scale, the complexity of the production processes, and the opportunities for the materials to be reintegrated within the biological loop at the end of the product's service life. There is no real case study applying food-waste-derived building materials in a single project or at a larger scale.

2.4.1 Rice husk and rice husk ash

Mixing the rice husk ash with cement can reduce the need for fillers. Rice husks can be found in huge quantities in milling plants (Figure 2.3). Rice husks can also

FIGURE 2.3 Rice husks can be found in huge quantities in milling plants.

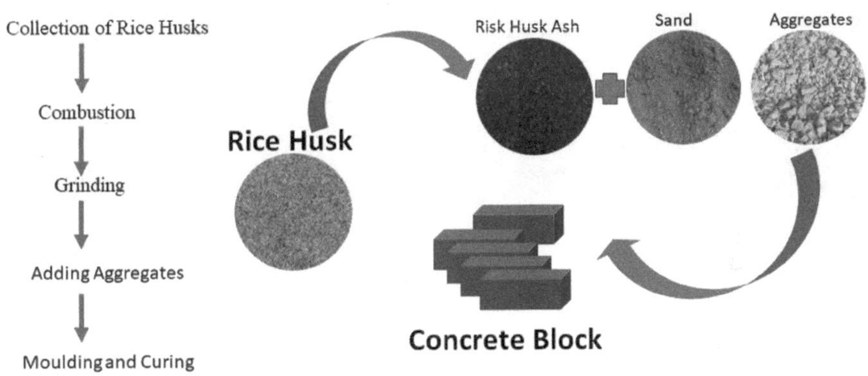

FIGURE 2.4 A simplified process flow diagram of replacing cement with rice hush ash for the construction of concrete blocks.

be used as a raw material for board production. Areas of application in construction include partial cement replacement, bricks, ceramic glazes, and insulation. The manufacturing process includes the combustion of husks and mixing with concrete as illustrated in Figure 2.4. The technical properties of using risk husk include (a) lower density compared to traditional concrete blocks, (b) better environmental footprint, and (c) highly durable.

Rice husk ash, a byproduct of risk husk, can be used as a partial replacement of cementitious materials. Figures 2.5 and 2.6 show sampling of RHA from a rice mill at northern region in Malaysia for characterisation test and rice husk ash stored at the milling plant, respectively.

FIGURE 2.5 Sampling of risk husk ash from a rice mill at northern region in Malaysia for characterisation test.

2.4.2 Potato peels

Mixing the peelings of potatoes, which are cleaned, pressed, and dried into the production of building material, can produce completely bio-based creations, i.e. materials derived entirely from biological resources, such as plants, agricultural residues, or organic waste. The main dietary fiber of potato peels are lignin, cellulose, hemicellulose, and proteins (Joshi et al., 2020). The final products could be used as insulation and acoustic absorbers, demonstrating a high degree of recyclability and remanufacturing potential due to their biodegradable and compostable qualities. Their properties can vary depending on the specific composition, processing methods, and additives used. By adjusting these parameters, manufacturers can tailor the properties of potato peel plywood for various applications, such as furniture, interior panels, or even structural components in sustainable construction. Each adjustment strikes a balance between performance, cost, and sustainability. Figure 2.7 suggested a schematic diagram for the production process of plywood made from potato peels, showing the stages of collection, hygienisation, grinding, dehydration, blending,

FIGURE 2.6 Rice husk ash stored at the milling plant.

moulding, and drying. Along the production process, proper temperature and moisture control at each stage is critical to ensure the correct consistency and properties of the plywood. The choice of blended materials (e.g. bamboo, recycled wood) and binders and additive (e.g., natural resins, starch, or natural fibres) can also influence the strength, water resistance, and environmental friendliness of the plywood (Ply Reporter, 2019).

The National and European (EU) Standards of traditional wood-based panels are standards that ensure an appropriate marketing of the plywood but not a specific standard exclusively developed for plywood made from food waste. The composite panels produced from agricultural residues were evaluated against existing standards for traditional wood-based panels to ensure they meet necessary performance and safety criteria. Table 2.2 shows the various specification standards applied for plywood. A similar table shall be prepared for all building and construction materials made from food waste. The EU standards provide a robust framework for ensuring the quality and safety of building materials. At this moment, most of the construction materials derived from food waste may not currently meet all EU standards for structural applications.

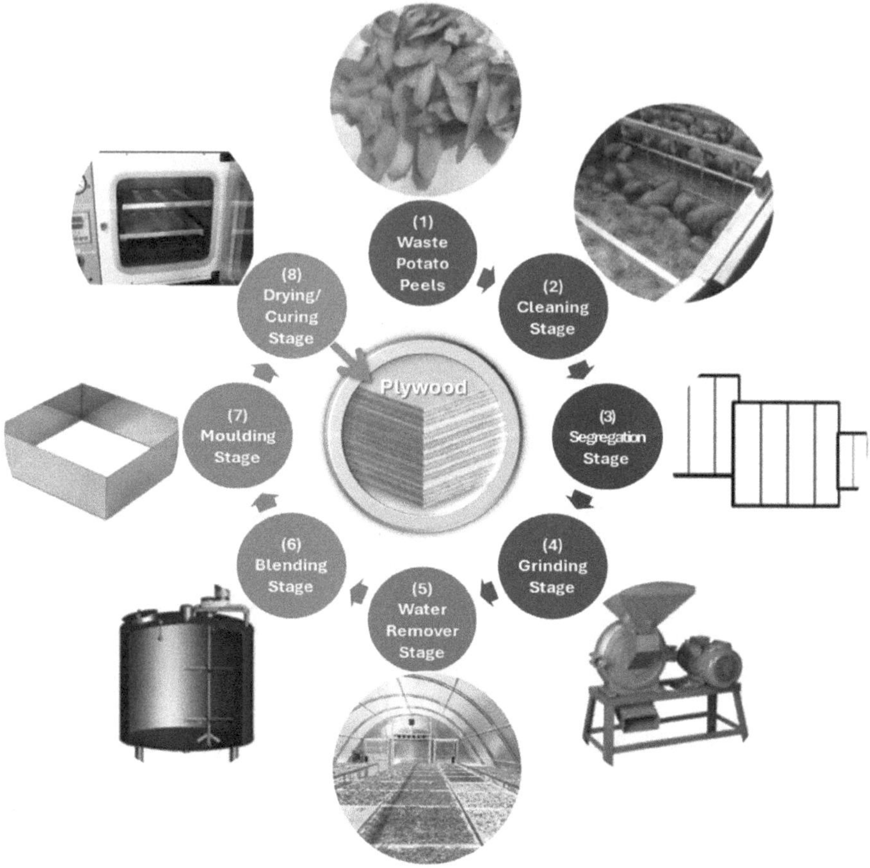

FIGURE 2.7 Schematic diagram of potato peels into plywood.

Table 2.3 presents two examples that utilise food waste in the production of construction materials, offering sustainable alternatives to traditional methods. Each project addresses specific challenges while aiming to contribute to an eco-friendly construction industry.

Rice husk-based building materials, including concrete, particleboard, insulation panels, and bricks, are sustainable alternatives that provide good thermal insulation, strength, and durability while reducing environmental impact through the reuse of agricultural waste. Potato-based insulation and acoustic absorbers offer excellent thermal conductivity and high sound absorption, are biodegradable, and suitable for various building applications. These projects demonstrate diverse approaches to repurposing food waste in construction materials, addressing environmental concerns and promoting sustainability. While challenges such as scalability, product consistency, durability, and meeting specific requirements persist, these initiatives represent significant steps towards a more sustainable future for the construction industry.

TABLE 2.2
Classification and specification standards for plywood

EU code	Standard	EU code	Standard
EN 313-1	Plywood – Classification and terminology – Part 1: Classification (June 1996)	EN 635-5	Plywood – Classification by surface appearance – Part 5: Methods for measuring and expressing characteristics and defects (May 1999).
EN 313-2	Plywood – Classification and terminology – Part 2: Terminology (May 1995). Revision published in 1999.	EN 636	Plywood – Specifications (2003): requirements for plywood for general purposes or structural application in dry, humid, or exterior conditions. It also gives a classification system based on bending properties.
EN 322	Wood-based panels – Determination of moisture content (June 1993). (Confirmed in November 1998).	EN 12369-2	Wood-based panels – Characteristic values for structural design – Part 2: Plywood (2004).
EN 635-1	Plywood – Classification by surface appearance – Part 1: General (April 1995).	ENV 14272	Plywood – Calculation method for the determination of some mechanical properties (2002).
EN 635-2	Plywood – Classification by surface appearance – Part 2: Hardwood (July 1995).	EN 13986	Harmonised standard -Wood-based panels for use in construction – Characteristics, evaluation of conformity and marking. (2004)
EN 635-3	Plywood – Classification by surface appearance – Part 3: Softwood (July 1995).	EN 12871	Wood-based panels – Performance specifications and requirements for load bearing boards for use in floors, walls, and roofs (Published in 2001).
EN 635-4	Plywood – Classification by surface appearance – Part 4: Parameters of ability for finishing, Guideline (December 1996).		

2.5 CONCLUSION

Integrating food waste scraps into building material production is a creative and multidimensional solution for the construction industry. This concept aligns with broader environmental objectives and represents a move towards a more responsible way of utilising resources in the construction industry. As the concept of food waste-derived construction materials is increasingly being explored as alternatives to traditional construction materials, this has created new revenue streams through the management of food waste and the sale of recycled materials.

Before implementing such ideas on a larger scale, it is advisable to conduct thorough research and concurrently consult with experts in materials science, engineering, and

TABLE 2.3
Selected case studies of utilisation of food waste in construction materials

Types of food waste source	• Agricultural waste: Rice husks • Building rubble and masonry waste • Forestry industry waste: Wood chips from waste wood	• Rice husks, peanut, banana, potato, and mushroom
Organisation/ country	• Fraunhofer Institute for Wood Research, Wilhelm-Klauditz-Institute WKI, Germany	• Arup, UK
Issue addressed	• High CO_2 emissions from cement production • Disposal of building rubble and agricultural waste • Need for sustainable and renewable raw materials in construction	• Arup notes that in the UK, construction accounts for 60% of raw materials consumed. • More than 40 million tonnes of dried organic waste from agriculture and forestry were produced in Europe in 2014.
Engineering solutions	• Use of rice husk ash as a substitute for cement • Production of recycled concrete using building rubble and agricultural waste • Development of sustainable insulation materials from plant residues • Creation of composite systems combining recycled concrete with timber	• Peanut shells can be used to produce low-cost materials such as partition boards that are resistant to moisture and fire • Banana fruit and leaves can make rugged textiles, as they contain high-strength fibres and have good acoustic absorption • Potato peel can be cleaned, pressed, and dried to create a low-weight, fire-resistant, and water-repellent insulating material • Arup has previously grown a building from fungus and agricultural by-products that produced no waste and can be composted after use
Expected benefits and link to Sustainable Development Goals (SDGs)	• Environmental benefits: Reduced CO_2 emissions from cement production • Effective use of agricultural and building waste • Economic benefits: Cost savings in material production • Sustainable alternatives to conventional materials • Social benefits: Improved building materials for better insulation and durability • Related SDGs: SDG 11 - Sustainable Cities and Communities, SDG 12 - Responsible Consumption and Production, SDG 9 - Industry, Innovation, and Infrastructure	• A circular system with building waste fed back into the biological cycle at the end of its service life with nutrients returned to the soil. • Capturing organic waste streams from cities and the countryside could provide the industry with lower-cost, lower-CO_2 building materials such as bricks, insulation, and partition boards.

(Continued)

TABLE 2.3 (Continued)

Case study link/ press release	• Rice husk ash as a cement substitute - https://biooekonomie.de/en/news/rice-husk-ash-cement-substitute)	• Waste potatoes used as insulation and acoustic absorbers - The Urban Bio-loop
Types of final product	• Concrete building blocks • Sustainable insulation materials from plant residues • Composite systems for use as floor slabs combining recycled concrete with laminated veneer lumber and cross-laminated timber.	• Insulation and partition boards

sustainability to ensure that the resulting products meet necessary safety and performance standards. Cross-sector partnerships between the food waste recycling chain, construction industry, sustainable product manufacturing, academia, as well as local policy makers, shall be established to create values involving multiple industries. The local and international standards for building materials such as The International Organization for Standardization (ISO), European Norms (EN) Standards, and British Standards (BS) shall be adhered to their structural and functional requirements, durability, environmental impact, fire resistance, and health and safety. Such regulatory requirements should not be overlooked or bypassed, as non-compliance could lead to legal issues or unsafe structures. Industries should also avoid food waste that contains harmful chemicals and contaminants without proper treatment in the valorisation process. It is important to establish standardised processes for food waste-derived building materials and incorporate these processes into the local policy and standards. This can avoid the production and marketing of unapproved material that is without proper testing for structural integrity, durability, and environmental safety. In addition, the production line should also balance this innovation with its economic viability to ensure practical adoption in the construction industry. As the idea involves high initial costs for research, development, and specialised equipment, by optimising the production techniques and scaling operations, the manufacturers can reduce costs and make the materials competitive with traditional options. Streamlining the food waste valorisation processes and using readily available technologies are also alternatives to enhance its cost-effectiveness while maintaining quality. With the right approaches, food waste-derived building materials will be accepted by the wider market and broadly adopted into mainstream practices for a more sustainable construction industry.

EXERCISES

1. Explain the multifaceted impacts brought by the idea of valorisation of food waste into building materials.
2. Discuss the potentials and benefits of using agricultural waste in building materials production.
3. Discuss the potential technical and regulatory challenges associated with utilising food waste as a construction material.

REFERENCES

Ahmadi, R., Souri, B., & Ebrahimi, M. (2020). Evaluation of wheat straw to insulate fired clay hollow bricks as a construction material. *Journal of Cleaner Production*, 254, 120043.

CEAP. (2020). Circular economy action plan. Retrieved from https://environment.ec.europa.eu/strategy/circular-economy-action-plan_en

Chen, J., Elbashiry, E. M. A., Yu, T., Ren, Y., Guo, Z., & Liu, S. (2018). Research progress of wheat straw and rice straw cement-based building materials in China. *Magazine of Concrete Research*, 70(2), 84–95.

Davey, R. (2023, July 24). Building with food waste. Retrieved from https://www.azobuild.com/article.aspx?ArticleID=8621

EPA. (2024, October 19). Funding opportunities about wasted food. Retrieved from https://www.epa.gov/sustainable-management-food/funding-opportunities-related-wasted-food

FAO. (2022). Food loss and waste database. Retrieved from https://www.fao.org/platform-food-loss-waste/flw-data/en/

Fathurahman, R., & Surjosatyo, A. (2022, June). Utilization of rice husks as a fuel for gasification–a review. In *IOP Conference Series: Earth and Environmental Science* (Vol. 1034, No. 1, p. 012065). IOP Publishing.

Garcia-Garcia, G., Woolley, E., Rahimifard, S., Colwill, J., White, R., & Needham, L. (2017). A methodology for sustainable management of food waste. *Waste and Biomass Valorization*, 8(6), 2209–2227.

Ghisleni, C. (2023, January 24). Edible cement: innovative material that uses food waste in civil construction. Retrieved from https://www.archdaily.com/995299/edible-cement-innovative-material-that-uses-food-waste-in-civil-construction

Jones, M., Bhat, T., Huynh, T., Kandare, E., Yuen, R., Wang, C. H., & John, S. (2018). Waste-derived low-cost mycelium composite construction materials with improved fire safety. *Fire and Materials*, 42(7), 816–825.

Joshi, A., Sethi, S., Arora, B., Azizi, A. F., & Thippeswamy, B. (2020). Potato peel composition and utilization. *Potato: Nutrition and Food Security*, 229–245.

Malsang, I. (2021, October 19). Concrete: the world's 3rd largest CO_2 emitter. Retrieved from https://phys.org/news/2021-10-concrete-world-3rd-largest-co2.html

Mounika, G., Baskar, R., & Sri Kalyana Rama, J. (2022). Rice husk ash as a potential supplementary cementitious material in concrete solution towards sustainable construction. *Innovative Infrastructure Solutions*, 7(1), 51.

Papargyropoulou, E., Lozano, R., Steinberger, J. K., Wright, N., & Ujang, Z. (2014). The food waste hierarchy as a framework for the management of food surplus and food waste. *Journal of Cleaner Production*, 76, 106–115.

Ply Reporter. (2019, April 26). MDF and particle board made of potato peel. Retrieved from https://www.plyreporter.com/article/50850/mdf-and-particle-board-made-of-potato-peel?utm_source=chatgpt.com

Sivakumar, A. A., Sankarapandian, S., Avudaiappan, S., & Flores, E. I. S. (2022). Mechanical behaviour and impact of various fibres embedded with eggshell powder epoxy resin biocomposite. *Materials*, 15(24), 9044.Vasudevan, G., & Wei, S. C. (2020, May). Utilisation of eggshell powder (ESP) as partial replacement of cement incorporating superplasticizer. In *IOP Conference Series: Materials Science and Engineering* (Vol. 840, No. 1, p. 012016). IOP Publishing.

Wang, L. K., Wang, M. H. S., Cardenas, R. R., Sabiani, N. H. M., Yusoff, M. S., Hassan, S. H.,... & Hung, Y. T. (2021). Composting processes for disposal of municipal and agricultural solid wastes. Solid Waste Engineering and Management: Volume 1, 399–523.

Xu, F., Li, Y., Ge, X., Yang, L., & Li, Y. (2018). Anaerobic digestion of food waste–challenges and opportunities. *Bioresource Technology*, 247, 1047–1058.

3 Fortification of meat-based products with up-cycled fruit and vegetables waste

Nesa Dibagar, Adam Figiel, and Moslem Namjoo

3.1 INTRODUCTION

Food waste serves as a reservoir of valuable substances and plays a crucial role in addressing the needs of a rapidly expanding global population for healthy products. For this reason, advancements in technology have enabled the recovery of valuable components from food waste, thus promoting their integration into the food supply chain in a sustainable manner. The concept of food-to-food fortification utilising natural fortifiers extracted from food waste has recently been developed within the food science and technology sector. The underutilised portions of fruits and vegetables, which are often discarded, contain high amounts of phytonutrients and valuable bioactive compounds that, if extracted, can have an immense added value in a wide range of applications, such as food additives, nutraceuticals, therapeutics, and cosmetics. Such value-added products isolated from fruit and vegetable waste (FVW) mainly involve polyphenols, proteins, lipids, polysaccharides, fibers, vitamins, essential minerals, fatty acids, volatiles, and pigments.

The incorporation of FVW into meat-based products formulations may significantly enhance their nutritional, sensory, and physical properties. Nutritionally, these alterations can increase fiber content, key vitamins (such as vitamins A, C, and E), and antioxidants, hence improving the health benefits of meat products. The presence of natural antioxidants in FVW can also improve the shelf life and stability of meat products by reducing lipid oxidation (Falowo et al., 2014). Sensory attributes, such as flavour, colour, and texture are augmented by the unique compounds in fruits and vegetables, providing a more complex and appealing taste profile. For example, the natural sugars and acids in fruits can provide an appealing flavour contrast to the savory ingredients of meat. Natural fibers in plant waste can help bind the meat mixture, resulting in a juicier and firmer texture. The natural pigments can also give the meat products a fresher and more vibrant appearance. Environmentally, the use of FVW promotes sustainability by diverting garbage from landfills, lowering the carbon footprint of waste disposal, and fostering a circular economy. This method

DOI: 10.1201/9781032718507-3

helps to create a more sustainable food system by optimising resource use while minimizing environmental effects.

The microbial quality of the reformulated meat products with FVW is a concern that must be considered. The microbiological quality of meat products can be affected by the addition of FVW. Longer shelf lives for meat products can be achieved by limiting the growth of pathogenic microbes and spoiling agents through the inherent antimicrobial qualities of some fruits and vegetables, such as the presence of organic acids, phenolic compounds, and essential oils. However, to avoid the introduction of new contaminants, FVW must be properly cleaned, sanitized, and handled. Furthermore, the moisture content and pH levels of the additional FVW can have an impact on microbial growth, with lower pH and less moisture generally reducing microbial multiplication. Proper processing of FVW is required to ensure microbiological safety and waste quality before introduction to different meat products.

Cost is a crucial factor in purchasing meat products. Reformulating meat-based products by incorporating FVW could be highly cost-effective. This strategy reduces reliance on expensive animal-derived ingredients by partially replacing them with more affordable plant waste. This not only lowers production costs but also adds value by repurposing by-products that would otherwise be discarded, promoting a more sustainable and cost-efficient manufacturing process. However, it is important to note that the initial expenses associated with developing and standardising new formulations might offset some of the cost benefits in the short term.

This chapter explores the use of FVW in meat product reformulation, with an emphasis on its nutritional, microbiological, sensory, and technological aspects. Despite FVW's enormous potential for usage in meat product compositions, current evidence is mostly lab-based and on a smaller scale. There is insufficient large-scale data from the food companies to fully comprehend and realize this promise. The industry lacks thorough data and large-scale studies to assess the practicality and scalability of adding FVW into meat products.

3.2 FRUIT AND VEGETABLE WASTE

Regrettably, almost half of the fruits and vegetables grown worldwide are wasted. Due to the high water content, the global loss of fruits and vegetables is notably high by weight, as depicted in Figure 3.1. Various calculations and statistical analyses have been conducted on the data shown in Figure 3.1. In the food industry, during the processing of fresh fruits and vegetables to obtain required products, underutilised fractions, including peel, seed, kernel, pulp, pomace, stem, stalk, leaves, roots, and bagasse, are discarded. These fractions are perishable and get rotten quickly, leading to environmental hazards mostly brought on by microbial deterioration.

3.3 VALORIZING FVW INTO VALUE-ADDED PRODUCTS

FVWs remain a significant source of vital nutrients, including polyphenols, carotenoids, unsaturated fatty acids, dietary fibers, organic acids, proteins, vitamins (in particular vitamin C), minerals, and various trace elements (Trigo et al., 2022). They are also abundant in antioxidants and possess antimicrobial, antibacterial,

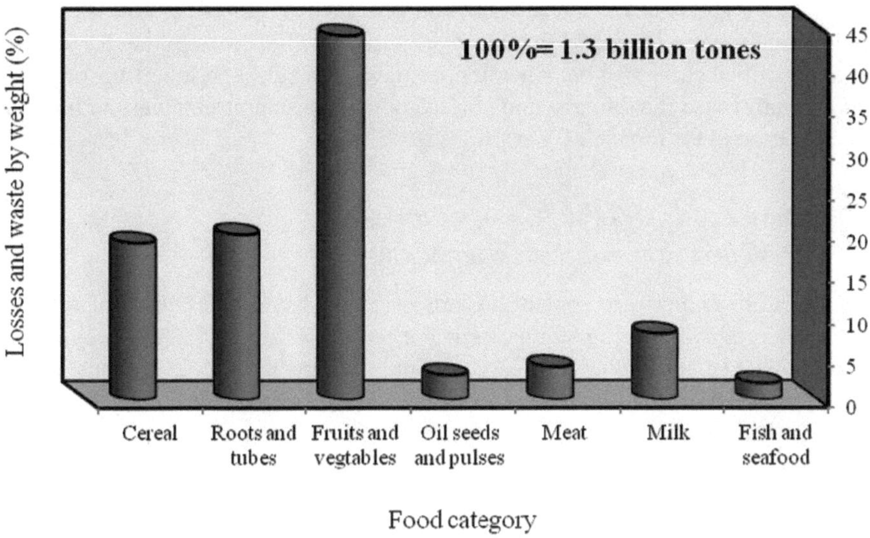

FIGURE 3.1 Global food loss and waste.

Source: Adapted by Zhu et al. (2023).

and antifungal properties (Skwarek and Karwowska, 2023). Scientists are actively exploring the chemical composition, properties, and potential applications of different types of FVW in food products with a focus on eco-friendly and health-enhancing approaches.

Innovative valorisation strategies aim to recover high-value ingredients from FVW, used as natural sources of bioactive compounds for functional food formulation. Nutrients extracted from FVW are indeed utilised in various value-added products. They are incorporated into bakery (such as different types of bread and cake) and confectionery (such as cookies and biscuits) items, noodles and pastries, dairy products (such as cheese, yogurt, salad dressing, and ice cream), and wine. They are also successfully used for the preparation of biogas, industrial oil, edible oil, pectin, essential oil, emulsifiers, bio-pigments and natural colourants, packaging films, vinegar, cosmetics, dietary supplements, and drugs.

3.4 NECESSITY OF PHYSICAL PRETREATMENT BEFORE UTILISATION OF FVW

Physical pretreatment plays a crucial role in transforming FVW into value-added ingredients. By pre-treating, an ingredient with microbial stability, minimal bioactive losses, and a wide range of health-improving advantages is developed. Several techniques are employed for FVW pretreatment: including drying, reducing particle size, and densifying. Due to the highly perishable nature, wet waste frequently begins to deteriorate chemically and biologically while being stored. In the absence of appropriate pretreatment, FVWs are susceptible to infestation by insects and larvae. Drying can help to remove the moisture content from the FVW to increase its stability. For huge samples, a size reduction process is required. If not, the drying duration will be very long, and the internal moisture will diffuse to the dry surface. Additionally, blanching and densification, achieved via baling, briquetting, or pelletizing, can minimise the damage and storage costs. These pretreatment methods are extremely crucial for fortified FVW.

3.4.1 Drying, size reduction, and densifying of FVW for meat products fortification

FVW due to high moisture content are prone to both microbial contamination and chemical degradation during storage or if not promptly handled. These wet materials accumulate in substantial volumes, particularly during specific seasons, posing challenges for large-scale treatment. As a result, their direct industrial application is not practical. Among waste processing techniques, drying stands out as a widely used method. An appropriate selection of a drying strategy with optimum working

FIGURE 3.2 A large-scale hybrid microwave-vacuum drying machine (HT-WZ2, Zhengzhou Hento Machinery Co., Ltd, China).

parameters effectively reduces microbiological contamination, prevents chemical deterioration, and minimise losses of bioactive ingredients. The reduction of chemical changes owing to water removal is one of the major benefits of drying that makes items more stable. Sun drying, solar drying, oven drying, convective air drying (cabinet drying, tunnel drying, spray drying, fluidised bed drying), drum drying, rotary drying, and hybrid drying methods, such as infrared-assisted convective drying, microwave-assisted convective drying, and microwave-assisted-vacuum drying (Figure 3.2) are among the FVW typical drying methods.

Sun drying, a natural food preservation method, uses sunlight and wind for dehydration. It is a convenient drying method involving just placing FVW in direct sunlight. However, the effectiveness of sun drying is weather-dependent, making it unreliable in rainy climates. Solar drying, on the other hand, improves sun drying using devices to capture more solar energy, thus increasing air temperature and speeding up the drying process.

Convective air drying is a widely used technique for dehydration due to its ease of application and readily available technology. It uses hot air for drying, either naturally or with forced circulation. Industrial settings often use forced air dryers for their efficiency in air circulation and uniform drying. The most applied temperatures for convective air drying of food materials fall within the range of 40 to 90°C, allowing products to dry and reach the safe water activity level of storage with no microbial proliferation ($a_w < 0.6$) (Tapia et al., 2020). Degradation mechanisms for bioactive compounds during convective drying reveal that vitamin C, phenols, flavonoids, and glycosides react with oxygen at high drying temperatures. This reaction rate leads to the degradation of these bioactive compounds due to their high reducibility. Research suggests that the optimal drying temperatures for preserving vitamin C, polyphenols, flavonoids, volatile compounds, and their antioxidant activity are as follows: 50–60°C, 55–60°C, 60–70°C, 40–50°C, and 50–70°C, respectively (ElGamal et al., 2023). For maintaining plant bioactive components, convective drying at relatively low temperatures is strongly recommended. Figure 3.3 illustrates a section of a pilot scale convective air dryer that can operate at four distinct drying temperatures. This configuration enables the identification of the optimal dehydration temperature for specific plant-based materials, ensuring the preservation of their biochemical profile and physical quality within acceptable standards.

Hybrid dryers have shown potential in improving drying speed and product quality, with a field-assisted drying method, which significantly reduces drying times and increases energy efficiency. These technologies are most effective early in the drying process when moisture content is high. Properly combining different drying methods can enhance drying performance and product quality.

Rotary dryers are the typical industrial machines used to minimize the moisture content of food waste. They involve a rotating drum or cylinder to heat and dry the material as it tumbles through the cylinder. Figure 3.4 demonstrates a schematic view of an industrial rotary dryer, which can be used for waste drying. A large-scale rotary dryer was used by Arjona and Ollero (2005) for olive oil mill waste drying.

Drying condition significantly affects the drying yield and FVW moisture content, water activity, bulk density, and physicochemical properties. To develop energy-efficient, cost-effective, and environmentally friendly FVW drying methods,

FIGURE 3.3 A pilot scale convective air dryer operating at four distinct drying temperatures.

research and development are required. Microwave drying has gained popularity as an alternative technique for various food products. In the food industry, microwave energy is widely used. Notably, microwave treatment significantly reduces drying time for biological products without compromising quality, making it an optimal drying method.

FVW generated by industry often arrives in sizes that are impractical for direct use and hinder efficient industrial extraction. Therefore, they need to undergo size reduction. This process involves breaking down the materials into smaller pieces. By applying various forces, particles of specific size and shape are created. Milling, a common physical pretreatment method, reduces particle size and increases the contact surface area. The extraction recovery is significantly influenced by particle size. Smaller particles of dehydrated waste material tend to yield better extraction results. This phenomenon occurs because reducing particle size ruptures cell walls, enhancing contact between fine particles and the extraction solvent. However, caution is necessary when using extremely small particle sizes, as they can lead to bed packing and channeling effects during extraction (Domínguez et al., 2020). Moreover, the reaction rate in biological and chemical processes is significantly accelerated because of the substantial specific surface area of these small particles. The optimized size reduction method can increase the yields of particularly costly value-added components extraction, so before the main extraction process, it is advised to perform this size reduction. The process of densification proves valuable in addressing specific limitations associated with biomass applications. These limitations include uneven particle size and shape,

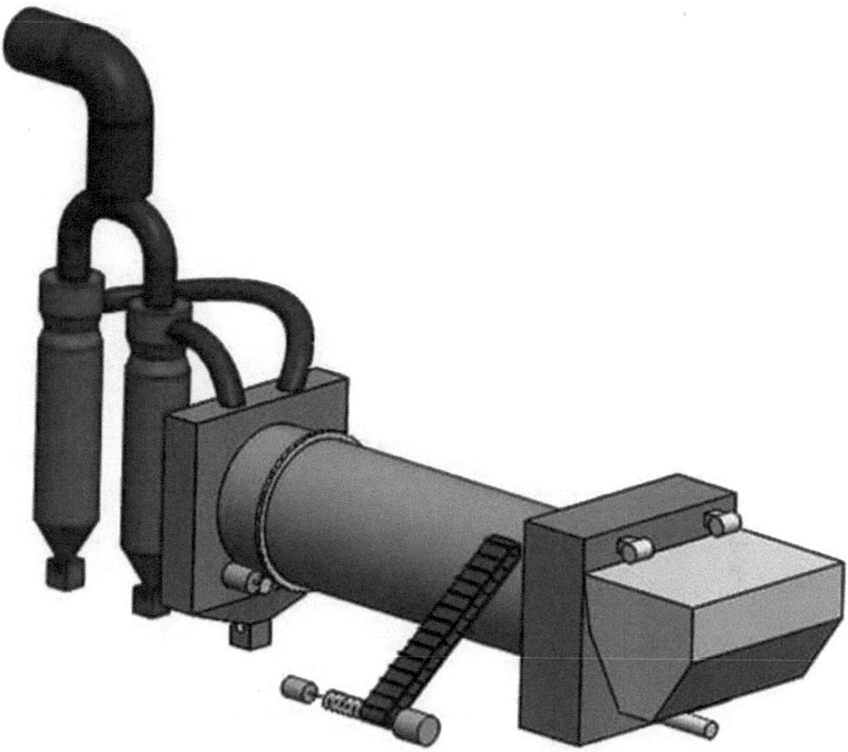

FIGURE 3.4 Schematic view of an industrial rotary dryer.

poor density, and high transportation costs (Sarangi et al., 2018). FVW, often clumpy and possessing low density, faces challenges due to seasonal fluctuations affecting their availability for storage and handling. These by-products are challenging to convert into value-added products, especially when they require transportation or extended storage in a transformation facility. Consequently, FVW densification by pelletization, briquetting, and log-forming methods becomes essential to enhance the extraction of precious substances. As an example, figure 3.5 shows the sea buckthorn fruit pomace pretreatment by drying, milling, and densifying processes. This pomace has been effectively utilized as a fortificant in feed formulations for lamb meat production, resulting in increased muscle mass, improved tenderness, enhanced water-holding capacity, and heightened antioxidative capacity in the meat (Qin et al., 2020).

3.5 RECOVERY OF BIOACTIVE COMPOUNDS FROM FVW TO FORTIFY MEAT-BASED PRODUCTS

3.5.1 Conventional extraction techniques

Recovering bioactive compounds from dried FVW involves employing various traditional and modern extraction approaches. The traditional methods are often simple

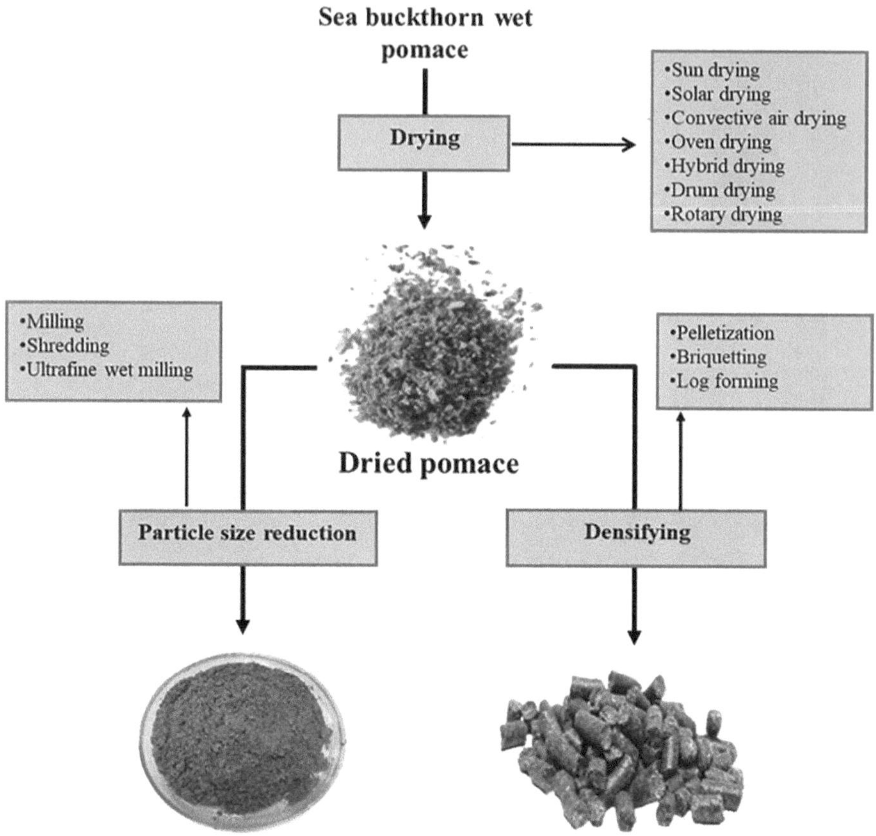

FIGURE 3.5 Sea buckthorn wet pomace pretreatment by drying, milling, and densifying processes.

with minimum equipment, such as maceration, decoction, infusion, cold pressing, and Soxhlet extraction. Maceration involves soaking waste material in a solvent, including water, alcohol, or oil to dissolve bioactive compounds. The quality and quantity of the extracted phytochemicals are influenced by the solvent, maceration duration, and process temperature (Kowalczewski and Zembrzuska, 2023). Decoction simmers plant materials in water for extended periods. The efficacy of decoction can fluctuate based on the type of waste material, boiling duration, and water-to-waste ratio (Manousi et al., 2019). Distillation is a common method for obtaining volatile compounds, but its effectiveness depends on the compound type and heat stability. Various distillation techniques, including simple distillation, short-path distillation, steam distillation, and vacuum distillation, can be used to extract bioactive compounds. The choice of method depends on the phytochemicals, their heat sensitivity, and the desired purity of the final product (Hasan et al., 2024). Figure 3.6 shows a large-scale extractor.

Cold pressing is a gentle extraction method that preserves natural flavours, aromas, and nutritional compounds. It is particularly effective for sensitive phytochemicals that can degrade at high temperatures (Hasan et al., 2024). Further to this, Soxhlet extraction is widely used for various compounds. It is especially useful when dealing with poorly soluble compounds or exhaustive extractions; however, it can be time-consuming and solvent-intensive (Bitwell et al., 2023). Cold pressing and Soxhlet extraction are preferred for preserving natural flavours and maximising yield, respectively (Hasan et al., 2024). The choice of method depends on specific requirements and available resources.

Despite their limitations, conventional extraction methods persist globally. Modern alternatives address these limitations by offering increased yields, shorter extraction times, and better process control.

FIGURE 3.6 A pilot-scale prototype of a three-chamber vacuum extractor operating on the principle of the Clevenger apparatus (Paszków Farma, Poland).

3.5.2 Modern extraction techniques

In modern practice, bioactive components from FVW are extracted and refined using sophisticated techniques and specialized equipment. The contemporary methodologies encompass supercritical fluid extraction (SFE), membrane separation techniques (MST), pressurized liquid extraction (PLE), pulsed electric field (PEF), ultrasound-assisted extraction (UAE), microwave-assisted extraction (MAE), and enzyme-assisted extraction (EAE).

SFE employs carbon dioxide (CO_2) as the main solvent. CO_2 has lower viscosity and higher diffusivity, allowing it to penetrate solid materials more effectively and enhance extraction yields. CO_2 ensures food-grade purity and safety (Gan and Baroutian, 2022). Hassas-Roudsari et al. (2009) found that SFE-extracted total phenol content and antioxidant activity of canola seed meal were highest at 160°C. Additionally, SFE has been utilised to extract β-carotene, polyphenols, and lycopene from apricot pomace, carrot pulp, and tomato waste (Wijngaard et al., 2012).

MAE effectively extracts phytochemicals using solvents and microwaves while preserving biological activity. Interestingly, MAE maintains the biological activity of extracts, improving antioxidant activity and colour quality, and preserving phenolic components. Important variables that significantly influence the extraction efficiency in MAE are temperature, extraction time, and microwave power (Bitwell et al., 2023). When extracting antioxidants from mango peel, MAE resulted in 1.5–6 times higher yields compared to the traditional method (Dorta et al., 2013).

UAE offers faster extraction times and is compatible with delicate substances by employing sound waves to dissolve cell walls. Advantages of UAE include lower energy and solvent use, compatibility with both thermolabile and unstable compounds, and quicker extraction durations (Mehta et al., 2022). When compared to conventional, microwave, and UAE methods, intermittent sonication resulted in higher pectin extraction from grapefruit (Bagherian et al., 2011).

PLE is a method that uses solvent extraction at elevated temperatures and pressures. In PLE, the solvent remains in a liquid state throughout the process. This technique offers several advantages, including higher extraction yields and reduced solvent consumption during sample preparation. PLE is highly versatile and can be employed with a wide range of solvents, with water being the most environmentally friendly choice. To optimize PLE, key parameters include extraction temperature, pressure, extraction time, and the type of extraction mode (dynamic, static, or a combination of both) (Hasan et al., 2024). For efficient extraction of procyanidins from red grape pomace using organic solvents, a temperature above 80°C was recommended by Monrad et al. (2010).

PEF is a non-thermal food preservation technology that uses high-voltage, low-energy pulses to disrupt the cellular structure of microorganisms in food. The electrical field causes permeabilization of the cell membrane, leading to cell death and the preservation of the food. PEF has been widely employed to enhance mass transfer and improve the extraction of phenols, betalains (pigments), and pectin from grape seed, red beetroot, and apple pomace. Additionally, this method can result in the inactivation of microorganisms, enhancing food safety (Zhu et al., 2023).

Furthermore, the EAE technique uses enzymes to release phytochemicals from aromatic and medicinal plant cells, including lactase, protease/lipase, lipase, phospholipase, pectinases, cellulases, and hemicellulases (Ebrahim, 2023). These plant cell wall and membrane-rupturing enzymes improve the release and extraction of active ingredients. This method is notably less energy and solvent-intensive than traditional extraction techniques, making it a more environmentally friendly and productive option (Zhu et al., 2023). Gaur et al. (2007) employed EAE to extract oils from mango kernels, soybean, and rice bran. The extraction rates were 98%, 86%, and 79%, respectively. The methods used to efficiently extract different bioactive substances may depend on the plant parts being processed, such as stems, leaves, peels, or pomace.

3.6 FVW VALORISATION INTO MEAT-BASED PRODUCTS

Meat is the primary protein source for many consumers, particularly young people and children who consume processed meat snacks as fast food. Meat faces significant challenges, including microbial spoilage and fat rancidity, which prompt the use of synthetic preservatives, such as nitrates and nitrites in derivative products. Unfortunately, these preservatives can have carcinogenic effects on the human body, posing a threat to overall health (Bhardwaj et al., 2022). Oxidation processes in meat lead to the breakdown of lipids and proteins. Consequently, this impacts the flavour texture, and colour of meat available for consumption. The production of meat products, in particular burgers with high-fat content, positions them as a significant contributor to chronic health conditions, such as obesity, atherosclerosis, and heart-related problems. In this regard, food-to-food fortification with organic and natural fortifiers can be a practical approach to alleviate the negative impacts of meat products. It may serve a dual purpose: safeguarding meat from corruption and rancidity while simultaneously boosting its nutritional and sensory values. FVW is a worthy candidate for studies on valorisation into natural ingredients for the fortification of meat and its products. Given that oxidative processes contribute to the degradation of lipids, proteins, and pigments in meat, incorporating FVW as a cost-effective and straightforward source of bioactive compounds helps maintain the nutritional and sensory quality of meat products. Moreover, FVW bioactives offer an opportunity to reduce the use of synthetic preservatives during meat snack production. This practice also contributes to reducing environmental impact. Figure 3.7 provides a graphical abstract illustrating the valorisation cycle of FVW into various meat products. Depending on the desired final product, FVW can be incorporated into meat products either as a powder or as an extract. This powder or extract serves as a bio-fortifier due to its potent bioactivity. The final meat product undergoes several analyses, including microbiome testing, physical and biochemical analysis, and sensory evaluation. The preservation of nutrients, convenience, safety, extended shelf life, enhanced bioavailability, and the variety and accessibility of fruit and vegetable products are critical factors that contribute to a healthy lifestyle. These factors not only ensure that the nutritional benefits of fruits and vegetables are retained and readily available but also facilitate the valorisation of their waste into meat products. This process of valorisation enhances the nutritional profile of meat products,

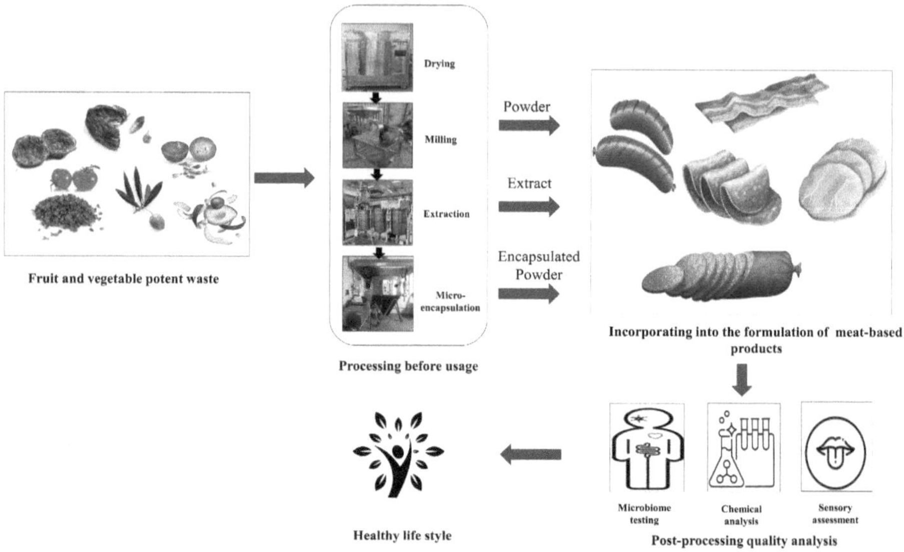

FIGURE 3.7 A schematic cycle of FVW valorisation into various meat products (Qin et al., 2020).

making them more beneficial for health. By integrating these processed fruit and vegetable components, the resulting meat products can offer improved bioactivity and nutritional value, thereby supporting overall well-being and promoting sustainable dietary practices.

3.7 VARIOUS APPLICATIONS OF FVW IN MEAT PRODUCTS

Integrating FVW into meat products can significantly enhance their nutritional, physical, and sensory qualities. This innovative approach involves utilising natural preservatives found in FVW, such as organic acids and phenolic compounds, to prolong the shelf life of meat products. This not only reduces the need for synthetic preservatives but also contributes to cleaner, more natural labels. Additionally, FVW can serve as a fat substitute, replacing animal fats with fibrous FVW to create juicier and healthier meat products.

The incorporation of FVW in meat analogs meets the growing demand for plant-based alternatives, appealing to both vegan and vegetarian consumers. By providing similar sensory and nutritional properties, these analogs offer a sustainable and nutritious option for those seeking to reduce their meat consumption. Moreover, the use of FVW can enhance the sensory attributes of meat products by adding unique flavours, colours, and textures, making them more appealing to consumers.

Reducing residual nitrites in meat products is another key benefit of using FVW. The natural antioxidants present in FVW can serve as a safer alternative to traditional preservatives, addressing health concerns while maintaining product safety and quality. From a cost perspective, utilising FVW can lower production expenses by reducing the dependence on expensive animal-based ingredients. This strategy not only makes meat products more affordable but also adds value to what would otherwise be a waste, promoting a circular economy and reducing environmental impact.

Nutritionally, FVW can boost the content of vitamins, minerals, fiber, and antioxidants in meat products, enhancing their health benefits. For example, the addition of carrot or beet pulp can increase fiber content and provide essential vitamins such as A and C. Furthermore, the environmental benefits of integrating FVW are significant, as it promotes sustainability by reducing food waste and lowering the carbon footprint associated with meat production. Figure 3.8 demonstrates a visual representation of the key benefits of incorporating FVW into meat –based derivatives.

3.7.1 Antioxidants fortification of meat products with FVW

Antioxidants play a critical role in the meat industry by reducing oxidative processes that cause deterioration and decreased shelf life. Biomolecules, including lipids, proteins, and DNA, are protected from oxidation by antioxidants, which neutralize free radicals (Manessis et al., 2020). The two main types of antioxidants used in the meat industry are synthetic and natural. Synthetic antioxidants, which are chemically produced, are commonly used to prevent spoilage and extend shelf life. However, concerns have been raised about their potential adverse health effects. For example, some studies suggest that prolonged consumption of chemical compounds could be linked to an increased risk of cancer

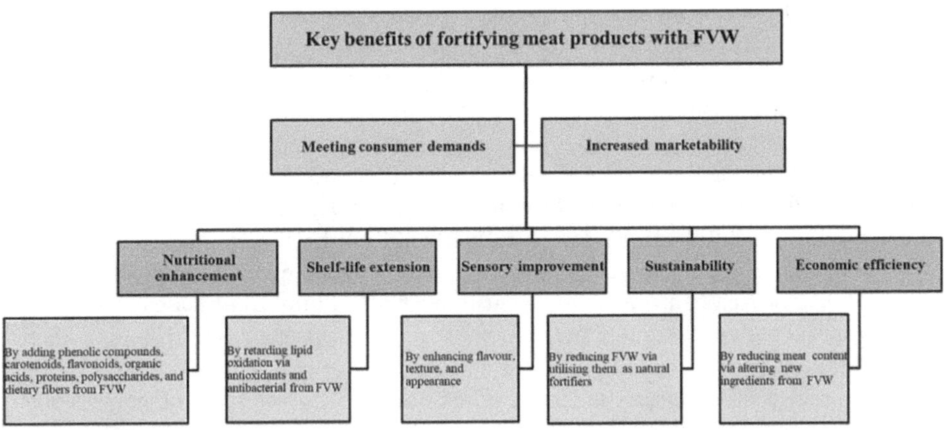

FIGURE 3.8 Key benefits of fortifying meat-based products with FVW.

and other health issues (Domínguez et al., 2019). Natural antioxidants are derived from sources, including fruits, vegetables, herbs, and spices. The natural compounds, such as polyphenols, tocopherols (vitamin E), and ascorbic acid (vitamin C), provide a more sustainable and health-conscious alternative to their synthetic counterparts. FVW can also stand out as an available, affordable, and sustainable source of natural antioxidants. The abundant bioactive compounds in FVW can offer an eco-friendly solution to enhance the antioxidant content in meat products. Examples of FVW with high antioxidant content include the peels and pulp of citrus fruits, such as oranges and lemons, which are rich in vitamin C and flavonoids. Similarly, the skins and seeds of berries, such as blueberries and strawberries, contain high levels of anthocyanins and other polyphenols. Additionally, the outer layers of onions and the pomace from grapes and pomegranates are also excellent sources of natural antioxidants. In fibrous FVW, dietary fibers can bind to the antioxidants, enhancing their stability and bioavailability (Haque et al., 2024). Together, these components can work synergistically to inhibit lipid peroxidation, reduce rancidity, extend shelf life, stabilise colour, and improve water holding capacity, emulsion stability, and cooking yield.

Researchers continue to explore novel natural sources from FVW, extraction methods, and nanotechnology for targeted delivery. Incorporating antioxidants and dietary fibers extracted from FVW into fresh meat and meat items offers several benefits. Table 3.1 demonstrates the antioxidant and dietary fiber fortification of meat products with FVW from various sources.

3.7.2 Antimicrobial fortification of meat products with FVW

Various types of meat are particularly prone to contamination by pathogenic and spoilage microorganisms. This susceptibility is due to their rich nutritional profile and ample moisture, which provide an ideal environment for microbial growth. The rapid perishability of meat, requiring prompt consumption, demands physical and chemical methods to extend shelf life and enhance flavour. These techniques, applied to fresh meat and its products, include following good manufacturing practices, employing low-temperature storage, using heat treatments, and adding synthetic and natural preservatives.

FVW exhibits inherent notable antimicrobial properties, which can be harnessed to enhance the safety and shelf life of meat products. Reformulating meat products by substituting natural antimicrobial compounds from FVW can inhibit the growth of spoilage microorganisms and pathogens, thereby reducing the risk of contamination. This approach can offer nutria-dense meat-based products by a cleaner label, mitigating negative consumer perceptions regarding synthetic additives and their potential health risks.

The precise mechanisms underlying the antimicrobial activities of polyphenols and other natural compounds derived from FVW remain uncertain. However, several potential modes of action have been proposed, including destabilisation of the cytoplasmic membrane, inhibition of certain metabolic pathways and enzymes, or chelation of essential metals for growth, such as zinc and iron (Morsy et al., 2018).

Citrus fruit peels, such as those from oranges and lemons, have been shown to contain flavonoids and essential oils that have antibacterial effectiveness against a range of infections (Alsakhawy et al., 2022). Similary, the skins and seeds of berries, such as strawberries and blueberries, are abundant in polyphenols and anthocyanins, which have strong antibacterial properties. The outer layers of onions and the pomace from grapes and pomegranates also exhibit antibacterial properties. Table 3.2 illustrates the antimicrobial fortification of raw meat and various meat snacks using different FVW.

3.7.3 Fortification of Meat Products with Fat Substitutes from FVW

The reduction of fat content in meat and meat products is driven by increasing consumer demand for healthier food options and the associated health concerns related to high-fat diets. However, reducing fat in meat poses several challenges. Fat plays a crucial role in determining the texture, flavour, tenderness, and overall sensory appeal of meat products. Low-fat versions often suffer from a less appealing appearance, reduced flavour, and loss of mouthfeel. Additionally, fat contributes to the juiciness and overall eating experience of meat products, and its reduction can lead to a less satisfying product. To address these drawbacks, innovative solutions are needed. One effective strategy is FVW incorporation into meat product formulations. These plant-based additions can improve the texture and moisture retention of low-fat meat products, compensating for the loss of fat. Dietary fibers included in FVW can help reduce animal fat content in meat products by functioning as fat replacers. The fibrous quality of FVW helps to retain moisture and improve the texture of meat products, compensating for the lower fat level. These fibers form a matrix that mimics the mouthfeel and juiciness of animal lipids. Furthermore, the fibers can bind with water and other components, increasing the overall stability and cohesiveness of the meat product during cooking.

Research has explored using banana peel and pulp as fat substitutes in hamburger meat, comparing them to other fiber-rich foods like apple peel and oatmeal. The results showed that meat products with banana by-products had higher yields: 88.98% with green banana pulp, 87.35% with green banana seed flour, and 86.71% with green banana pulp flour (Stragliotto et al., 2022). These banana fibers improved the meat yield at a 3% addition rate without extra animal fat. Interestingly, compared to a control group with no added animal fat and another group with 9% fat, the meat with banana waste was easier to chop after cooking and shrank less during cooking (Stragliotto et al., 2022). Flours derived from certain vegetables and tubers, such as potato pulp flour and yellow pea flour, are rich in dietary fiber and starch. When incorporated into sausages and ham, they enhance water retention capacity, leading to improved texture in terms of cohesiveness, gumminess, and hardness (Kaack and Pedersen, 2005). Oat fiber, abundant in β-glucans, forms intricate three-dimensional networks and exhibits remarkable water and fat retention capacity (Liu et al., 2015). When incorporated into hamburger beef, 13.45% gel oat fiber content facilitated the reduction of water and fat loss during cooking. Simultaneously, it increased the burger's softness without compromising its sensory properties or shelf life (Pinero et al., 2008).

TABLE 3.1
Examples of meat products fortification with antioxidants and dietary fiber derived from FVW

FVW	Employed technique	Infusion level	Meat product	Key findings	References
Ash (*Benincasa hispida*) gourd peel	Powder: peel drying in a hot air oven at a 50°C, milling, and sieving	4%, 8%, and 12%	Buffalo meat sausage	Increase in TPC, antioxidant capacity, and crude fiber by about 62.5, 50, and 800% at the concentration of 12%.	Haque et al. (2024)
Sugar beet (*Beta vulgaris var. saccharifera* L.) molasses	Non-crystallised syrup and powder	Appropriate amount	Fermented Turkish sucuk	Strong antioxidant power of molasses, decreasing oxidation when they were used as the same dose of sodium nitrate (100 ppm).	Dilek and Karakaya (2022)
Mosambi peel	Powder: peel salting to remove the bitterness and drying by a tray drier at 65°C for 24 h, milling	2%, 4%, 6%, and 8%	Buffalo sausages and patties	Total dietary fiber content of enriched sausages and patties with 8% mosambi peel powder was about 16 times more than control.	Younis et al. (2021)
Date (*Phoenix dactylifera* L.) pits	Powder: pre-vacuum drying at 50°C, milling, sieving, and finish drying for 2 h	1.5%, 3%, and 6%	Beef burgers	A dose of 3% avoided lipid oxidation.	Sayas-Barberá et al. (2020)
Mexican oregano (*o Lippia berlandieri Schauer*) essential oil	Not mentioned	0.1%	Beef patty	3.55% increase in antioxidant capacity	Perales-Jasso et al. (2018)
Sage (*Salvia officinalis* L.) essential oil	Extraction by hydrodistillation and supercritical fluid extraction techniques	0.05, 0.075, and 0.1 μL/g	Pork sausage	Highest inhibitory potential (0.24 mg malondialdehyde/kg) against lipid oxidation in 0.1 μL/g	Šojić et al. (2018)
Olive leaves	Extract: leaves drying at 40°C, extraction by maceration method using ethanol 95%	500 ppm	Frankfurter-type sausage	26% reduction in lipid oxidation in comparison to the control	Alirezalu et al. (2017)

Japanese plum (*Prunus salicina*) peel	Extract: Freeze drying, extracting with ethanol	1.6%	Raw chicken breast patty	Approximately 46.8% decrease in lipid oxidation vs. the control	Basanta et al. (2018)
Jaboticaba (*Plinia jaboticaba*) peel	Milling, drying, extracting with ethanol at 95% for 1 h	0.25%, 0.5%, 0.75%, and 1%	Bolognatype sausages	Decreased lipid oxidation in a concentration-dependent manner. The maximum reduction was by 26.14%.	de Almeida et al. (2015)
Grape (*Bogazkere*) seed flour	Method not reported	0.5–5%	Vacuum-packed Frankfurters	Decreased lipid oxidation in a dose-dependent manner. Maximum decreased by 35.29%.	Özvural and Vural (2014)
Date (*Phoenix dactylifera* L.) palm	Thawing, blanching, pitting, peeling, and blending	10% and 15%	Campagnetype pork liver paté	Decrease by 73.77% in lipid oxidation vs. the control	Martín-Sánchez et al. (2013)
Onion peel	Drying for 8 h at 50°C in a hot air dryer, milling, extracting with 100 mL of 50% ethanol at 20°C	1% and 2%	Raw minced pork	Reduced lipid oxidation by 40% vs. control (TBARS), 4.95% (peroxide values) and 4% (conjugatedienes)	Shim et al. (2012)
Avocado (Persea americana var Hass and Fuerte) peel	Homogenization of peel with 30 mL of acetone/water (70:10), centrifugation at 2500 rpm at 4°C for 3 min and filtration	Hass: 60.8 ± 14.4 mg GAE/g $_{fm}$. Fuerte: 69.1 ± 16.9 mg GAE/g $_{fm}$.	Porcine patties	Hass: reduction by 76.9% in lipid oxidation and 30% in protein oxidation. Fuerte: Reduction by 92.3% in lipid oxidation and 20% in protein oxidation.	Rodríguez-Carpena et al. (2011)
Pomegranate (*Punica granatum*) rind	Powder: Dried at 60°C for 48 h and ground	2%	Ground goat meat patties	69% decrease in lipid oxidation vs. the control	Devatkal and Naveena (2010)
Pomegranate (*Punica granatum*) seed	Powder: Dried in a tray and ground	2%	Ground goat meat patties	80% decrease in lipid oxidation vs. the control	Devatkal and Naveena (2010)
Kinnow rind (*Citrus reticulate*)	Powder: Dried at 60°C for 48 h and ground	2%	Ground goat meat patties	49% decrease in lipid oxidation vs. the control	Devatkal and Naveena (2010)

(Continued)

TABLE 3.1 (Continued)

FVW	Employed technique	Infusion level	Meat product	Key findings	References
Pomegranate peel (*Punica granatum*)	Freeze-dried, ground, extracted with ethanol (40 g powder + 1 Lethanol) for 24 h, filtered and concentrated	100 mL with 20 g meat	Raw pork	Decreased lipid oxidation by 60% vs. the control	Shan et al. (2009)
Citrus peel (*Citrus unshiu*)	Frozen at –40°C, ground and extracted with 70% ethanol at 20°C for 72 h.	0.1%	Raw and cooked beef loin, pork loin, and salmon filet	Lipid oxidation decrease by 28% in raw beef, 11.7% in raw pork, 42% in, raw salmon, 11.1% in cooked pork, 82% in, cooked chicken, 37.5% in cooked salmon	Jo et al. (2004)

vs.: versus, fm.: fresh material, TPC: total polyphenol content.

TABLE 3.2
Examples of different meat products fortification with antimicrobials derived from FVW

FVW	Employed technique	Infusion level	Target meat product	Key findings	References
Ash (*Benincasa hispida*) gourd peel	Powder: drying in a hot air oven at a 50°C	4%, 8%, and 12%	Buffalo meat sausage	Total plate count, yeast and mold count reduced from 2.87 to 2.12 log CFU/g and 1.93 to 1.38 log CFU/g, respectively.	Haque et al. (2024)
Tomato (*Solanum lycopersicum* L.) pomace	Freeze drying at −50°C, milling,	0.5%, 1%, and 1.5%	Raw fermented sausages	The lowest Enterobacteriaceae counts (1.74 log CFU/g) were observed in higher amounts of pomace.	Skwarek and Karwowska (2022)
Olive leaf extract	Ventilated oven drying at 120°C for 8 min, milling, ultrasound-assisted extraction	200, 400, 800 mg/kg	Ripened pork sausages	Inhibitory effect against sulfite-reducing Clostridia, Coliforms, *Escherichia coli* and *Staphylococcus coagulase* positive.	Difonzo et al. (2022)
Mosambi peel powder	Salting to remove the bitterness and drying by a tray drier at 65°C for 24 h, milling	2%, 4%, 6%, and 8%	Buffalo sausages and patties	Thiobarbituric acid, yeast, mold and total plate count showed that the incorporation of Mosambi peel powder increased the shelf life of sausages and patties.	Younis et al. (2021)
Pomegranate peel nanoparticles (*Punica granatum* L.)	Ball-milling process of lyophilized pomegranate peel	1% and 1.5%	Raw beef meatballs on plastic trays sealed with wrapping film for 15 days at 4°C	On day 9, decreased by about 50% of total viable bacteria and psychrophilic and lipolytic bacteria.	Morsy et al. (2018)
Grape seed extract powder	Not reported	1%	Mutton slices	Overall psychrophilic counts reduced vs. control.	Reddy et al. (2013)
Sage (*Salvia officinalis* L.) essential oil	Extraction by hydrodistillation and supercritical fluid extraction techniques	0.05, 0.075, and 0.1 μL/g	Pork sausages	Reduced the number of aerobic mesophilic bacteria by about 7% at the concentration of 0.1 μL/g vs. control.	Šojić et al. (2018)

(Continued)

TABLE 3.2 (Continued)

FVW	Employed technique	Infusion level	Target meat product	Key findings	References
Pomegranate peel (*Punica granatum*)	Extracting with 80% methanol, 2nd extraction with 40% ethanol, 40% methanol and 20% water, and 3rd extraction with 80% methanol	7.5%	Meat paté inoculated with 4 log CFU/g of Listeria monocytogenes at 4, 7, and 12°C for 46 days	A 4 log decrease in the growth of Listeria at 4°C, 1 log at 7°C, and no changes at 12°C vs. the control.	Hayrapetyan et al. (2012)
Grape seed (*Vitis vinifera* L.)	Freeze drying, milling, extracting by ethanol (40 g powder+ 1 L ethanol) for 24 h, filtration and concentration	100 mL with 25 g meat	Raw pork meat in aerobic packaging at 20°C. Inoculated with 105 CFU of Listeria monocytogenes, S. aureus, and S. enterica	Reduced growth of Listeria by 17.5%, S. aureus by 14%, and Salmonella by 20% vs. the control.	Shan et al. (2009)
Pomegranate peel (*Punica granatum*)	Freeze drying, milling, extracting by ethanol (40 g powder+ 1 L ethanol) for 24 h, filtration and concentration	100 mL with 25 g meat	Raw pork in aerobic packaging at 20°C. Inoculated with 105 CFU of Listeria monocytogenes, S. aureus, and S. enterica	Reduced growth of Listeria by 20.5%, *S. aureus* by 17.5%, and Salmonella by 20% vs. the control.	Shan et al. (2009)

vs.: versus, fm.: fresh material.

Another intriguing fat substitute is the hazelnut husk, a by-product of the roasting process. With a substantial 59% dietary fiber content, it has been successfully used in beef burgers. The hazelnut husk effectively reduces cooking losses and prevents the decrease in meat diameter and size. However, its incorporation impacts juiciness, appearance, and overall acceptability, with effects varying based on the concentration used (Turhan et al., 2005). Table 3.3 summarizes the findings from lab-scale research on incorporating various FVW into different meat-based products as substitutes for animal fat.

3.7.4 Meat Analogs and Extenders

As alternatives to animal-based products, meat analogs and meat extenders are important components of many dietary patterns, particularly flexitarian, vegetarian, and vegan diets. These products satisfy dietary requirements and preferences while offering important proteins and nutrients. Meat substitutes are essential for vegetarian and vegan diets because they mimic the flavour, texture, and nutritional profile of meat.

They are produced from plant-based substances, such as soy, wheat gluten, and mushrooms. Meat extenders are advantageous to consumers who are health-conscious and flexitarians since they lessen the need for actual meat while maintaining flavour and texture. These products enhance environmental sustainability by encouraging plant-based consumption and lowering the carbon footprint associated with meat production. They also help consumers adopt healthier eating habits and consume less saturated fat.

Vegetable proteins are often considered meat analogs because they exhibit similar characteristics to meat proteins. These shared properties include water retention, gelation, and emulsification, among others. When incorporated into food products, vegetable proteins can yield a final product with sensory and texture attributes comparable to those of real meat. Various sources of vegetable proteins, such as cereal, legume, fungal, and algal proteins, exhibit functional properties like meat proteins. Specifically, they share similarities in terms of texture, gel formation, cutting force, and water retention (Jones, 2016).

One intriguing example is tomato pulp, an industrial by-product. Tomato pulp contains skin, insoluble solids, and seeds, along with a high protein content (ranging from 30% to 39%) and lipids (27%). Additionally, it contains other components like fiber and pectin. When tomato pulp residue is incorporated into meat-free sausages at different concentrations (1%, 3%, 5%, and 7%), several positive effects are observed, including an increase in yield by increasing the water absorption capacity of the system, brightness and yellow colour in a concentration-dependent manner, as well as the adhesiveness and hardness (Savadkoohi et al., 2014). It has good sensory acceptability and offers a red colour that is remarkably comparable to sausage meat, even at quantities lower than 5%. Burgers with a 12.9% reduction in raw beef and no changes to diameter, texture, or acceptance after cooking benefited from the addition of 0.5–1.5% oat fiber (1.31% protein, 0.06% fat) combined with pea fiber (5.43% protein, 0.71% fat) (Besbes et al., 2008).

In beef patties, incorporating a 1:1 mixture of dried pumpkin pulp and seed (with 25.92% protein and 16.45% fat) as a meat substitute at 5% increased the water retention capacity without changing texture or acceptability compared to the control with a full amount of meat (Serdaroğlu et al., 2018). Additionally, yield-enhancing extenders are non-meat ingredients with high protein or carbohydrate content. These extenders serve to lower the product cost, lower the meat content, improve the product's final sensory qualities, or minimize cooking-related changes (Dekkers et al., 2018).

Fiber and its derivatives are commonly used as extenders in food products. Lemon albedo (fiber) raises the water content of products it is added to, like fermented sausages, because of its high pectin content (25%) which boosts the product's final yield (Aleson-Carbonell et al., 2003). Aleson-Carbonell et al. (2005) found that if albedo is combined with other compounds that have a higher water retention capacity, the product's humidity rises and the cooking losses of water and fat in emulsified products, like sausages, are decreased.

3.7.5 Texture and colour fortification of meat products with FVW

A food product's texture and colour, particularly when it comes to meat snacks, are characteristics that indicate both its quality and the consumer's acceptability and

TABLE 3.3
Examples of different meat products fortification with substitutes of fat derived from FVW

FVW	Employed technique	Target meat product	Infusion level	Key findings	References
Artichoke outer and inner bracts, choke, and stem	Immersing in 0.5% citric acid solution, blanching at 100ºC for 3 min, drying at 50°C for 18 h in an electric air oven, milling	Beef sausage	25%, 50%, 100%	Improving cooking loss, cooking yield and shrinkage. Decreasing contents of total fat (maximally by 54.58%). Increased moisture and crude fiber	Abdel-Rahman et al. (2023)
Lemon albedo	Powder: albedo drying in a hot air oven at 50°C for 48 h, grounding into fine powder for 2–3 min	Chicken patties	50%	Lower fat and cholesterol content along with a significant increase in cooking yield, moisture content, fat retention and moisture retention values	Chappalwar et al. (2021)
Sunflower seed flour	Not reported	Pork frankfurters	50%	Improvements in the lipid content of protein, and minerals, colour darkening, increased cohesiveness, springiness, and chewiness	Grasso et al. (2020)
Tomato peel	Powder: Drying and crushing by conventional method to 0.5 mm	Beef and pork sausages	15%	Increased free water. Structure is denser and more compact. Overall sensory acceptability	Wang et al. (2017)
Pineapple peel and pomace	Thermal treatment: inactivation of bromelain (100°C, 2 h)	Beef burger	50%	In raw burgers, reduced pH, redness and lightness. In cooked burgers, reduced redness and cooking loss, and increased hardness	Selani et al. (2016)
Green banana flour	Dehydration (60°C for 12 h) of banana slices. Cooked pork skin (80°C for 60 min)	Bolognatype sausages	20–100%	Decreased cooking loss; increased emulsion stability, no changes in colour, sensory acceptability, texture parameters and microbiology stability, increased resistant starch	Alves et al. (2016)
Apple pomace fiber (*Malus domestica cv. Tsugaru*)	Dehydrated (55°C) and starch remotion (gelatinized-95°C), washed with hot water (100°C) and ethanol (preheated to 60°C) and dried at 55°C	Uncured, reduced fat chicken sausages	17–33%	Lower cooking loss, increased emulsion stability, increased a* colour value, and decreased pH value. Improved texture profile of sausages compared with fat reduction control	Choi et al. (2016)

tendency to purchase. The amount of fat, the quality and quantity of fiber or hydrocolloid used, the type of packaging, and/or conservation technologies, among other variables, all affect meat products' texture and colour. However, subjectivity in consumer acceptability needs to be objectively reflected; therefore, the instrumental evaluation of texture and colour changes has always been the main focus of the meat industry.

Enhancing meat products with FVW can significantly improve both their texture and colour. The dietary fibers inherent in FVW augment the texture by increasing water retention, thereby providing a meat-like mouthfeel, which is particularly advantageous in low-fat meat formulations. Furthermore, the natural pigments found in FVW, including anthocyanins and carotenoids, can enhance the colour of meat products, resulting in a more visually appealing appearance.

For instance, citrus fruit albedo, the white, fibrous portion of the peel, has a particular composition that may modify the colour and texture of meat products. As a result of the high fiber content in albedo, the meat retains more water, improving texture by making the meat more cohesive and juicy. This is particularly beneficial for low-fat meat formulations, which can otherwise be dry and less appealing. Also, citrus albedo contains natural pigments, such as carotenoids and flavonoids. These pigments not only contribute to the meat's antioxidant capabilities, which serve to increase its shelf life, but they also improve the visual appeal by changing the hue. Carotenoids, for example, can enhance the red and yellow hues in meat products, making them more visually appealing. Adding lemon albedo (fiber) has enhanced the cohesiveness and chewiness of both fermented and unfermented sausages (Aleson-Carbonell et al., 2003). Fiber from different citrus fruits has changed the luminosity of bologna sausages, regardless of concentration. It has changed the intensity of the red and yellow colours and probably reduced reactions of myoglobin and nitrosomyoglobin due to the carotenoid content and its antioxidant action (Fernández-Ginés et al., 2003; Viuda-Martos et al., 2010).

Adding cooked or dried lemon albedos to hamburger beef has also improved its ability to retain water and fat, increasing cooking yields as well as its dietary fiber content. However, this harmed other parameters, such as hardness and the gumminess of the finished product. The final texture of a meat product can vary depending on the amount of fiber added. For instance, citrus fiber concentrations below 1.5% (0.5% and 1%) increase hardness, while at 2%, hardness decreases, likely due to changes in the product's water retention capacity. Conversely, higher fiber concentrations lead to a decrease in elasticity, cohesion, and chewiness. Rodríguez-Carpena et al. (2011) found that adding avocado seed and peel extracts to raw and chilled pork burgers for 15 days decreases the product's loss of red colour and brightness. According to Jo et al. (2004), adding freeze-dried citrus peel extract to raw and cooked beef and pork has raised the red (a*) colour.

Conservation methods, such as radiation and altered atmospheres, commonly change the meat product's final colour characteristics. The addition of specific organic substances, such as carotenoids, prevents the modification of colour changes during this process.

Burgers that had been exposed to radiation retained their red colour (a*) when dried tomato peel was added to their formulation. Tomato waste, including peels and seeds, has huge potential to significantly enhance the colour and texture of different meat products due to its rich composition of bioactive compounds (Stajić et al.,

2024). Given the substantial amount of tomato waste produced globally, leveraging these by-products is sustainable and economical. The high levels of lycopene and β-carotene in tomato waste provide vibrant red pigments that naturally align with the red colour of meat products, enhancing their visual appeal. These natural pigments also help preserve the red colour during storage and processing, reducing the reliance on synthetic colour additives.

In terms of texture, the dietary fibers in tomato waste can improve water retention and binding properties in meat products, leading to a juicier and more cohesive texture, particularly beneficial for low-fat formulations. Additionally, these fibers help stabilize emulsions and enhance the firmness of meat products, resulting in a more desirable mouthfeel. The popularity of using tomato waste for meat product fortification stems from its effectiveness in improving both sensory qualities and nutritional value while promoting sustainability by repurposing a significant food industry by-product. As an example, incorporating tomato extracts into sheep meat through sprinkling has lessened the loss of red colour in meat after 7 days at 2°C (Andres et al., 2017). In contrast, the red colour of the sausage was maintained for 21 days by adding dehydrated tomato peel; however, this also increased the intensity of the yellow colour (b*) and lowered the luminosity of the product. The addition of tomato peel has preserved the red colour of meat products better than pulp due to its high content of lycopene and β-carotene. Hardness and cutting force increased, while cohesiveness decreased (Calvo et al., 2008). This phenomenon can be attributed to insoluble fiber, which can increase the hardness of the products.

3.8 CASE OF STUDIES

3.8.1 Overview

Existing studies have primarily focused on the impact of incorporating FVW into different meat products from a biochemical, textural, and sensory perspective. These investigations explore aspects, such as antioxidant and antimicrobial properties, shelf life extension, and consumer acceptance. However, detailed information on large-scale implementation and cost analysis is relatively scarce. Implementing FVW in meat processing at an industrial scale involves complex logistics, supply chain considerations, and regulatory compliance. The variability in FVW composition (depending on the specific fruits and vegetables) adds further challenges. Regarding the cost aspect, incorporating FVW into meat products can be cost-effective. By utilising by-products from fruit and vegetable processing industries, meat quality features (such as physicochemical properties, microbial safety, and sensory attributes) can be enhanced while providing health benefits. However, detailed cost assessments specific to these enriched meat products are not widely available in the literature. This overview emphasizes the need for more real-world case studies.

Despite the growing awareness of sustainability and the need to reduce food waste, there remains a significant lack of companies that incorporate FVW into their meat products. While some pioneering companies in Europe and Asia have begun to integrate fruits, vegetables, and herbs into their meat products, these examples are still relatively rare. This highlights the potential for further innovation and expansion in the industry to embrace more sustainable practices and enhance the nutritional value of meat products through the valorisation of food waste.

3.8.2 Incorporating fruit and vegetables for enhanced sustainability in meat products: the Rügenwalder Mühle model

Rügenwalder Mühle (https://www.ruegenwalder.de/de), based in Bad Zwischenahn, Germany, is a renowned food company celebrated for its innovative approach to meat product preparation. The company emphasizes sustainability and the integration of plant-based ingredients into their offerings. They incorporate a variety of fruits and vegetables, such as apples, onions, chives, mushrooms, and olives, to enhance the flavour, texture, nutrition, and sustainability of their products. For instance, Rügenwalder Mühle addresses the fortification of pork spreadable sausage by adding apples and onions to their Pomeranian Liver Sausage. This product consists of 40% pork, 30% pork liver, 9% onions, and 6% apples, along with spices, spice extracts, antioxidants, emulsifiers, and preservatives. The inclusion of apples and onions not only enhances the taste with a balance of sweetness, tartness, and savory notes but also boosts the nutritional profile by adding dietary fiber, vitamins, and antioxidants.

Valorizing apple and onion peels into this product can significantly reduce waste and provide economic benefits. Apple and onion peels are rich in bioactive compounds, such as polyphenols, flavonoids, and dietary fiber, which have numerous health benefits, including antioxidant, anti-inflammatory, and antimicrobial properties. By utilising these peels, the company can create value-added products, reducing the environmental impact of food waste. This approach can lower waste disposal costs and generate revenue from high-value products like dietary supplements and natural preservatives. Additionally, it aligns with sustainable practices and enhances the nutritional profile of food products, offering additional health benefits to consumers.

In another attempt, Rügenwalder Mühle produces Pomeranian Liver Sausage Chives with high-quality pork liver (46%), pork meat (27%), and fresh, mild flavour chives (6%), creating a smooth and spreadable sausage. Chive addition into sausage enhances the taste by providing a mild onion-like taste and freshness and also boosts nutritional value, providing essential proteins, fats, iron, and vitamins, such as A and B12.

The company's products align with Sustainable Development Goals (SDGs). These processes support SDG 12 (Responsible Consumption and Production) and SDG 3 (Good Health and Well-being) by promoting sustainability and providing health benefits. This case study highlights Rügenwalder Mühle's commitment to sustainability and innovation in food production.

3.8.3 Innovative use of fruits, vegetables, and nuts in meat products: the DEMES Company approach

DEMES Company (http://www.demes.ir), established in Shiraz, Iran, is a leading manufacturer of a wide range of high-quality meat products, including sausages, bologna, kebabs, and hamburgers. Over the years, DEMES has grown significantly since its inception, expanding its product range to include both red and white meats, such as beef, sheep, chicken, and turkey. This company has also expanded its product line to fortify its meat products, including various fruits, vegetables, and herbs, such as carrots, mushrooms, olives, onions, garlic, parsley, and coriander.

In one of the chicken bologna sausages, DEMES incorporates 60% chicken, 13% mushrooms, 13% carrots, soya, ascorbic acid, flavourings, and other permitted additives in the product formulation (Figure 3.9). The mushrooms are blanched, pasteurized, and chopped before being added, while individually quick-frozen (IQF) carrot slices are included in the mix. Adding mushrooms and carrots to meat products enhances their nutritional value by providing essential vitamins, minerals, and antioxidants. Mushrooms contribute a savory umami flavour and a meaty texture, while carrots add a subtle sweetness and moisture, improving the overall taste and juiciness. These vegetables are low in calories and high in fiber, supporting digestive health and weight management.

In cattle meat bologna sausages, 55% meat, 15% olive, and 5% carrot are considered as the main ingredients (Figure 3.10). Olive is rich in healthy fats, particularly monounsaturated fats and improves the nutritional profile of bologna. It adds a rich, savory, and slightly tangy flavour, which can complement the meat's taste. The natural oils in olives also contribute to a smoother texture. Furthermore, the inclusion of olives adds a pleasant, slightly chewy texture, enhancing the overall mouthfeel of the product. The vibrant

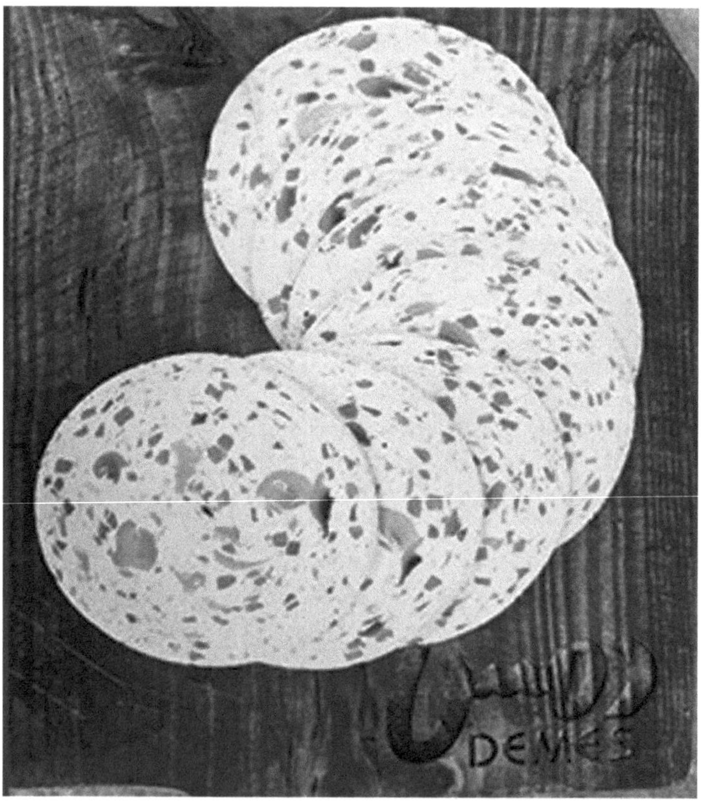

FIGURE 3.9 Incorporating finely chopped mushrooms and carrots into chicken bologna sausage at DEMES Company.

Source: **http://www.demes.ir.**

FIGURE 3.10 DEMES bologna sausage enriched with olive and carrot.

Source: http://www.demes.ir.

colours of olives and carrots can make the bologna more visually appealing, which can be particularly attractive in deli displays or when served as part of a charcuterie board.

DEMES Company has introduced an innovative idea of adding pistachios to their meat products (Figure 3.11). The incorporation of pistachios into meat products represents a novel approach to enhancing both the sensory and nutritional qualities of meat products. Pistachios contribute a distinctive nutty flavour that compliments the savory profile of meat, while their rich content of healthy fats, proteins, fibers, and essential vitamins and minerals, such as vitamin B6 and phosphorus, significantly boosts the nutritional value of the final product. Additionally, the inclusion of pistachios improves the texture by adding a pleasant crunch and enhances visual appeal with their vibrant green colour, making the products more attractive to consumers. This innovative addition differentiates DEMES Company's offerings in a competitive market and aligns with the growing consumer demand for gourmet and health-conscious food options.

FIGURE 3.11 DEMES meat products with pistachios.

Source: **http://www.demes.ir.**

The engineering solutions at DEMES involve the meticulous preparation and integration of these plant-based ingredients into their meat products, resulting in a balanced and flavourful final product. Incorporating vegetables, such as mushrooms and carrots, can reduce the overall cost of meat products by extending the volume without compromising on quality. The company's dedication to sustainability and innovation exemplifies how traditional meat products can be improved with up-cycled FVW.

3.9 CONCLUSION

The food waste components are bioactive substances with immense potential. It can be used to enhance the nutritional and functional qualities of newly developed meat products. Incorporating various forms of FVW, whether in powder or extract form, can profoundly enrich these meat products. These agro-industrial by-products are a treasure trove of dietary fiber, antioxidants, and health-promoting compounds. Their inclusion not only improves sensory attributes and shelf life but also addresses critical health concerns. Meat products, often rich in fat (especially saturated fat) and lacking in dietary fiber, can contribute to gastrointestinal and cardiovascular disorders. Herein lies the promise of natural preservatives derived from FVW, peels, pomace, seeds, and more. These natural alternatives can be used to replace or at least reduce reliance on synthetic additives. The ecological footprint associated with food-grade waste disposal is mitigated and consumer apprehensions about artificial additives and health implications are addressed. Industrial-scale fortification of meat products with FVW still lacks comprehensive information. While the economic incentive of utilising low-cost FVW-derived products is promising, stakeholders in the meat processing industry must ensure quality and safety standards.

EXERCISES

Question 1

a) What are the optimal extraction methods for bioactive compounds from fruit and vegetable waste for meat product fortification?
b) How do different extraction techniques (e.g., solvent-based, enzymatic) impact compound yield?
c) Is extraction efficiency enhanced by considering waste particle size and composition?
d) What is the stability of extracted compounds during processing and storage?

Questions 2

a) What are the potential benefits of fortifying different meat products with micronutrients, considering both nutritional enhancement and public health impact?"
b) How is the fortification process optimized to ensure uniform distribution of micronutrients in different meat matrices?
c) What specific micronutrients (e.g., iron, vitamin B12, omega-3 fatty acids) are most relevant for fortification in meat products?
d) How do sensory attributes (taste, texture) and stability (shelf life) of fortified meat products vary based on the type of meat and fortification strategy?

Question 3

a) What sensory attributes are impacted by fruit and vegetable waste fortification in meat products, and how can we optimize acceptability?"
b) How does waste inclusion affect colour, flavour, and texture?
c) Can formulation adjustments (e.g., masking agents, flavour enhancers) improve sensory profiles?
d) What consumer preferences and cultural factors influence acceptance?

Question 4

a) How can sustainable packaging solutions complement fruit and vegetable waste fortification in meat products?"
b) What waste-derived materials (e.g., biopolymers, edible coatings) can enhance packaging functionality?

c) How do these materials impact product safety, freshness, and environmental impact?
d) Can waste-based packaging align with circular economy principles?

REFERENCES

Abdel-Rahman, A. M., Ahmed, R. A-K. "Utilization of Artichoke Processing Wastes as fat replacer in beef sausage." *Egyptian Journal of Food Science* 51, no. 1 (2023): 13–31.

Aleson-Carbonell, L., J. Fernández-López, E. Sayas-Barberá, E. Sendra, and J. A. Pérez-Alvarez. "Utilization of lemon albedo in drycured sausages." *Journal of Food Science* 68, no. 5 (2003): 1826–1830.

Aleson-Carbonell, L., J. Fernández-López, J. A. Pérez-Alvarez, and V. Kuri. "Characteristics of beef burger as influenced by various types of lemon albedo." *Innovative Food Science & Emerging Technologies* 6, no. 2 (2005): 247–255.

Alirezalu Kazem, Javad Hesari, Mohammad Hadi Eskandari, Hadi Valizadeh, and Mohammad Sirousazar. "Effect of green tea, stinging nettle and olive leaves extracts on the quality and shelf life stability of frankfurter type sausage." *Journal of Food Processing and Preservation* 41, no. 5 (2017): e13100.

Alsakhawy, Sara. A, Baghdadi, Hoda H, El-Shenawy, Moustafa A, and El-Hosseiny, Lobna S. "Antibacterial activity of silver nanoparticles phytosynthesized by citrus fruit peel extracts". BioNanoScience 12, no. 4 (2022): 1106–1115.

Alves dos Santos, Larissa Aparecida Agostinho, José Manuel Lorenzo, Carlos Antonio Alvarenga Gonçalves, Bibiana Alves Dos Santos, Rosane Teresinha Heck, Alexandre José Cichoski, and Paulo Cezar Bastianello Campagnol. "Production of healthier bologna type sausages using pork skin and green banana flour as a fat replacers." *Meat Science* 121 (2016): 73–78.

Andres, A. I., M. J. Petron, J. Delgado-Adamez, M. Lopez, and M. Timon. "Effect of tomato pomace extracts on the shelf-life of modified atmosphere-packaged lamb meat." *Journal of Food Processing and Preservation* 41, no. 4 (2017): e13018.

Arjona, R., and P. Ollero. "Automation of an olive waste industrial rotary dryer." *Journal of Food Engineering* 68, no. 2 (2005): 239–247.

Bagherian Homa, Farzin Zokaee Ashtiani, Amir Fouladitajar, and Mahdy Mohtashamy. "Comparisons between conventional, microwave-and ultrasound-assisted methods for extraction of pectin from grapefruit." *Chemical Engineering and Processing: Process Intensification* 50, no. 11–12 (2011): 1237–1243.

Basanta Maria Florencia, Sergio Anibal Rizzo, Natalia Szerman, Sergio Ramón Vaudagna, Adriana Maria Descalzo, Lia Noemi Gerschenson, Carolina Daiana Pérez, and Ana Maria Rojas. "Plum (Prunus salicina) peel and pulp microparticles as natural antioxidant additives in breast chicken patties." *Food Research International* 106 (2018): 1086–1094.

Besbes Souhail, Hamadi Attia, Claude Deroanne, Sakander Makni, and Christophe Blecker. "Partial replacement of meat by pea fiber and wheat fiber: effect on the chemical composition, cooking characteristics and sensory properties of beef burgers." *Journal of Food Quality* 31, no. 4 (2008): 480–489.

Bhardwaj Kanchan, Agnieszka Najda, Ruchi Sharma, Renata Nurzyńska-Wierdak, Daljeet Singh Dhanjal, Rohit Sharma, Sivakumar Manickam, Atul Kabra, Kamil Kuča, and Prerna Bhardwaj. "Fruit and vegetable peel-enriched functional foods: potential avenues and health perspectives." *Evidence-Based Complementary and Alternative Medicine* 2022, no. 1 (2022): 8543881.

Bitwell Chibuye, Singh Sen Indra, Chimuka Luke, and Maseka Kenneth Kakoma. "A review of modern and conventional extraction techniques and their applications for extracting phytochemicals from plants." *Scientific African* 19 (2023): e01585.

Calderon-Oliver, Mariel, and Luis Humberto Lopez-Hernandez. "Food vegetable and fruit waste used in meat products." *Food Reviews International* 38, no. 4 (2022): 628–654.

Calvo, M. M., M. Luisa García, and M. Dolores Selgas. "Dry fermented sausages enriched with lycopene from tomato peel." *Meat Science* 80, no. 2 (2008): 167–172.

Chappalwar Anita M., Vikas Pathak, Meena Goswami, Arun Kumar Verma, and Vincent Rajkumar. "Efficacy of lemon albedo as fat replacer for development of ultra-low-fat chicken patties." *Journal of Food Processing and Preservation* 45, no. 7 (2021): e15587.

Choi Yun-Sang, Young-Boong Kim, Ko-Eun Hwang, Dong-Heon Song, Youn-Kyung Ham, Hyun-Wook Kim, Jung-Min Sung, and Cheon-Jei Kim. "Effect of apple pomace fiber and pork fat levels on quality characteristics of uncured, reduced-fat chicken sausages." *Poultry Science* 95, no. 6 (2016): 1465–1471.

de Almeida Patrícia Leal, Silvério Nepomuceno de Lima, Luciene Lacerda Costa, Cintia Cristina de Oliveira, Karina Aparecida Damasceno, Bibiana Alves dos Santos, and Paulo Cezar Bastianello Campagnol. "Effect of jabuticaba peel extract on lipid oxidation, microbial stability and sensory properties of Bologna-type sausages during refrigerated storage." *Meat Science* 110 (2015): 9–14.

Dekkers Birgit L., Remko M. Boom, and Atze Jan van der Goot. "Structuring processes for meat analogues." *Trends in Food Science & Technology* 81 (2018): 25–36.

Devatkal Suresh K., and B. M. Naveena. "Effect of salt, kinnow and pomegranate fruit by-product powders on colour and oxidative stability of raw ground goat meat during refrigerated storage." *Meat Science* 85, no. 2 (2010): 306–311.

Difonzo Graziana, Michela Pia Totaro, Francesco Caponio, Antonella Pasqualone, and Carmine Summo. "Olive leaf extract (OLE) addition as tool to reduce nitrate and nitrite in Ripened Sausages." *Foods* 11, no. 3 (2022): 451.

Dilek Nazik Meziyet, and Mustafa Karakaya. "Natural alternative curing agent in fermented sucuk production: sugar beet (Beta vulgaris var. saccharifera L.) molasses." *Selcuk Journal of Agriculture and Food Sciences* 36, no. 3 (2022): 380–386.

Domínguez Rubén, Mirian Pateiro, Mohammed Gagaoua, Francisco J. Barba, Wangang Zhang, and José M. Lorenzo. "A comprehensive review on lipid oxidation in meat and meat products." *Antioxidants* 8, no. 10 (2019): 429.

Domínguez Rubén, Patricia Gullón, Mirian Pateiro, Paulo ES Munekata, Wangang Zhang, and José Manuel Lorenzo. "Tomato as potential source of natural additives for meat industry. A review." *Antioxidants* 9, no. 1 (2020): 73.

Dorta Eva, M. Gloria Lobo, and Mónica González. "Improving the efficiency of antioxidant extraction from mango peel by using microwave-assisted extraction." *Plant Foods for Human Nutrition* 68 (2013): 190–199.

Ebrahim M. S. "Plant-based bioactive compounds: a comprehensive review of conventional and novel extraction techniques." *The Pharma Innovation Journal* 12, no. 2 (2023): 2609–2615.

ElGamal Ramadan, Song, Cheng, Rayan, Ahmed M., Liu, Chuanping, Al-Rejaie, Salim, and ElMasry, Gamal. (2023). Thermal degradation of bioactive compounds during drying process of horticultural and agronomic products: A comprehensive overview. Agronomy, 13(6), 1580.

Falowo Andrew B., Peter O. Fayemi, and Voster Muchenje. "Natural antioxidants against lipid–protein oxidative deterioration in meat and meat products: a review." *Food Research International* 64 (2014): 171–181.

Fernández-Ginés, J. M., J. Fernández-López, E. Sayas-Barberá, E. Sendra, and J. A. Pérez-Alvarez. "Effect of storage conditions on quality characteristics of bologna sausages made with citrus fiber." *Journal of Food Science* 68, no. 2 (2003): 710–714.

Gan Andrew, and Saeid Baroutian. "Current status and trends in extraction of bioactives from brown macroalgae using supercritical CO_2 and subcritical water." *Journal of Chemical Technology & Biotechnology* 97, no. 8 (2022): 1929–1940.

Gaur Ruchi, Aparna Sharma, S. K. Khare, and Munishwar Nath Gupta. "A novel process for extraction of edible oils: enzyme assisted three phase partitioning (EATPP)." *Bioresource Technology* 98, no. 3 (2007): 696–699.

Grasso, S., T. Pintado, J. Pérez-Jiménez, C. Ruiz-Capillas, and A. M. Herrero. "Potential of a sunflower seed by-product as animal fat replacer in healthier frankfurters." *Foods* 9, no. 4 (2020), 445.

Haque Abdul, Saghir Ahmad, Mohd Adnan, Mohammad Idreesh Khan, Syed Amir Ashraf, and Z. R. A. A. Azad. "Fortification of conventional Buffalo meat sausage with ash gourd peel enhances shelf life, nutritional, functional and microstructural characteristics." *NFS Journal* 35 (2024): 100179.

Hasan Md Mehedi, Md Rakibul Islam, Ahmed Redwan Haque, Md Raihan Kabir, Khursheda Jahan Khushe, and SM Kamrul Hasan. "Trends and challenges of fruit by-products utilization: insights into safety, sensory, and benefits of the use for the development of innovative healthy food: a review." *Bioresources and Bioprocessing* 11, no. 1 (2024): 10.

Hassas-Roudsari Majid, Peter R. Chang, Ronald B. Pegg, and Robert T. Tyler. "Antioxidant capacity of bioactives extracted from canola meal by subcritical water, ethanolic and hot water extraction." *Food Chemistry* 114, no. 2 (2009): 717–726.

Hayrapetyan Hasmik, Wilma C. Hazeleger, and Rijkelt R. Beumer. "Inhibition of Listeria monocytogenes by pomegranate (Punica granatum) peel extract in meat paté at different temperatures." *Food Control* 23, no. 1 (2012): 66–72.

Jo, C., H. J. Kang, M. Lee, N. Y. Lee, and M. W. Byun. "The antioxidative potential of lyophilized citrus peel extract in different meat model systems during storage at 20C." *Journal of Muscle Foods* 15, no. 2 (2004): 95–107.

Jones Owen Griffith. "Recent advances in the functionality of non-animal-sourced proteins contributing to their use in meat analogs." *Current Opinion in Food Science* 7 (2016): 7–13.

Kaack Karl, and Lene Pedersen. "Application of by-products from industrial processing of potato flour and yellow peas as ingredients in low-fat high-fibre sausages." *European Food Research and Technology* 221 (2005): 313–319.

Kowalczewski Przemysław Łukasz, and Joanna Zembrzuska. "Advances in biological activities and application of plant extracts." *Applied Sciences* 13, no. 16 (2023): 9324.

Liu Rui, Nan Wang, Qian Li, and Min Zhang. "Comparative studies on physicochemical properties of raw and hydrolyzed oat β-glucan and their application in low-fat meatballs." *Food Hydrocolloids* 51 (2015): 424–431.

Manessis Georgios, Aphrodite I. Kalogianni, Thomai Lazou, Marios Moschovas, Ioannis Bossis, and Athanasios I. Gelasakis. "Plant-derived natural antioxidants in meat and meat products." *Antioxidants* 9, no. 12 (2020): 1215.

Manousi Natalia, Ioannis Sarakatsianos, and Victoria Samanidou. "Extraction techniques of phenolic compounds and other bioactive compounds from medicinal and aromatic plants." In Alexandru Mihai Grumezescu and Alina Maria Holban (Eds.), *Engineering tools in the beverage industry*, pp. 283–314. Woodhead Publishing, 2019."

Martín-Sánchez Ana María, Gelmy Ciro-Gómez, Estrella Sayas, José Vilella-Esplá, Jamel Ben-Abda, and José Ángel Pérez-Álvarez. "Date palm by-products as a new ingredient for the meat industry: application to pork liver pâté." *Meat Science* 93, no. 4 (2013): 880–887.

Mehta Nitin, Pavan Kumar, Akhilesh Kumar Verma, Pramila Umaraw, Sunil Kumar Khatkar, Anju Boora Khatkar, Devendra Pathak, Ubedullah Kaka, and Awis Qurni Sazili. "Ultrasound-assisted extraction and the encapsulation of bioactive components for food applications." *Foods* 11, no. 19 (2022): 2973.

Monrad Jeana K., Luke R. Howard, Jerry W. King, Keerthi Srinivas, and Andy Mauromoustakos. "Subcritical solvent extraction of anthocyanins from dried red grape pomace." *Journal of Agricultural and Food Chemistry* 58, no. 5 (2010): 2862–2868.

Morsy Mohamed K., Enas Mekawi, and Rasha Elsabagh. "Impact of pomegranate peel nanoparticles on quality attributes of meatballs during refrigerated storage." *LWT* 89 (2018): 489–495.

Özvural Emin Burçin, and Halil Vural. "Which is the best grape seed additive for frankfurters: extract, oil or flour?" *Journal of the Science of Food and Agriculture* 94, no. 4 (2014): 792–797.

Perales-Jasso Yessica J., Stephannie A. Gamez-Noyola, Juana Aranda-Ruiz, Carlos A. Hernandez-Martinez, Guadalupe Gutierrez-Soto, Alejandro I. Luna-Maldonado, Ramon Silva-Vazquez, Michael E. Hume, and Gerardo Mendez-Zamora. "Oregano powder substitution and shelf life in pork chorizo using Mexican oregano essential oil." *Food Science & Nutrition* 6, no. 5 (2018): 1254–1260.

Pinero, M. P., K. Parra, N. Huerta-Leidenz, L. Arenas De Moreno, M. Ferrer, S. Araujo, and Y. Barboza. "Effect of oat's soluble fibre (β-glucan) as a fat replacer on physical, chemical, microbiological and sensory properties of low-fat beef patties." *Meat Science* 80, no. 3 (2008): 675–680.

Qin, Xuze, Zhang, Ting, Cao, Yuxin, Deng, Buhao, Zhang, Jianxin, and Zhao, Junxing. "Effects of dietary sea buckthorn pomace supplementation on skeletal muscle mass and meat quality in lambs." Meat science 166, (2020): 108141.

Reddy, G. V., Bhaskar, A. R. Sen, Pramod N. Nair, K. Sudhakar Reddy, K. Kondal Reddy, and N. Kondaiah. "Effects of grape seed extract on the oxidative and microbial stability of restructured mutton slices." *Meat Science* 95, no. 2 (2013): 288–294.

Rodríguez-Carpena Javier German, D. A. V. I. D. Morcuende, and M. A. R. I. O. Estévez. "Avocado by-products as inhibitors of colour deterioration and lipid and protein oxidation in raw porcine patties subjected to chilled storage." *Meat Science* 89, no. 2 (2011): 166–173.

Sarangi Prakash Kumar, Sonil Nanda, and Pravakar Mohanty, eds. *Recent advancements in biofuels and bioenergy utilization*. Vol. 232. Springer, 2018.

Savadkoohi Sobhan, Henk Hoogenkamp, Kambiz Shamsi, and Asgar Farahnaky. "Colour, sensory and textural attributes of beef frankfurter, beef ham and meat-free sausage containing tomato pomace." *Meat Science* 97, no. 4 (2014): 410–418.

Sayas-Barberá Estrella, Ana María Martín-Sánchez, Sarra Cherif, Jamel Ben-Abda, and José Ángel Pérez-Álvarez. "Effect of date (Phoenix dactylifera L.) pits on the shelf life of beef burgers." *Foods* 9, no. 1 (2020): 102.

Selani Miriam M., Giovanna AN Shirado, Gregório B. Margiotta, Erick Saldaña, Fernanda P. Spada, Sonia MS Piedade, Carmen J. Contreras-Castillo, and Solange G. Canniatti-Brazaca. "Effects of pineapple byproduct and canola oil as fat replacers on physicochemical and sensory qualities of low-fat beef burger." *Meat Science* 112 (2016): 69–76.

Serdaroğlu, M., H. S. Kavuşan, G. A. M. Z. E. İpek, and B. U. R. C. U. Öztürk. "Evaluation of the quality of beef patties formulated with dried pumpkin pulp and seed." *Korean Journal for Food Science of Animal Resources* 38, no. 1 (2018): 1.

Shan Bin, Yi-Zhong Cai, John D. Brooks, and Harold Corke. "Antibacterial and antioxidant effects of five spice and herb extracts as natural preservatives of raw pork." *Journal of the Science of Food and Agriculture* 89, no. 11 (2009): 1879–1885.

Shim So-Yeon, Yun-Sang Choi, Hack-Youn Kim, Hyun-Wook Kim, Ko-Eun Hwang, Dong-Heon Song, Mi-Ai Lee, Ju-Woon Lee, and Cheon-Jei Kim. "Antioxidative properties of onion peel extracts against lipid oxidation in raw ground pork." *Food Science and Biotechnology* 21 (2012): 565–572.

Skwarek Patrycja, and Małgorzata Karwowska. "Fatty acids profile and antioxidant properties of raw fermented sausages with the addition of tomato pomace." *Biomolecules* 12, no. 11 (2022): 1695.

Skwarek Patrycja, and Małgorzata Karwowska. "Fruit and vegetable processing by-products as functional meat product ingredients-a chance to improve the nutritional value." *LWT* 189 (2023): 115442.

Šojić Branislav, Branimir Pavlić, Zoran Zeković, Vladimir Tomović, Predrag Ikonić, Sunčica Kocić-Tanackov, and Natalija Džinić. "The effect of essential oil and extract from sage (Salvia officinalis L.) herbal dust (food industry by-product) on the oxidative and microbiological stability of fresh pork sausages." *LWT* 89 (2018): 749–755.

Stajić Slaviša, Patrycja Skwarek, Sanja Đurđević, Małgorzata Karwowska, Boris Pisinov, Igor Tomasevic, and Vladimir Kurćubić. "Tomato pomace powder as a functional ingredient in minced meat products—influence on technological and sensory properties of traditional Serbian minced meat product Ćevapi." *Processes* 12, no. 7 (2024): 1330.

Stragliotto, Leticia Kahler, Ferrari, Gabriel Tonin, Campagnol, Paulo Cezar Bastianello, Strasburg, Virgílio José, Zandonadi, Renata Puppin, and de Oliveira, Viviani Ruffo, "Green banana by-products on the chemical, technological and sensory quality of meat products". International Journal of Gastronomy and Food Science 30, (2022): 100614.

Tapia María S., Stella M. Alzamora, and Jorge Chirife. "Effects of water activity (aw) on microbial stability as a hurdle in food preservation." *Water Activity in Foods: Fundamentals and Applications* 60 (2020): 323–355.

Trigo João P., Elisabete MC Alexandre, Jorge A. Saraiva, and Manuela E. Pintado. "High value-added compounds from fruit and vegetable by-products–characterization, bioactivities, and application in the development of novel food products." *Critical Reviews in Food Science and Nutrition* 60, no. 8 (2022): 1388–1416.

Turhan Sadettin, Inci Sagir, and N. Sule Ustun. "Utilization of hazelnut pellicle in low-fat beef burgers." *Meat Science* 71, no. 2 (2005): 312–316.

Viuda-Martos, M., Y. Ruiz-Navajas, J. Fernández-López, and J. A. Pérez-Álvarez. "bologna-type sausage." *Meat Science* 85, no. 3 (2010): 568–576.

Wang Qiang, Hongbin Wu, Yuejie Xie, Haijun Chang, Xue Li, Chengjiang Liu, and Zhengwei Xiong. "Effects of tomato peel as fat replacement on the texture, moisture migration, and sensory quality of sausages with varied fat levels." *CyTA-Journal of Food* 15, no. 4 (2017): 582–591.

Wijngaard Hilde, Mohammad B. Hossain, Dilip K. Rai, and Nigel Brunton. "Techniques to extract bioactive compounds from food by-products of plant origin." *Food Research International* 46, no. 2 (2012): 505–513.

Younis Kaiser, Saghir Ahmad, and Mudasir Ahmad Malik. "Mosambi peel powder incorporation in meat products: Effect on physicochemical properties and shelf life stability." *Applied Food Research* 1, no. 2 (2021): 100015.

Zhu Yingdan, Yueting Luan, Yingnan Zhao, Jiali Liu, Zhangqun Duan, and Roger Ruan. "Current technologies and uses for fruit and vegetable wastes in a sustainable system: a review." *Foods* 12, no. 10 (2023): 1949.

4 Valorisation of food waste into bioactive compounds

*Haripriya Ravikumar, Lee Hong Tee,
Yin Hui Chow, Soorya Haridas, Athira Mundassery,
and Janci Rani Ramaswamy*

4.1 INTRODUCTION

Food waste (FW) can be categorized into two major groups which are plant-based and animal-based FW, including expired food. This chapter will explore how FW can be transformed into valuable bioactive compounds. This approach addresses the challenge of FW and opens opportunities to extract high-value compounds for use in various industries, including food, nutraceuticals, pharmaceuticals, and cosmetics. Agriculture or plant-based FW refers to any edible parts of plants discarded or wasted at any stage of the food supply chain, from production and processing to consumption. This includes fruits, vegetables, grains, legumes, nuts, seeds, and other plant-derived foods that are often wasted due to various factors. Plant-based FW contributes significantly to overall FW globally. One of the significant contributors to agricultural FW is fruit and vegetable trimming, such as peelings, stems, cores, and seeds, which are frequently discarded despite containing valuable bioactive compounds with considerable nutritional benefits. For example, citrus peels, including orange, lemon, and lime are often rich in flavonoids, essential oil, limonene, and vitamin C. In addition, grape seeds and skins are known to possess bioactive compounds such as resveratrol, tannins, flavonoids, and phenolic acid. Further to this, vegetables waste such as tomato skins contain significant amounts of lycopene, polyphenols, and dietary fiber. Another prevalent source of plant-based FW is the unsold products in the retail due to excess inventory, issues with storage, cosmetic imperfections, or minor flaws as well as size discrepancies. Spoiled or expired products further contributed to the FW when fruits and vegetables exceeded their optimal freshness due to improper storage and handling and thus were deemed no longer suitable for consumption.

In industrial beverage processing, the peels and pulps discarded after juice extraction are often rich in nutrients and fibers, presenting an untapped potential for valorisation into bioactive compounds. During the industry process, by-products generated from the canning, trimming, freezing, and drying process of the fruit and vegetables offer a wealth of bioactive compounds that can be harnessed for various

DOI: 10.1201/9781032718507-4

applications. For example, guava pomace generated from the guava juice industry is a key source of bioactive compounds such as phenolic compounds, lycopene, and dietary fibers. Field losses, another major source of plant-based FW, occur when crops are left unharvested or damaged due to pests, adverse weather, mechanical harvesting issues, or overproduction.

Some of the avoidable FW, especially peel, skin, and twigs of fruits, contain a greater number of bioactive compounds than the edible portions do. In addition, it is also reported that food processing waste is often rich in protein, lipids, and carbohydrates, thus exhibiting valorisation potential in producing animal feed, cosmetics, chemicals, prebiotics as well as valuable nutraceuticals. Conversion of FW or by-products or undervalued biomass into value with the effective application is known as valorisation of the waste.

4.2 METHODS FOR RECOVERY OF BIOACTIVE COMPOUNDS FROM FVW

Recovering bioactive compounds from food and vegetable waste involves three major stages: pre-treatment, extraction, and stabilisation/encapsulation. Each stage is crucial to maximizing the yield of the compounds extracted and preserving the integrity and functionality of the bioactive compounds. The valorisation of fruit and vegetable waste (FVW) through the three major stages is illustrated in the flowchart in Figure 4.1.

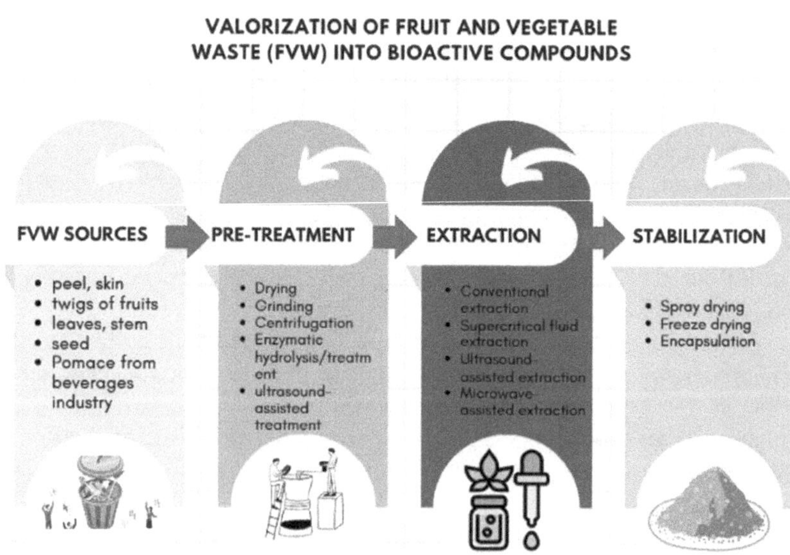

FIGURE 4.1 Valorisation of fruit and vegetable waste (FVW) into bioactive compounds.

4.2.1 Pre-treatment of the FVW

The first step of the valorisation of the FVW is a pre-treatment process which is designed to enhance the efficiency of the subsequent extraction process by breaking down the cell structures and increasing the accessibility of bioactive compounds. Pre-treatment processes, often considered as part of sample preparation, play an important role in the recovery of bioactive compounds. These processes are essential as they can induce physiochemical modifications that can impact on the chemical structure of the target compounds, thereby enhancing the extractability and overall yield of the targeted bioactive compounds. Methods such as drying, grinding, freezing, and milling are commonly used. Besides, advanced techniques such as enzyme-assisted treatment, ultrasound-assisted treatment, and microwave pre-treatment can also disrupt the plant cell walls of FVW, thus enhancing the release of bioactive compounds.

Extraction and micro-encapsulation of bioactive compounds. Paulo and Santos (2021) reported that valorisation of phenolic compounds present in olive oil by-products (OOBP) can be done via extraction and micro-encapsulation approaches. The OOBP was collected and conserved in a liquid nitrogen environment, dried using freeze drying, or stored in a moisture-free environment. The OOBP includes olive leaves and olive stones like olive oil pomace left over from olive oil production. As an alternative, the OOBP was acidified until pH 2–3 to preserve the OOBP from oxidation and polymerization reaction. Low pH also promotes the precipitation of protein, the unwanted component thus increasing the amount of free phenolic compounds for extraction in the later stage. During the sample preparation before the extraction process, OOBP often had high water content (approximately 60%) and centrifugation was used to separate the solid residue from the vegetative waters. Vegetative waters consist of water used during olive oil production and the natural water content found within the olive drupes (Suárez et al., 2009). After reducing the water content, it is typical to remove the lipid fraction of the OOBP samples. The removal of non-polar compounds such as fatty acids, triacylglycerols, fat-soluble pigments, and vitamins can help in the quantification and enrichment of polar bioactive compounds such as phenolic acid. The defatting process normally is achieved through extraction using a solvent like hexane or petroleum ether.

4.2.2 Extraction of bioactive compounds from FVW

Extraction techniques are among the most important processes for characterizing and recovering the bio-accessibility and availability of bioactive compounds from FVW. Several extraction methods are discussed here to recover bioactive compounds from FVW.

4.2.2.1 Conventional extraction (CE)

Various conventional extraction (CE) methods have been employed to extract bioactive compounds from fruits and vegetable by-products. These methods include maceration, Soxhlet extraction, hydrodistillation, percolation, and digestion. The amount of solvent used, equipment utilised, and extracted compounds vary in these methods.

Solvents such as water, ethanol, methanol, hexane, and petroleum ether are used in these methods. Bioactive chemicals are usually extracted using solvents with high polarity while certain pigments like chlorophyll and carotenoids and lipophilic compounds can be extracted using solvents with low polarity such as petroleum ether and chloroform.

Bioactive substances can be extracted from FW through maceration, with or without heat treatment. Safdar et al. (2017) found that the extraction of polyphenols from the peel of the kinnow (*Citrus reticulate* L.) using methanol, ethanol, acetone, and ethyl acetate by maceration for 20 h. Using 80% methanol resulted in the highest overall polyphenol concentration (28.40 ± 0.33 mg GAE/g extract), whereas 80% ethyl acetate produced the lowest polyphenol level (8.64 ± 0.20 mg GAE/g extract). In addition, Nepote et al. (2005) found that the highest amount of polyphenol was observed at 50% ethanol concentration. The ethanol concentration has increased by over 70%. The polarity and concentration of the solvent, whether it is aqueous or absolute, affect the number of polyphenols that can be extracted. Aqueous alcohols such as ethanol and methanol are commonly used to extract phenolic chemicals from plant sources. The alcohol increases the solubility of the solute, while water aids in the solute's desorption. Compared to other traditional extraction methods, this method is more cost-effective because it does not require advanced equipment. However, it is time-consuming and inefficient since the extraction process is mainly controlled by diffusion.

Soxhlet extraction is recommended to be used for bioactive compounds sensitive to heat. This method is simple and efficient, allowing for continuous production processes and the potential reuse of the extraction solvent after distillation. Caldas et al. (2018) found that the Soxhlet extraction method yielded a higher number of phenolic compounds (catechin, rutin, and epicatechin) with a greater yield of 80 mg of gallic acid equivalent (GAE)/g compared to non-conventional methods. The extract obtained using non-conventional methods like ultrasonic, microwave, and ohmic had higher-purity. Kapoor et al. (2020) compared two methods for extracting guava seed oil: Soxhlet extraction and supercritical fluid extraction. The results showed that solvent extraction using n-hexane for six hours at 40–60 °C increased the extracted oil yield by 13.63% compared to supercritical carbon dioxide extraction with ethanol as a co-solvent at 50 °C and 45 MPa. It was discovered that the extracted guava seed oil was high in polyunsaturated fatty acids, especially linoleic acid (75.25%), one of the essential fatty acids, while oleic acid made up most of the monounsaturated fatty acids. Although Soxhlet extraction is simple and efficient, it requires a considerable amount of hazardous and flammable organic solvents, which can result in the emission of harmful gases during the extraction process, posing a threat to the environment.

Hydrodistillation is a technique used to extract volatile organic compounds, such as essential oils, from plant matrices. In this method, water is used as an extractant, which leads to a higher yield of oil compared to traditional extraction methods. Shaw et al. (2023) managed to recover yields as high as 9.98% and 9.70% from the peels of kinnow and orange, respectively. The essential oil composition of kinnow and orange peel is mainly comprised of D-limonene, terpenoids, and sesquiterpenes. These constituents increase the medicinal value of essential oils. Fagbemi et al. (2021)

conducted a study to compare the hydrodistillation method and the Soxhlet extraction method for extracting volatile compounds from Tamarind seeds. The hydrodistillation methods can extract more oxygenated compounds, which contribute to the fragrance of the oil, compared to Soxhlet extraction. However, hydrodistillation requires more distillation cycles and more heating energy, making it economically unfeasible.

4.2.2.2 Supercritical fluid extraction (SFE)

Supercritical fluid extraction (SFE) has been remarked as a green and tunable technology for extracting highly sensitive bioactive compounds from fruit waste. SFE offers several benefits such as single initial capital cost, low operating cost, shorter extraction duration, high selectivity, and purity. Selective extraction of a bioactive fraction enriched with desired compounds can be achieved via SFE without additional purification steps. The stability of bioactive compounds, which would normally be vulnerable to deterioration during solvent extraction, is better preserved by SFE due to the absence of light and air. Hence, the extracts maintain a greater concentration of antioxidants. By subjecting to the pressure and temperature of the extraction solvent close to or above its critical values, the mobile phase solvating power and diffusivity are enhanced while polarity is reduced, facilitating rapid penetration of the cell wall by the solvent. Such conditions also enable more selective extraction of bioactive compounds with different polarities as the density and transport properties can be adjusted between a gaseous and a liquid state with small pressure variations. The mechanism of SFE involves three sequential steps: (a) dissolution of bioactive compounds by SF; (b) sweeping of bioactive compounds; (c) trapping of bioactive compounds. When supercritical fluid is injected into the extraction chamber, it diffuses through the plant matrices, extracting bioactive compounds based on their solubility in the fluid. The supercritical fluid is allowed to return to its gaseous state by decreasing the pressure after extraction, which separates and concentrates the bioactive chemicals in the extract.

The selection of suitable supercritical fluid is crucial for the extraction of bioactive compounds from fruit by-products. Supercritical water ($Tc = 374$ °C, $Pc = 22.06$ MPa) and carbon dioxide ($Tc = 31$ °C, $Pc = 7.38$ MPa) are the most often used supercritical fluids because they are readily available, safe for the environment, non-explosible, non-toxic, inexpensive, and easily removed. Supercritical carbon dioxide (SC-CO_2) is a more cost-effective option for extraction of non-polar to moderately polar compounds since, in comparison to other solvents, it has a lower critical point (Tc and Pc) and thus requires less energy to compress and heat. It can be used to extract heat-sensitive bioactive compounds without inducing degradation or decreasing their bio-activity. It has been applied for the extraction of polyphenols, pigments, lipids, and vitamins that are soluble in fat or water from fruit wastes. Nevertheless, the low critical temperature limits its application to the extraction of bioactive compounds with low molecular weights of less than 500 Da, which can be overcome by increasing the operating pressure. Co-solvent can be added to SC-CO_2 to improve the solubility and extraction yield of polar compounds. The addition of small amounts of co-solvent can alter the polarity and interactions without appreciably altering the properties of the SC-CO_2. More hydrophilic compounds can be extracted using SC-CO_2 by adding polar organic solvents such as methanol, ethanol, and ethyl acetate. Ferrentino et al. used SF-CO_2 at 30 MPa, 45 °C to study the recovery of antioxidants and phenolic compounds from apple pomace, both in the presence and absence of 5% ethanol as a co-solvent. In comparison to extracts prepared from SFE without co-solvent (3.12 ± 0.15 mg Trolox equivalent antioxidant (TEA)/g of extract), the extracts obtained with 5% ethanol produced a greater antioxidant activity (5.63 ± 0.10 mg TEA/g of extract) (Ferrentino et al., 2018). For the extraction of polyphenols, natural pigments, and polysaccharides, supercritical water exhibits greater solvation power and selectivity (Marzorati et al., 2020; Ostolski et al., 2021). Hydrocarbons such as ethane, ethylene, propane, methanol, ethanol, dimethyl ether, and toluene have been used to extract volatile compounds, polyphenols, and pigments as well.

Extraction of bioactive compounds from fruit by-products, including the peels of apple, banana, mango, and pomegranate, can be set at the operating pressure of 20 to 35 MPa. Increased density and solubility of SCF are associated with an increase in pressure. Additionally, as pressure rises, the average intermolecular distance between the solutes decreases, strengthening the solute-solvent interaction. As a result, the extraction yield rises with pressure. Extremely high-pressure reduces the mass transfer coefficient, which raises internal and external mass transfer resistances and hinders the extraction of bioactive compounds from plant matrices. The performance of SFE is also greatly influenced by temperature. When the temperature increases, the density of the supercritical fluid decreases, which negatively impacts the yield. Conversely, the solute's vapor pressure increases as the temperature rises, which has a positive effect on the yield. Energy is required for the solute to desorb from the plant matrices into the supercritical fluid. Increasing the operating temperature can increase the solute's vapor pressure (i.e., facilitating vaporization of solute into SF), enhance the diffusivity and mass transfer coefficient, decrease both internal and external dispersion resistances, speed up the solute desorption process, and ultimately improve the efficiency of the extraction process. This counteracts the impact of the density decrease brought on by the temperature rise. High temperatures might damage the structure and bio-activity of targeted bioactive compounds, such as carotenoids. A study by Franco-Arnedo et al. that optimized the amount of ethanol co-solvent, pressure, and temperature discovered that raising the temperature enhanced the extraction of total flavanone content (TFC) and total phenolic content (TPC) from tangerine peels (Franco-Arnedo et al., 2020). The highest yield of TPC (112.42 mg/g extract) and TFC (3.57 mg/g extract) was achieved 22.0 MPa, 80 °C, and 5.0% ethanol. Therefore, finding the ideal operating conditions that offer the optimum trade-off between yield and economic performance in SFE requires careful consideration of temperature and pressure optimization.

Other factors influencing the SFE of bioactive compounds from fruit by-products include the flow rate of SF. During the initial stage of extraction, the amount of solute in the extracted stream increased rapidly due to the extraction of surface-attached solute. Concurrently, a higher flow rate reduces the mass transfer resistance, improving the initial rate of extraction. At a later stage, the extraction is mainly governed by solute diffusion from the plant cell matrices. The maximum yield is then reached at an equilibrium state. The flow rate and the residence time needed to extract bioactive compounds are interdependent. Increasing the solvent flow rate does not always translate into a higher extraction yield in a shorter time. A rapid increase in the flow rate of the solvent can significantly reduce the residence time for effective contact between solvent and solute, potentially causing channeling effect, which lead to lower yields and solvent wastage. Hence, the competing factors of time and solvent costs must be taken into consideration while determining the flow rate.

4.2.2.3 Ultrasound-assisted extraction (UAE)

Ultrasound could serve as a key technology to achieve green extraction. Compared to CE methods such as maceration, hydro distillation, and solvent extraction, high extraction yield can be achieved in a shorter duration with high reproducibility, reduced consumption of solvent, and lower energy requirement while retaining the quality of extract through ultrasound-assisted extraction (UAE). Several classes of food components such as aromas, pigments, antioxidants, and other organic and mineral compounds have been extracted, analysed, and formulated efficiently from a variety of matrices (mainly animal tissues, microalgae, yeasts, food, and plant materials).

In UAE, solvents and ultrasound energy are used to extract target compounds from cells and plant matrices. Ultrasound is a form of mechanical energy conducted at a frequency greater than 20 kHz. These waves can travel through solid, liquid, or gaseous media and are made up of a sequence of cycles of compression and rarefaction that cause molecules to shift and become dislodged from their initial positions. When the mechanical sound wave intensity is strong enough during rarefaction, the molecules are pulled apart, resulting in cavitation bubbles in the liquid. These bubbles can grow by coalescence or rectified diffusion. Upon reaching a critical size, these bubbles burst during the compression phase and produce a transient hot spot with extreme conditions that could reach up to a temperature of 5,000 K and pressure of 1,000 atm. Thus, the medium's biochemical reactions are accelerated at the generated hot spot.

The primary mechanism underlying UAE is acoustic cavitation. The plant's cellular matrix may experience fragmentation, localized erosion, sonoporation, shear force, and detexturation following the implosion of cavitation bubbles and sound waves. These phenomena can occur independently or in combination. As cavitation bubbles collapse in the solvent, they create inter-particle collisions and shockwaves that can cause friable solid particles from the plant matrices to fragment. Fragmentation of these solids decreases their size, which increases the effective surface area and mass transfer rates in the solid matrix's boundary layer. This results in improved solubility of the target compound in the solvent and leads to higher extraction yield. Ultrasound has been shown to cause erosion in UAE-treated samples, as observed through Scanning Electron Microscope (SEM). When the cavitation bubbles collapse on the surface of the plant matrix, the plant tissues and structures are eroded. This erosion allows better access of the solvent to the plant matrix, leading to greater solubility and increased yield during extraction. Additionally, the rapid oscillation and collapse of cavitation bubbles in the solvent and near the surface of the plant matrix create shear force and turbulence in the solvent. This local shear stress and turbulence can cause damage to the plant structures, which helps to release bioactive substances. Following UAE, plant structures have been observed to undergo detexturization, whereby cell structures are damaged and transformed into undefined shapes. Cavitation can cause sonoporation. Reversible or irreversible pores that form in the cell membranes during cavitation release the bioactive substance contained in the plant matrices through sonoporation. Ultrasound also enhances the water absorption capacity of plant matrices, improving the solvent's accessibility and diffusivity of bioactive substances. Besides that, ultrasound can increase the swelling index of plant matrices, which makes it easier for solutes to desorb and diffuse, ultimately improving the efficiency of extraction. In most cases, the increase in extraction yield by UAE is due to the cumulative impact of all mechanisms, rather than any one mechanism acting alone.

Ultrasonic power is a significant factor that impacts process costs, particularly when operating on an industrial-scale. The power used during UAE has been reported as an amplitude percentage in the range of 0 to 100%, with 100% amplitude being the rated power of the equipment. It has been demonstrated that, for amplitudes ranging from 30 to 80% of maximum power, there is a linear relationship between ultrasonic amplitude and output power. Depending on the component to be extracted and the type of plant matrix, a power range of 20 to 700 W could be used to extract bioactive compounds from fruit and vegetable by-products. Maran and Priya (2014) reported that the production of polysaccharides from the fruit peel of *Nephelium lappaceum* L. increased with an increase in power up to 100 W and decreased with an increase in power over 110 W. Kaur et al. (2023) found that as the amplitude grew from 20 to 40%, the yield of TPC from *Citrus reticulata* (kinnow mandarin) peels increased; however, the TPC yield decreased at higher amplitudes. The yield of the UAE rises as power increases, peaks, and subsequently falls. As the resonant bubble size increases with the power of the ultrasonic wave, the cavitation bubbles implode violently, producing high shear and microjets that ensure higher cellular degradation, improve rate of molecular mixing, and increase solvent penetration, thereby improving the extraction yield. Increasing the power during an ultrasonic extraction process can lead to a higher number of bubbles being generated. A greater number of bubbles can result in larger inter-bubble collisions, distortion, and nonspherical collapse, which reduces the impact of the bubble implosion. Additionally, the high ultrasonic power can degrade the bioactive substance, further reducing the yield. This phenomenon has been reported for UAE of polysaccharides, where when ultrasonic power surpasses a certain threshold, it triggers a cavitation or chemical effect. This effect weakens the site of polysaccharide bonding energy and breaks the glycosidic bonds, ultimately leading to a reduction in yield. Thus, when using ultrasonic to extract polysaccharides, its output power intensity was often kept at a modest level to minimise the deterioration effect of the technology.

The cavitation effect during extraction is significantly affected by the frequency. It affects the chemical and physical consequences following the collapse of cavitation bubbles. Frequencies within the range of 20–100 kHz are commonly used in UAE of bioactive compounds from fruit and vegetable by-products. This range of frequencies regulates the physical consequences, such as mechanical shear force brought on by the cavitation bubble collapsing, while the frequencies that control the chemical effects of cavitation bubbles are between 200 and 500 kHz. High-frequency ultrasound can break down polyphenols, produce large amounts of free radicals and reduce their biological activity. Moreover, if the compression rarefaction cycle is too short, the cavitation bubble will not develop and grow. There is a negative correlation between the ultrasonic frequency and the rarefaction phase duration. Thus, the cavitation bubble does not have sufficient time to increase in diameter at high frequencies, impeding the implosion effect. Studies have shown that pectin (Chua et al., 2020; Guandalini et al., 2019; Hosseini et al., 2019; Oliveira et al., 2016), antioxidants (Pan et al., 2012), anthocyanins (Corrales et al., 2008), malic and tartaric acid (Palma and Barroso, 2002) can be extracted from fruit peels and by-products at low frequencies ranged from 20 to 40 kHz.

An increase in temperature affects both the solute and solvent. An increase in temperature affects the solute in two ways. Firstly, increasing the temperature affects the molecular motion, transport and interaction of solutes, improving the solute's solubility and desorbability in the solvent. Secondly, increasing the temperature leads to a decrease in surface tension and viscosity of the solvent while increasing its vapor pressure and diffusivity in the plant tissue matrices. As the vapor pressure increases, more solvent vapors enter the bubble cavity, causing the bubbles to collapse less violently. As a result, the sonochemical effects resulting from the collapse of cavitation bubbles may be reduced at higher temperatures. Al-Dhabi et al. (2017) claimed that the yield of phenolic chemicals extracted from spent coffee grounds waste in the UAE increases as the temperature is raised from 30 to 45 °C, but decreases when the temperature exceeds 45 °C. Phenolic compounds from orange peel (Shehata et al., 2021), polysaccharide from rambutan fruit peel (Maran and Priya, 2014) and pectin from dragon fruit peel (Chua et al., 2020) have all shown similar trends in extraction. Moreover, a higher temperature may also lead to the degradation of plant cells and the transportation of unstable compounds. Lower temperatures promote sonochemical effects, thus temperature control is typically used to prevent temperature rise. It is crucial to choose an appropriate extraction temperature for the target compound of the extraction to prevent thermolabile chemicals from degrading. Therefore, temperature optimization and control are necessary.

Various hypotheses have been proposed to explain the impact of UAE temperature on the yield of bioactive substances. High temperatures increased solvent vapor pressure, the cavitation bubble created during the rarefaction cycle is then occupied by the solvent, which lowers the pressure difference between the gas-liquid phases (i.e., bubbles and solvent). Therefore, even though there are more cavitation bubbles at high temperatures, the cells are damaged less severely, leading to a lower yield. The second hypothesis suggests that an increased shear stress caused by large number of cavitation bubbles that grow at high temperatures and collapse later degrades the desired component. For the third hypothesis, the solvent's reduced surface tension at higher temperatures is the primary reason for the decreased intensity of the collapsing cavitation bubble and the consequent reduction in mass transfer of the component that needs to be extracted.

A longer sonication duration leads to a higher yield initially, which then decreases with subsequent increments of time. During the initial increase in sonication duration, the ultrasonic waves produce a cavitation effect, which intensifies the swelling, pore enlargement, and disintegration of plant cell walls. More contact time between the solute and the extraction solvent is made possible by these improvements with longer sonication times, leading to more efficient extraction. Prolonged exposure to ultrasonic radiation can cause unfavorable reactions such as oxidation or deterioration of the solute's structural integrity, which lowers the extraction yield. According to Moorthy et al. (2017), the solute concentration in plant cells and the solvent have reached equilibrium, and the maximum extraction yield of pectin from jackfruit peel was obtained at 30 min. After 30 min, the pectin yield decreases as overexposure to ultrasonic treatment could have caused structural disintegration in pectin. The extraction of pectin from fruit peels of pomegranate (Moorthy et al., 2015), rambutan (Maran and Priya, 2014), banana (Maran et al., 2017), and the extraction of phenolic compounds from black chokeberry waste (Ramić et al., 2015) have shown similar results. The optimal sonication duration for pectin extraction and phenolic component extraction varies from 10 to 60 min and 10 to 90 min, respectively, depending on the specific fruit or vegetable waste.

4.2.2.4 Microwave-assisted extraction (MAE)

Microwave-assisted extraction (MAE) is a green and emerging extraction method that combines microwave and solvent extraction. This method is superior to CE methods due to its high extraction yield, reduced solvent usage, and shorter extraction time. The MAE method enables faster extraction of bioactive compounds from plant by-products compared to Soxhlet extraction and UAE. A microwave is an electromagnetic wave that combines electric fields and magnetic forces. Microwaves oscillate between 300 MHz and 300 GHz and have a wavelength that ranges from 1.0 cm to 1.0 m (Alvi et al., 2022). The most common frequency and wavelength bands used in MAE are 915–2,450 MHz and 12–20 cm, respectively. Polar materials, such as bound water in plant matrices, absorb microwave radiation, which is then converted into heat via ionic conduction and dipole rotation mechanisms. The heating caused the water molecules to evaporate leading to an increase in intracellular pressure and rapid swelling in plant cells. As a result, the cell walls ruptured, facilitating the diffusion of solvent into the plant matrices and the leaching of bioactive compounds from the plant matrices into the solvent. To optimize the MAE process, several factors need to be considered, including extraction temperature and time, microwave power, solid-to-solvent ratio, and solvent.

The MAE process is significantly influenced by the selection of solvents. The choice of solvent is based on various factors such as the solubility of the targeted bioactive compounds, the solvent's penetration ability, the interaction between plant matrices and solvent, and the solvent's dielectric constant (ε'). Preferably, the solvent should exhibit a high selectivity toward the targeted bioactive compounds. It is also important to select extraction solvents based on their ability to absorb microwave radiation and heat up, rather than employing the same solvent used in CEs. A solvent with a higher dielectric constant has a greater capacity to absorb microwave energy, leading to a faster rate of solvent heating. Ethanol ($\varepsilon' = 24.3$) is one of the most often utilized solvents in MAE, either by itself or in conjunction with water ($\varepsilon' = 78.3$), due to its good capabilities in solubilizing phenolic compounds and absorbing microwave energy. Ethanol or water can be added to ineffective microwave absorbers, like hexane ($\varepsilon' = 1.89$) to increase the extraction efficiency. Nevertheless, it is crucial to conduct an optimization study for the extraction of phenolic compounds from each FW material because the phenolic compounds' polarity and chemical structure also affect the extractability and efficiency of the solvent. In a study on the MAE of carotenoids from Citrus clementina peel, Kadi et al. (2022) discovered that the mixture of hexane and acetone, which contained 68.97% hexane, produced the best total carotenoids content when compared to the results obtained by hexane and acetone ($\varepsilon' = 20.7$) separately. Besides that, choosing the right solvent-to-solid ratio is also crucial because an excessive amount of extracting solvent would need more time, energy and cost for its removal and purification of the bioactive compounds in subsequent steps. Moreover, nonuniform irradiation and distribution may cause MAE to yield poorer recoveries. Adequate solvent volume is required to ensure complete submersion of the sample in the solvent during the MAE process, particularly for plant matrices that swell during the process. A solvent-to-solid ratio of 10:1 to 20:1 (ml/mg) was optimal in many applications (Veggi et al., 2013).

Compared to CE methods, the extraction time for MAE is much shorter, usually ranging from a few min to half an hour. For instance, phenolic compounds from banana peels were extracted in 6 min using MAE (960 W/g, a S/F ratio of 2:100 g/mL), which was shown to be significantly more efficient than Soxhlet extraction, which took 9 to 10 h (Vu et al., 2019). This helps to avoid possible thermal deterioration and oxidation, which is important for thermally targeted labile compounds that can be affected by the solute-solvent system overheating. Extraction times for MAE can vary depending on the plant parts, so a thorough study of extraction time optimization is essential. In general, the extraction yield tends to rise as the extraction time increases, but the yield only increases slightly as the time is further increased. The solvent's dielectric properties also affect the irradiation duration. Extended exposure of high-dielectric solvents such as ethanol and methanol ($\varepsilon' = 32.6$) to microwaves, when diluted with water, can significantly heat them up, posing a risk to the stability of thermally labile targeted compounds (Veggi et al., 2013). When a longer extraction time is required, samples can be extracted in multiple steps using sequential extraction cycles. In each cycle, fresh solvent is added to improve the extraction yield and prevent solvent saturation by solute, as well as overheating from using the same volume of solvent. Silva et al. (2021) conducted a sequential MAE at 87 °C for 30 min per cycle using 70% ethanol. They recovered 27.17 ± 0.9 mg of piceatannol/mg extract from the residual cake of passion fruit seed oil extraction. In contrast, only 13.03 ± 0.4 of piceatannol/mg extract was obtained using Soxhlet extraction for 120 min.

The amount of energy delivered to the sample, and transformed into heat in the dielectric material, is determined by the intensity of the incident microwave power. MAE ruptures the plant cell wall and tissues by localized heating, increasing the interaction between the solute and solvent (Prakash Maran et al., 2013). Thus, raising power will often result in a higher extraction yield and shorter extraction time. Lasunon and Sengkhamparn (2022) conducted a study to examine how different microwave powers and times affect the MAE of pectin from industrial tomato waste. Their findings showed that when the extraction time was kept constant at 3 min, increasing the microwave power from 300 to 600 W led to an increase in pectin yield from 9.43 to 22.85%. There was a slight decrease in the percentage of pectin extracted when the extraction time was increased from 5 to 10 min at a high microwave power of 600 W/g, resulting in a yield drop from 29.67 to 28.28%. Prolonged application of high microwave power leads to thermal degradation which can rupture the cell wall rapidly at high temperature, causing impurities to leach into the solvent along with targeted solutes. Moreover, the application of MW power above a threshold value may result in considerable losses of bioactive compounds and their chemical structure in plants by-products due to enzymatic and thermal reactions of oxidation (Rifna et al., 2023), whereas low power levels cause gradual cell wall rupture, enabling selective MAE. Since MAE is carried out in closed vessels, the temperature may rise significantly over the solvent's boiling point. As a result, it's important to select the microwave power carefully to prevent overheating, which might cause degradation of selected bioactive compounds and internal vessel overpressure.

Temperature and microwave power are interrelated. Increasing microwave power can raise the system's temperature and increase the extraction yield until it either reaches a plateau or decreases. Higher temperature increases the desorption of solutes from the active sites of plant matrices, resulting in greater extraction efficiency. This occurs because enhanced intermolecular interactions and molecular motion within the solvent brought on by a higher temperature reduces the viscosity and surface tension solvent, and increases the solvent's solubilization capacity, thereby improving the matrix penetration and wetting (Martínez-Abad et al., 2020). Additionally, as the temperature rises, there is an increase in cellular pressure that results in cell rupture and opening of the plant cell matrices, increasing the availability of solutes for extraction into the solvent (Rivadeneira et al., 2020). In some circumstances, extraction efficiency increases with temperature until an optimal temperature is reached, at which point it begins to drop as temperature increases further. Skenderidis et al. (2020) optimized the conditions for the industrial-scale vacuum microwave-assisted aqueous extraction of bioactive compounds (total phenols) from pomegranate peel. They discovered that when temperature increased, the extraction performance was improved until it reached the optimal temperature (61.5 °C), beyond which any additional temperature increase reduced the extraction efficiency as the stability of the phenolic compounds was affected. Similar findings were reported by Thu Dao et al. (2021) for pectin extraction from the peels of passion fruit, white and red flesh dragon fruits. They discovered that raising the temperature above the optimal point (i.e., 76 °C for passion fruit peels and 84 °C for the peels of white and red flesh dragon fruits) reduces the pectin yield and accelerates the depolymerization process, leading to low-polymerization pectin that does not precipitate in ethanol.

4.2.3 Preservation or Stabilisation of the Bioactive Compounds from FVW

After extraction, stabilisation of the bioactive compound is crucial to maintaining the integrity and bio-activity of the extract often sensitive to light, heat, and oxygen. Common techniques such as spray drying, freeze drying, and encapsulation are used to protect the bioactive compounds from degradation during transportation or storage. The spray drying technique is a common industrial encapsulation technique and has been used to prepare food additives and flavours in the food industry. In spray drying, liquid feed is automised into tiny droplets using a spray nozzle and these droplets are exposed to hot gas in a drying chamber. As these droplets are suspended in the hot air, moisture evaporates quickly, leaving behind the fine powder. For lipophilic extracts such as carotenoids or fat-soluble vitamins or pigments, the spray drying process is more straightforward than lipophilic extract as easily emulsified in an aqueous containing water-soluble polymers. A considerable variety of natural and synthetic polymers like Arabic gum, chitosan, maltodextrin, sodium caseinate, and soybean oil were used as encapsulating materials. Spray drying has been used to encapsulate the bioactive compounds extracted from olive leaves with different materials like insulin, sodium caseinate-soy lecithin, and chitosan. For the encapsulation of olive oil pomace (OOP) extract, the spray drying technique also stands out as the most applied technique with maltodextrin as the encapsulating agent (Aliakbarian et al., 2018). Supercritical-assisted atomization was proposed to be an alternative encapsulation technique for OOP in maltodextrin polymeric matrix (Aliakbarian et al., 2017). The hydrophilic extracts that are not favored by spray drying techniques often alternative approaches like emulsification in polymers soluble in organic solvents to obtain extract-loaded microparticles. Spray drying and freeze drying were both employed to encapsulate the extract from olive oil wastewater (OOWW) (Petrotos et al., 2012). Encapsulation methods combined with extraction techniques using generally recognized as safe (GRAS) materials are anticipated to have a positive impact on environmental issues while also enhancing or reinventing value-added products for the food and nutraceutical industries.

4.3 FIBERS FROM DIFFERENT FOOD WASTE AND EXPIRED FOOD

Dietary fibers consist of carbohydrate polymers, including cellulose, hemicellulose, lignin, and pectin. Dietary fibers are classified as soluble dietary fiber (SDF) or insoluble dietary fiber (IDF) according to their solubility in water. The dietary soluble fiber categories consist of pectin derived from whole grain and legume sugars, polysaccharides derived from bean and legume sugar monomers, mucilage found in aquatic plants, cacti, aloe vera, and okra, and glycoproteins found in food additives. In contrast, polysaccharides, β-glucans, mucilage, oligosaccharides, and inulin, insoluble fibers consist of cellulose, hemicellulose, chitin (and its derivatives such as chitosan), and chitin is insoluble fiber. Figure 4.2 represents all kinds of fibers extracted from FW and expired food. For example, cellulose derived from fruits, root vegetables, and other sources of glucose monomers, hemicellulose complex carbohydrates found in cereal bran and grains, and lignin aromatic alcohols derived from vegetables.

Fibers can be extracted using chemical, microbial, dry and moist processing, enzymatic gravimetric, and enzymatic gravimetric (with certain limitations) techniques. The extraction methods are steam extraction, water extraction, ethanol extraction, ultrasonic-assisted extraction, and high hydrostatic pressure-assisted extraction. Table 4.1 shows details of various methods used for the extraction of fibers from FW and expired FW. The FW and expired FW need to go through processes like dry milling, wet milling, ultrasound-assisted extraction, and particle size reduction. Dry and wet milling techniques are suggested to enhance the extractability

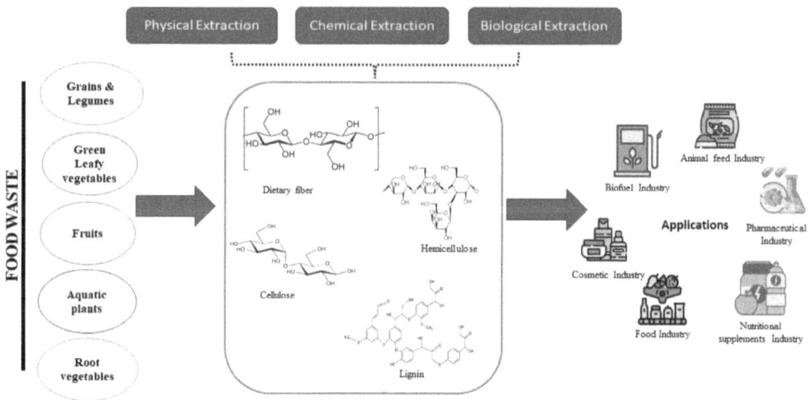

FIGURE 4.2 Fibers extracted from food waste.

of their economic nature, capacity to generate fiber of superior purity, and utilisation of diminished quantities of chemicals and water in comparison to conventional methods (Khanpit et al., 2021). The utilisation of ultrasonic treatment has the potential to increase the extraction yield of fibers from various food by-products while simultaneously reducing the extraction time (Plakantonaki et al., 2023). Particle size is invariably a critical factor in physical separation. The primary factor that propels the dry processing method is the disparity in density. Dietary fiber (DF) of reduced density is isolated through air elutriation from starch and protein. In addition to density, several other physical characteristics of fruit peel, including terminal velocity, particle size (equivalent spherical diameter), particle shape (sphericity), and moisture content, influence the separation of DF via dry processing (Kaur et al., 2023). To achieve maximum yield with maximum purity (TDF), the peel particle size and terminal velocity of air are critical parameters optimized during elutriation (Chockchaisawasdee et al., 2023; Khanpit et al., 2023). There are different chemical extraction methods using water, alkali, alcohol, and acid digestion methods. The different extraction techniques help to determine the physiological and prebiotic properties of the sample (Fidriyanto et al., 2023). As Baker explained, the alkaline process in conjunction with high shearing blending produced the greatest apparent yield of SDF in olive pomace (Baker et al., 2023). Chemical methods are among the diverse pre-treatment procedures that employ chemicals to dissolve hemicelluloses and break the bond between cellulose and lignin (Moradi and Fathi, 2023). The efficacy of cellulose and lignin in the renewable resource sector for effluent sterilization has been demonstrated (Ganguly et al., 2020). Similarly, the glycosidic linkages in DF may be disrupted under intense alkaline conditions, which may affect the ratio between IDF and SDF (Zhang et al., 2020). The alcohol-insoluble residue was digested using acids which resulted in pectin which was used as an emulsifying agent in the food industry (Petkowicz and Williams, 2020). Through enzymatic hydrolysis, compounds that function as specific growth-promoting substrates for lactic acid bacteria in the human intestinal tract are generated naturally. Microorganisms present in the human gastrointestinal tract readily ferment DF components (Radenkovs et al.,

TABLE 4.1
Physical, chemical and biological methods used in the extraction of fibers from various food waste

S.No	Food waste	Method	Control parameters	Application	References
			Physical methods		
1	Sugarcane (*Saccha rum officina rum*) bagasse	Wet milling for total, soluble and insoluble dietary fiber	Sugarcane bagasse was steam-cooked to reduce microbes followed by oven drying and milling into powder.	Noodles - Food product development	(Lau et al., 2022)
2	Papaya (*Carica papaya L.*) peel	Hot air convection & microwave-assisted dehydration for Dietary fiber concentrates	Forced convection drying: 40°C	Baked products & Noodles - Food Industry	(Nieto-Calvache et al., 2019)
3	Pearl millet (*Pennis etumglaucum*) bran	Ultrasound-assisted extraction for dietary fiber	0.08 M phosphate buffer, Ultrasonication: 50 °C, 10 min Solvent: ethanol	Thickeners - food industry	(Kaur et al., 2023)
4	Orange (*Citrus sinensis L.*) peel	Hot air oven drying for dietary fiber	Temperature: 40°C for 48 h, Sieve mesh size: 150 mm	Biscuits - Food product development	(Obafaye and Omoba, 2018)
5	Tomato (*Solanum lycoper sicum*) waste, Bell pepper (*Capsic um*) placent a & seeds	Dry milling for dietary fiber	EAC 30-Lab pasta dryer: 30 –40 °C, relative humidity: 25–75%	Pasta - Cereal based industry	(Teterycz and Sobota, 2023)
6	Amla (*Emblic a officina lis*) pomace	Wet drying for total dietary fiber	Amla pomace is blanched in boiling water, cooled, dried, and ground to fine powder for sieve.	Biscuits - Food product development	(Kohli et al., 2019)
7	Blueberry (*Vaccini um sect. Cyanococcus*) pomace	Dry milling for total dietary fiber	Temperature: 45 ± 2°C, Sieve size: 1 mm	Cookies - Food industry & Functional food	(Tagliani et al., 2019)
8	Lettuce (*Lactuc a sativa*)	Dry milling for total dietary fiber	Temperature: 70± 0.5 °C (15 h), Equilibrium Relative Humidity 55–65%, Sieve size: 125 μm	Bread - Bakery industry	(Plazzotta, Sillani and Manzocco, 2018)

(Continued)

TABLE 4.1 (Continued)

S. No	Food waste	Method	Control parameters	Application	References
9	Date (*Phoenix dactylif era L.*) seeds	Air drying and grinding for dietary fibers	Date seeds were washed, soaked, and air-dried before being ground into a fine powder	Functional food	(Kamal et al., 2023)
10	Onion (*Allium cepa L.*) peel	Dry milling for dietary fiber	Temperature: 70 °C (10 h)	Bread - Bakery industry	(Ateeq et al., 2023)
11	Spent barley (*Horde um vulgare*) grains	Grinding for Dietary fiber	Grinding speed: 6,000 rpm	Noodles - Food Product Development	(Shi et al., 2023)
12	Seedless dates (*Phoenix dactylif era*)	Dry milling for Dietary fiber	Temperature: 70 °C (15 min), lyophilisation 60 or 70 °C, Milling: 7,000 rpm (1 min)	Muffins - Bakery Industry	(Hamza et al., 2021)
13	Faba beans (*Vicia faba L.*)	Heating and filtering for crude fiber & lignin	Temperature: 70 °C (15 min) and the fiber extracts dried at 100 °C	Animal feed - Agricultural industry	(Krenz et al., 2023)
14	Peanut (*Arachi s hypoga ea L.*) skin	Decortication, dry grinding for dietary fiber	Temperature: 160 °C (4–5 min) Grinding: 1,800–2,300 rpm (2 min)	Peanut butter - Food Product Development	(Dikkala et al., 2023)
15	Peanut (*Arachi s hypoga ea L.*) seed coat & residue	High Energy Media milling for dietary fiber	High energy media mill (7.35% w/w) mixed with Zirconia beads (0.6–0.8 mm) in a 3:1 ratio (w/w)	Food gels - Food industry	(Li et al., 2024)
16	Faba bean (*Vicia faba L.*) husks	Particle size reduction for dietary fiber	Stainless steel sieves: 500 µm and 250 µm.	Bread - Bakery Industry	(Chockchaisawasdee et al., 2023)
17	Sweet lime (*Citrus limetta*) peel	Elutriator for total dietary fiber	Grinding velocity: 1 to 2.5 m/s	Fat replacer & yogurt preparation - Food industry and Prebiotics - Nutraceuticals	(Khanpit et al., 2023)
18	Prickly pear (*Opunti a ficus indica cv. Verde Villanu eva*) peel	Drying for soluble, insoluble and total dietary fiber	Temperature: 55 °C	Prebiotics	(Garcia-Amezquita et al., 2018)
19	Mango (*Mangif era indica*) peel	Ultrasound-assis ted extraction for dietary fiber concentrates	Ultrasonicated for 9 min at 25 °C. Solvent: Ethanol.	Food Processing Industry	(Kaur et al., 2021)

(Continued)

TABLE 4.1 (Continued)

S.No	Food waste	Method	Control parameters	Application	References
			Chemical methods		
1	Apple (*Malus pumila*) pomace	Alkali, and acid Extraction for Crude Fiber & Dietary Fiber	Alkali: NaOH at 45 °C (4 h). Acid extraction: sulfuric acid at 80 °C (4 h)	Functional food	(Fidriyanto et al., 2023)
2	Olive (*Olea europaea*) pomace	Alkaline extraction for insoluble & soluble dietary fiber	Solvent: Sodium hydroxide	Dietary supplements	(Baker et al., 2023)
3	Peanut (*Arachi s hypoga ea L.*) shells	Alkaline extraction for cellulose & lignin	Solvent: 0.5 M Sodium hydroxide at 90 °C	Renewable resource industry	(Ganguly et al., 2020)
4	Sugarcane (*Saccha rum officina rum*) bagasse	Alkaline extraction for cellulose & lignin	Solvent: NaOH	Functional food	(Paroha, 2020)
5	Palmyra fruit (*Borass us flabellif er*) peduncle	Alkaline, and cyanamide-activated hydrogen peroxide treatment for cellulose microfibrils	Solvent: CH_2O_2/ CH_3COOH /H_2O at 85 °C (4 h)	Reinforcing agents - Medical, electronic, food packaging, and cosmetics industry	(Balasubramani et al., 2024)
6	Tomato (*Solanu m lycoper sicum*) pomace	Alkaline treatments for lignin and hemicellulose	Solvent: 2% sodium hydroxide at 95 °C (4 h)	Emulsions - Food Industry	(Moradi and Fathi, 2023)
7	Orange (*Citrus sinensis*) peel	Alkaline hydrogen peroxide hydrolysis for soluble, insoluble and total dietary fiber	Solvent: 1.0% hydrogen peroxide, solid-liquid ratio (1:20), pH at 11.5 at 60 °C (4 h)	Emulsifiers - food industry	(Zhang et al., 2020)
8	Broccoli (*Brassica oleraceaa*) stalk	Acid extraction for pectin	Acid: 0.1 M nitric acid	Thickening & emulsifying agent - Food industry	(Petkowicz and Williams, 2020)
			Biological methods		
1	Grape fruit (*Citrus paradisi*) peel pomace	Enzyme digestion for soluble & insoluble dietary fiber	Lipase (3%) at 50 °C (2 h)	Food packaging	(Chen et al., 2023)

(Continued)

TABLE 4.1 (Continued)

S.No	Food waste	Method	Control parameters	Application	References
2	Grape (*Vitis vinifera*) must lees, apricot (*Prunus armeni aca*) and apple (*Malus pumila*) pomace, rice (*Oryza sativa*) husk, corn (*Zea mays*) and sunflower (*Helianthus annuus*) pulp, spent grains	Enzymatic-gravimetric procedure for dietary fiber	Amylase, amyl glucosidase, and protease	Ice creams - Food product development	(Ayar et al., 2018)
3	Mango (*Mangifera indica*) peel	Enzymatic extraction technique for dietary fiber concentrates	α-amylase, protease, and amyl glucosidase	Food Processing Industry	(Kaur et al., 2021)
4	Mango (*Mangifera indica*) peel	Ultrasound-assisted enzymatic extraction technique for dietary fiber concentrates	Phosphate buffer (0.08 M, 50 mL)	Food Processing Industry	(Kaur et al., 2021)
5	Soy (*Glycine max*) waste	Aerobic fermentation for soluble, insoluble and total dietary fiber	*Bacillus thuringiensis*	Biofuel & Biopesticides	(Ballardo et al., 2016)
6	Mango (*Mangifera indica*) & orange (*Citrus sinensis*) peel	Enzymatic and gravimetric method for soluble, insoluble and total dietary fiber	Enzymes mimicking human small intestine digestion	Functional food	(Mosa and Kkalil, 2015)
7	Potato (*Solanum tuberos L.*) residue	Ultrasound-assisted enzymatic extraction technique for soluble and insoluble fiber	α-amylase	Functional food	(Ma et al., 2022)

(*Continued*)

TABLE 4.1 (Continued)

S.No	Food waste	Method	Control parameters	Application	References
8	Grape (*Vitis vinifera*) marc	Enzyme- catalysed treatment for cello-oligosaccharides	10 mM sodium acetate buffer & endo-1,4-β-d-glucanase (EG) enzyme	Prebiotics	(Liu, Hu and Haritos, 2024)
9	Rice (*Oryza sativa*) straw	Enzyme hydrolysis for cellulose	Xyn11 enzyme and Tris-HCl	Food and packaging industry	(Cabrera et al., 2023)
10	Spent green tea (*Camellia sinensis*) leaves	Enzymatic treatment for soluble, insoluble and total dietary fiber	Enzyme loadings varied from 0 to 25 U/g at 50 °C for 0–2 h	Cookies - Bakery Industry	(Nguyen et al., 2022)
11	Date fruit (*Phoenix dactylif era L.*) pomace	Enzymatic and gravimetric method for total dietary fiber	Amylase, amylglucosidase, and protease	Biorefinery industry	(Ali et al., 2022)

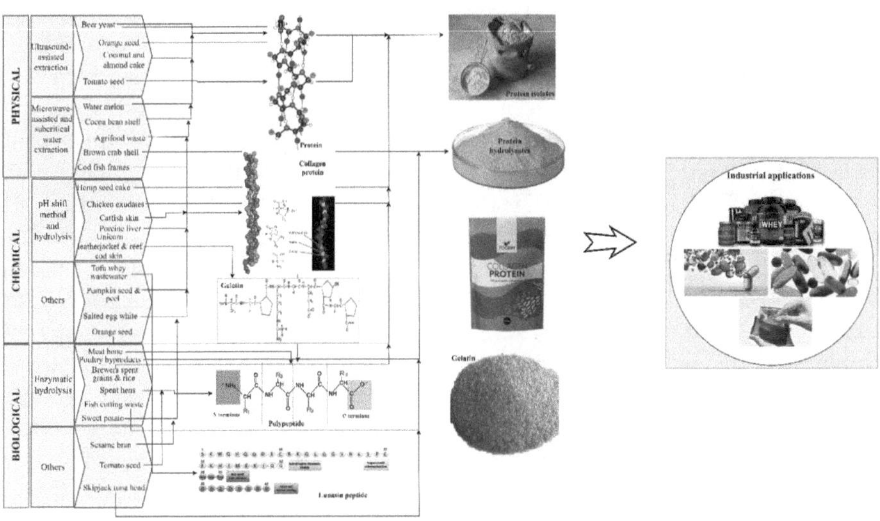

FIGURE 4.3 Proteins extracted from food waste.

2018). Delignification of lignocellulosic substances is a critical application of enzymatic hydrolysis, a secure and environmentally friendly process (Chen et al., 2023). The enzymatic method combined with ultrasonication could enhance pectin production and quality while reducing the reliance on hazardous solvents and extremely corrosive conditions during pectin recovery (Mosa and Kkalil, 2015). Under the biological method, the enzyme gravimetric method helps to extract the SDF and IDF from orange and mango peels. The enzymatic gravimetric method is most frequently used to ascertain TDF (Ali et al., 2022). The enzymatic treatment proved to be a highly effective approach in altering the composition of DF (Nguyen et al., 2022).

4.4 PROTEINS FROM DIFFERENT FOOD WASTE AND EXPIRED FOOD

FW and expired food consist of protein. Different types of extraction technologies, including enzymatic hydrolysis, alkali hydrolysis, and ultrasound extraction, can be used to extract protein are shown in Table 4.2. In addition, the detailed operating conditions including inedible parts such as seeds, peels, and kernels are shown in Table 4.2. For example, the beer industries produce tons of beer yeast (*Saccharomyces cerevisiae*), around 10–15 tons from 10^4 hL of beer, that is currently used for animal feed and as an enzyme source (Adesogan et al., 2020). Whereas protein and polypeptides can be extracted from the same sources with good antioxidant activity toward 2,2-diphenylpicrylhydrazyl (DPPH) and 2,2′-azino-bis (3-ethylbenzothiazoline-6-sulfonic acid) (ABTS). In addition, agri-FW consists of 85% of brewer's spent grain, 50% spent coffee grounds, and 70% kale (*Brassica oleracea*). Around 92–95% of protein with higher antioxidant properties was obtained from this waste by microwave-assisted alkali and hydrothermal extraction (Barrios et al., 2022). The protein also found to have 64–68% purity and good sources of essential and non-essential amino acids (Jaeger et al., 2020; Tarna et al., 2021; Zhu et al., 2022; Ummels et al., 2023). Figure 4.3 shows the protein forms from different FWs and expired food.

Similarly, the protein from orange seed, tomato seed, watermelon seed, chicken exudates, chicken bones, and fish by-products showed enhanced solubility, gelling, emulsifying properties, oil and water holding capacity, greater dispersibility, foaming capacity, and improved surface hydrophobicity (Bezerra et al., 2019; Jiang et al., 2019; Özyurt, Tetik and Ötleş, 2021; Rosas Ulloa et al., 2023). Whereas greater antioxidant activity, anti-inflammatory properties, and angiotensin-converting enzyme (ACE) 1 inhibitory properties were shown in agri-FWs, fish by-products, chicken wattles, and combs (Moayedi et al., 2018; Tkaczewska, Bukowski and Mak, 2018; Kumar et al., 2021; Dao et al., 2022). Iron-binding peptides also extracted from shrimp by-products, showed 17.5 μmol/g of protein iron-binding capacity and 1–6 kDa peptide molecular weight (Zhou et al., 2023). Spent hens and chicken exudates are the waste generated after losing their egg-lying capacity and waste generated during the storage and de-frozen stage. The polypeptides from spent hens have low molecular weight peptide fractions and good amino acid profiles such as Val, Met, Lys, Leu, Ile, Glu, Asp, Arg, and Ala which can be used for dietary supplements. The protein isolates from the chicken exudates can be used as additives in the food industry due to their higher gelation and emulsion properties (Kumar et al., 2021; Jiang et al., 2022).

Enzymatic hydrolysis (EH) was the widely adopted technology for the extraction of proteins in the form of hydrolysates and degrades complex proteins to peptides or polypeptides (Yao et al., 2020; Melgosa et al., 2021; Dao et al., 2022). Electrochemical hydrolysis, electro-activation, and MAE are the physical methods effectively used for the extraction of proteins in different forms with a yield of 90–95% (Barrios et al.,

TABLE 4.2
Physical, chemical and biological methods used in the extraction of proteins from various food waste

S.No	Food waste	Methods of extraction	Control parameters	Applications	References
			Physical methods		
1	Beer yeast (*Saccharomyces cerevisiae*)	UAE for water- soluble protein	Extracting agent: distilled water; 300 W ultrasound power; 8.5 pH; 12% solid- liquid ratio	Food, feed, and pharmac eutical industries	(Jaeger et al., 2020; Tarna et al., 2021; Zhu et al., 2022)
	Electro-activation for protein concentrates and hydrolysates	Electric current density: 20 mA/cm^2			
3	Fish cutting waste (scale and skin)	Electrochemical hydrolysis for Protein and polypeptide hydrolysates	1% aqueous NaCl solution at pH 12.2 and Eh 920 mV	Peptide supplements, functional foods, and Protein-mineral supplement	(Kuprina et al., 2023)
4	Hemp (*Cannabis sativa*) seed cake	High-pressure processing combined with UAE for protein	200 Mpa for 4–8 min	Food industry	(Cabral et al., 2023)
5	Agri-food waste [brewer's spent grain, spent coffee grounds, and kale] (*Brassica oleracea*)]	MAE alkali and hydrothermal extraction for protein	50 Hz with 10% solid-to-liquid ratio (w/v) of 0.5 M NaOH at 110 °C for 10 min	Food, feed, and pharmaceutical industries	(Barrios et al., 2022)
6	Orange (*Citrus sinensis*) seed	UAE for protein isolates	42 kHz; 130 W for 15 and 30 min; pH: 6–12	New product development in the food industry	(Rosas Ulloa et al., 2023)

(Continued)

TABLE 4.2 (Continued)

S.No	Food waste	Methods of extraction	Control parameters	Applications	References
7	Coconut (*Cocus nucifera*) and almond (*Prunus dulcis*) cake	UAE for protein	18 kHz; 4.5 kW; oscillation amplitude: 45– 60 µm at 60 °C; residence time: 2 min	Additives in the food industry	(Aiello et al., 2022)
8	Tomato (*Solanum lycopersicum*) seed	UAE for Protein isolates	Liquid: solid ratio: 52 ml/g; ultrasound duration: 2 min 18 s at 24 kHz; extraction time: 60 min	Food industry	(Özyurt, Tetik and Ötleş, 2021)
9	Watermelon (*Citrullus lanatus*) seed	MAE for protein	50 W; 1:30 solid-to-solvent ratio; pH 10 for 2 min	Functional ingredient in the food industry	(Behere, Patil and Rathod, 2021)
10	Brown crab (*Cancer pagurus*) shell	Subcritical water extraction for Protein hydrolysates	Temperature: 200 °C; solid- liquid ratio: 1:15 g/ml; heating rate: 6 °C/min	Nutraceutical, pharmaceutical, and food industry	(Rodrigues, Matias and Paiva, 2021)
11	Cocoa (*Theobroma cacao*) bean shell	MAE for protein	Temperature: 97 °C; pH: 12 for 5 min	Food additives	(Mellinas, Jiménez and Garrigós, 2020)
12	Codfish (*Gadus morhua*) frames (backbone with residual meat)	Subcritical water extraction for collagen protein	Temperature: 90–190 °C; pressure:100 bars	Food and nutraceuticals	(Melgosa et al., 2021)
13	Chicken bone	Ultrasonic treatment with enzymatic extraction for protein	400 W for 120 min at 45 °C; enzyme used: papain at pH 6.9	Functional ingredient in the food industry	(Dong et al., 2019)
14	Tilapia (*Oreochromis* sp.) fish scale	Extrusion-hydro- extraction for collagen protein	Single screw extrusion method; Screw speed: 360 rpm; barrel temperature: 135 °C; hydro-extraction temperature: 25–50 °C	Cosmetic, medical, and food industry	(Huang et al., 2016)

15	Clown featherback (*Chitala ornata*) skin	UAE for collagen protein	750 W and 20 kHz single frequency pulse mode and 80% amplitude for 10 min	Food industry	(Petcharat et al., 2021)
16	Banana peel	Osborne fractionation for albumin, glutelin, prolamin, and globulin	Treating with deionized water in a 1:10 ratio (W/V) at pH 7 and treated with an alkali solution	NM	(Deb, Kumar and Saxena, 2022)
			Chemical methods		
1	Wastewater from soya bean (*Glycine max*) oil body processing	Alkaline extraction- isoelectric-point- precipitation for protein	Various pH levels	Edible packaging	(Song et al., 2023)
2	Hemp (*Cannabis sativa*) seed cake	pH shift method for protein isolates	pH: 10.5 for 1 h at room temperature	Plant-based meat or dairy analog for human consumption	(Helstad et al., 2022)
3	Chicken exudates	pH shift method for protein isolates	NaOH-HCl/$Ca(OH)_2$-citric acid/ $Ca(OH)_2$-Glucono-δ-lactone	Additives in food industry	(Jiang et al., 2022)
4	Tofu whey wastewater	Ethanolic-isoelectric precipitation and gel filtration chromatography for lunasin peptide	Alcoholic precipitation by 30% ethanol (v/v); pH: 4.5; chromatography by mobile phase: 20 mM Tris-HCl buffer + 0.15 M NaCl, pH: 7.6 vs. 100 mM ammonium bicarbonate; pH: 7.8	Dietary supplements against degenerative and chronic disease	(Nieto-Veloza et al., 2021)
5	Pumpkin (*Cucurbita* g.) seed and peel	Solvent extraction for protein	Temperature: 32.7 °C; pH: 9.51 for 16.06 min	Edible film in food industry	(Lalnunthari et al., 2020)
6	Catfish (*Ictalurus punctatus*) skin	Acid hydrolysis for collagen protein	HCl at pH 2.4, Acetic acid at pH 2.7	Food industry	(Tan and Chang, 2018)
7	Salted egg white	Aqueous two-phase flotation for ovalbumin protein	28 ml of 28% of ammonium sulfate and 5 ml of 80% of polyethylene glycols; flotation time: 30 min; flow rate: 35 ml/min	NM	(Jiang et al., 2019)

(Continued)

TABLE 4.2 (Continued)

S.No	Food waste	Methods of extraction	Control parameters	Applications	References
8	Unicorn leatherjacket (*Aluterus monoceros*) and reef cod (*Epinephelus diacanthus*) skin	Chemical hydrolysis for gelatin protein	0.2 M NaOH for 2 h at 4 °C and 0.05 M phosphoric acid for 2 h	Food industry	(Renuka et al., 2019)
9	Golden pompano (*Trachinotus blochii*) skin and bone	Acetic acid-pepsin method for pepsin soluble collagen	1:40 (w/v) solid: solvent ratio of acetic acid solution containing 1% pepsin for 48 h.	Food and biomedical industry	(Cao et al., 2019)
10	Porcine liver	Autohydrolysis for protein	Autolysis incubation at 45 °C and pH 4.8 under anaerobic conditions after treating with acetic and ascorbic acid, and then treating in alkaline conditions	Natural bioactive pigment in various industries	(López-Pedrouso et al., 2023)
11	Orange (*Citrus sinensis*) seed	Alkaline extraction and isoelectric precipitation method for protein isolates	42 kHz and 130 W ultrasonic treatment for 15– 30 min	New product development in food industry	(Rosas Ulloa et al., 2023)
			Biological methods		
1	Meat bone	EH for polypeptides	Substrate concentration: 9%; lipase: 0.08%; temperature: 40 °C; pH: 7.5; time: 4 h.	Medicine, health care and food	(Yao et al., 2020)
2	Brewers spent grains [barley (Hordeum vulgare) and rice (Oryza sativa) protein]	EH for protein isolates	NM	Skeletal muscle synthesis, product development to give sensory characteristics and textural properties	(Ummels et al., 2023)

3	Spent hens	EH for polypeptides	pH: 6.5; temperature: 55 °C; Flavoenzyme (endopeptidase and exopeptidase from *Aspergillus oryzae*): 2%	Dietary supplements	(Kumar et al., 2023)
4	Fish cutting waste (scale and skin)	EH for protein and polypeptide hydrolysates	0.25% Alcalase 2.5 L at 50 °C for 6 h; pH: 7.0	Peptide supplements, functional foods and Protein- mineral supplement	(Kuprina et al., 2023)
5		Enzymatic-thermal hydrolysis for protein and polypeptide hydrolysates	Temperature treatment at 130 °C for 1h after enzymatic hydrolysis		
6	Rice (Oryza sativa) bran and soybean (Glycine max) cake	EH for protein	5% carbohydrase treatment at optimum time and temperature	Supplements	(Scarabattoli et al., 2023)
7	Sweet potato (Ipomoea batatas) waste	EH for protein	0.5% alcalase; temperature: 55 °C pH: 13 for 0.5 h	Foaming and stabilizing agent in food industry	(Gao et al., 2023)
8	Rohu (Labeo rohita) fish waste (head, fins, scale, swim bladder)	EH for peptides	1.08% alcalase; temperature: 52.10 °C; hydrolyzing time: 129.18 min	Nutrient supplements, functional foods and pharmaceutical	(Kumar et al., 2021)
9	Sesame (Sesamum indicum) bran	Microwave-assisted enzymatic extraction for protein	1.94 AU/100 g enzyme alcalase concentration; temperature: 49 °C for 98 min	Functional ingredient in food industry	(Görgüç, Özer and Yılmaz, 2020)
10	Carp fish by- products [(Cyprinus carpio) head, skin and skeleton]	EH for collagen protein	2% alcalase at 55 °C for 3 h	Functional food and pharmaceuticals	(González-Serrano et al., 2022)

(Continued)

TABLE 4.2 (Continued)

S.No	Food waste	Methods of extraction	Control parameters	Applications	References
11	Tomato (Solanum lycopersicum) seed	Fermentation by Bacillus subtilis A14h and lyophilized hydrolysate for peptides	2% (v/v) culture for 24 h at 37 °C	Functional foods and nutraceuticals	(Moayedi et al., 2018)
12	Shrimp shell	EH for protein hydrolysates	Pepsin at pH 2; trypsin at pH 8; chymotrypsin at pH 8; and alcalase at pH 8; Hydrolyzed for 1 h at 50 °C	Supplements, pharmaceuticals for blood pressure	(Ambigaipalan and Shahidi, 2017)
13	Carp (Cyprinus carpio) fish by-products	EH (protamex enzyme mixture) for Protein (gelatin) hydrolysates and peptides	2% (w/w) enzyme culture at pH 7 and temperature 50 °C for 3 h	Functional food and supplements	(Tkaczewska, Bukowski and Mak, 2018)
14	Skipjack tuna (Katsuwonus pelamis) head	In vitro gastrointestinal digestion for protein hydrolysates	Hydrolyzed by pepsin pH 1.5 and trypsin pH 7.0 for 2 h at 37 °C.	Health care products	(Zhang et al., 2019)
15	Catfish (Ictalurus punctatus) skin	EH for collagen protein	Solid-to-liquid ratio: 1:30; 2.4 HCl; 23.6 KU/g pepsin	Food industry	(Tan and Chang, 2018)
16			Solid-to-liquid ratio: 1:20; 2.4 HCl; 5.90 KU/g pepsin		
17	Perilla seed (Perilla frutescens) meal	EH for peptides	25 units of trypsin at pH 8.0 for 3 h at 37 °C	Food industry	(Kim, Liceaga and Yoon, 2019)

18	Poultry by-products (feathers, viscera, and guts), swine mucosa, and liver.	EH for protein hydrolysates	NM	Fish food	(Alves et al., 2019)
19	Skin and scales of sole fish (Cynoglossus arel)	EH for gelatin hydrolysates	1% alcalase at 55 °C for 2 h	Nutraceutical and functional food ingredients	(Viji et al., 2019)
20	Chicken wattles and combs	EH for protein hydrolysates	Treated with 5% alcalase for 240 min	Biomedical industry	(Bezerra et al., 2019)
21	Pig bones	EH for protein	Alcalase: 2.3% at 56 °C for 138 min; pH: 7.5	Functional foods	(Dao et al., 2022)
22	Shrimp by-products	EH for iron-binding peptides	80% degree of hydrolysis with alcalase enzyme at pH 7.8 and 55 °C.	Additives in the food industry	(Huang, Ren and Jiang, 2011)
23	Kuruma shrimp (Marsupenaeus japonicus) heads	EH for protein hydrolysates	Papain enzyme at pH 7.0 and temperature 50 °C, and enzyme/substrate is 10,000 U/g	Food and pharmaceutical industries	(Zhou et al., 2023)

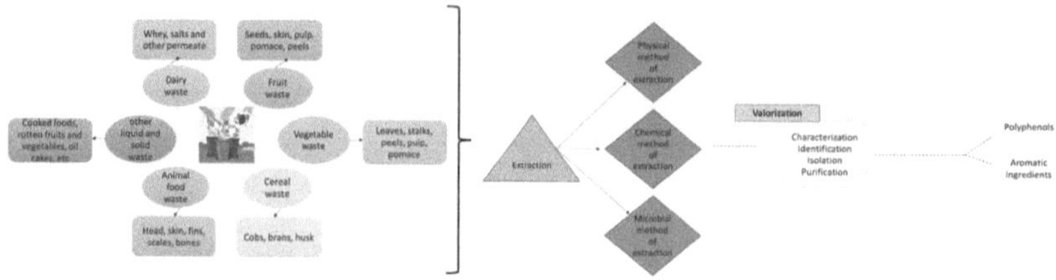

FIGURE 4.4 Polyphenols identified from food waste.

2022). Meanwhile, UAE and high-pressure processing are innovative technologies that have been used for the same with greater yield of protein. These are widely applied green and sustainable technologies due to their reduced solvent utilisation, time efficiency, and non-thermal properties. Conventional methods like acid and alkali hydrolysis, solvent extraction, and pH shift methods were used to extract gelatin, collagen, and protein isolates. pH shift method yielded 90.3% purity of protein content with the reduced phytic acid profile (Jiang et al., 2019; Renuka et al., 2019; Petcharat et al., 2021; Helstad et al., 2022). Meanwhile, the physical methods are found to generate protein with a higher yield around 60–95% as compared to other methods (Tarna et al., 2021; Cabral et al., 2023; Kuprina et al., 2023)

4.5 POLYPHENOLS AND AROMATIC INGREDIENTS FROM DIFFERENT FOOD WASTE AND EXPIRED FOOD

Polyphenols are a class of naturally occurring antioxidants found in various plant-based foods and aromatic compounds like essential oils are concentrated hydrophobic liquids containing volatile aroma compounds from plants (Saba et al., 2023). Figure 4.4 shows polyphenol and aromatic compounds identified from FW or expired food. They are commonly used in various applications, including aromatherapy, perfumery, and natural medicine (Yang et al., 2023). These bioactive compounds have antioxidants, anti-inflammatory, and antimicrobial properties that contribute to health and wellness (Raunhan et al., 2023). The entire process of converting FW into polyphenols and other ingredients involves sourcing discarded or waste food materials from various points in the food supply chain, such as farms, processing facilities, or consumer households, including parts of fruits, vegetables, and other plant-based materials that are typically discarded (Patsalou et al., 2020). The different extraction methods for essential oils from FW, particularly from sources like citrus peels and herbs offer polyphenols and aromatic compounds are shown in Table 4.3. The detailed operating parameters can be used as references for different types of polyphenol or aromatic compounds. The subsequent step includes extracting polyphenols or other target compounds from the collected FW using methods like physical, chemical, microbial and other biological methods (Ravikumar et al., 2021). Following extraction, the crude polyphenol or aroma compound extracts will be purified and concentrated to isolate and enhance the concentration of polyphenolic compounds to obtain higher quality and more potent polyphenol products (Krishnan et al., 2022). Referring to Table 4.3, citrus peels, for example, are a common source

TABLE 4.3
Physical, chemical and microbial methods used to extract polyphenols and aromas from food waste

S.No	Food waste	Method	Control parameters	Application on one health	References
			Physical methods		
1	Orange peel (*Citrus sinensis*), Lime peel (*Citrus limon*)	Ultrasound-assisted extraction (UAE) for extraction of bioactive compounds	Solvent: water: methanol, water: acetone, water: ethanol (1:20 *w/v*), Frequency: 38.5 kHz, Power: 50.93 W, Time: 30 min, Temperature: 40 °C	Human health	(Singanusong et al., 2015)
2	Pomegranate peel (*Punica granatum*)	Ultrasound-assisted extraction (UAE) for extraction of antifungal property identification	Solvent: water: ethanol, Frequency: 20 kHz, Power: 400 W, Time: 5, 10 and 15 min, Temperature: 50 °C	Human and animal health	(Sharayei et al., 2019)
3	Rambutan peel (*Nephelium lappaceum*)	Ultrasound-assisted extraction (UAE) for extraction of bioactive compounds	Solvent: water, Frequency: 20 kHz, Power: 60 W, Time: 20 min, Temperature: 60 °C	Biological health	(Prakash Maran et al., 2017)
4	Orange peel (*Citrus sinensis*)	Ultrasound-assisted extraction (UAE) for extraction of phytochemicals	Solvent: water: ethanol, Frequency: 20 kHz, Power: 150 W, Time: 40 min Temperature: 50 °C	Biological health	(Wang et al., 2018)
5	Jackfruit peel (*Artocarpus heterophyllus*)	Ultrasound-assisted extraction (UAE) for extraction of pectin	Solvent: ethanol, Frequency: 70 kHz, Power: NA, Time: 15– 30 min, Temperature: 50–70 °C	Biological health	(Moorthy et al., 2017)

(Continued)

TABLE 4.3 (Continued)

S.No	Food waste	Method	Control parameters	Application on one health	References
6	Sweet orange peel (*Citrus limetta*)	Ultrasound-assisted extraction (UAE) for extraction of polyphenols	Solvent: water: ethanol, Frequency: 25 kHz, Power: 150 W, Time: 10–40 min Temperature: 40 °C	Environmental health	(Khan et al., 2010)
7	Grape peel (*Vitis vinifera*)	Ultrasound-assisted extraction (UAE) for extraction of pectin	Solvent: water: ethanol: citric acid, Frequency: 55 kHz, Power: 100 W, Time: 60 min, Temperature: 75 °C	Biological health	(Minjares-Fuentes et al., 2014)
8	Dragon fruit peel (*Selenicereus undatus*)	Ultrasound-assisted extraction (UAE) for extraction of pectin	Solvent: water: ethanol, Frequency: 37 kHz, Power: 400 W, Time: 75 min (retrieved every 5 min), Temperature: 50, 60 and 70 °C	Biological health	(Lin, Kai and Ali, 2018)
9	Hawthorn peel (*Crateagus monogyna*)	Ultrasound-assisted extraction (UAE) for extraction of phytochemicals	Solvent: water: ethanol, water: methanol, aqueous, Frequency: 37 kHz, Power: 400 W, Time: 30 min, Temperature: 50 °C	Human health	(Ravikumar, Lin and Hui, 2022)
10	Pomelo peel (*Citrus maxima*)	Ultrasound-microwave-assisted acid extraction (UMAE) for extraction of pectin	Solvent: water: ethanol: citric acid, pH: 1.80, Sonication time: 27.52 min, Microwave time: 6.40 min, Irradiation power: 643 W	Biological health	(Liew et al., 2016)
11	Ripe mango peel (*Mangifera indica*)	Microwave-assisted extraction (MAE) extraction of polyphenols	Solvent: lactic acid/sodium acetate deep eutectic mixtures, Power: 100–800 W, 2450 MHz	Human health	(Pal and Jadeja, 2020)

(Continued)

TABLE 4.3 (Continued)

S.No	Food waste	Method	Control parameters	Application on one health	References
12	Orange peel (*Citrus sinensis*)	Microwave-assisted extraction (MAE) extraction of pectin	pH: 1.2, Time: 7 min, Solid: liquid ratio: 21.5 (*v/w*)	Biological health	(Su et al., 2019)
13	Orange peel (*Citrus sinensis*)	Soxhlet apparatus for extraction of essential oil	Solvent: acetone, chloroform, diethyl ether, ethanol, ethyl acetate, and n-hexane, Temperature: 70 °C, Time: 8 h	Environmental health	(Battista et al., 2020)
14	Orange peel (*Citrus sinensis*)	Solid/liquid extraction for extraction of essential oil	Solvent: Distilled water (1:5), Rotary Evaporator temperature: 70 °C Rotation: 40 rpm and 750 mm of Hg, Time: 1 h	Industrial use	(Jha et al., 2019)
15	Orange peel (*Citrus sinensis*)	Solid/liquid extraction for extraction of essential oil	Solvent: Cyclopentyl Solid: liquid ratio: 1:10, Time: 50 min, Temperature: 70 °C, Rotation: 900 rpm	Industrial use	(Ozturk, Winterburn and Gonzalez-Miquel, 2019)
16	Orange peel (*Citrus sinensis*)	Steam distillation for extraction of essential oil	Solvent: aqueous ethanol, Particle size: 1– 4 mm, Distillation apparatus: Oenological Distillation unit “DE-1626”	Environmental health	(Tsouko et al., 2020)
17	Citrus peel (*Citrus limon*)	Steam explosion for extraction of essential oil	Solid: liquid ratio: 10:13 (*w/v*), Temperature: 170 °C, Pressure: 8 bar, Time: 240 s	Industrial use	(Golmohammadi et al., 2018)
18	Citrus peel (*Citrus limon*)	Hydro distillation for extraction of essential oil	Solid: liquid ratio: 100:2 (*w/v*), Time: 4 h	Industrial use	(Golmohammadi et al., 2018)

(Continued)

TABLE 4.3 (Continued)

S.No	Food waste	Method	Control parameters	Application on one health	References
19	Grape waste (*Vitis vinifera*)	Lyophilisation for extraction of polyphenols	Pre-freezing: 48 h (−20 °C), Solvent: methanol/ethyl acetate, Centrifuge: 10,000 rpm, Time: 15 min, Evaporating temperature: 40 °C	Human health	(Liu et al., 2023)
		Chemical methods			
1	Solid food waste	Precipitation polymerization for extraction of polyphenols	Minimum working volume: 80 ml, Size of feed vessel: 500 ml Membrane active surface: 200 cm^2 Operating pressure: 1–40 bars	Biological health	(Mansour et al., 2018)
2	Citrus peel (*Citrus limon*)	Extraction of essential oils using a bioreactor	Autoclave temperature and time: 121 °C and 15 min, pH: 6.8 ± 0.05, Feed constant temperature: 37 °C, Rate of CO_2 supply: 0.5 vvm	Industrial use	(Patsalou et al., 2020)
		Microbial methods			
1	Cereal waste	Anaerobic digestion for extraction of polyphenols	Fermentation: (pre- treatment) absence of O_2	Biological health	(Cudjoe et al., 2022)
2	Chopped vegetable waste	Anaerobic digestion for extraction of polyphenols	Fermentation: (pre-treatment) absence of O_2 in the presence of *T. mechernichensis TL1*	Biological health	(Raunhan et al., 2023)
3	Cottonseed hull (*Gossypium hirsutum*)	Solid-state fermentation to analyse the antifungal and antibiotic properties	Sample 2: autoclaved CSH, sampling time: 0, 14, 21, 28 and 35 days, Triplicate analyses pattern: 2 × 5 × 3		

(Continued)

TABLE 4.3 (Continued)

S.No	Food waste	Method	Control parameters	Application on one health	References
			= 30 bottles Storage temperature: −80 °C	Human and animal health	(Du et al., 2022)
4	Cauliflower leaves (*Brassica oleracea*)	Enzyme-assisted fermentation for extraction of polyphenols	Solvent: aqueous enzyme solution Incubation temperature: 40 °C pH 4.0, Enzyme concentration: 0–5% E/S, Storage temperature: −18 °C	Industrial use	(Huynh et al., 2014)

of essential oils such as orange, lemon, or grapefruit oil. Other food by-products, like herbs and spices, can also yield essential oils through distillation or cold-press extraction methods. This process uses FW or expired food to produce natural fragrances and flavouring (Pandey and Rizvi, 2009). The extracted essential oils can find applications in aromatherapy, perfumery, and the food and beverage industry (Pavlić et al., 2023).

4.6 CASE STUDY OF DIFFERENT FW

4.6.1 Dragon Fruit By-Products and Extraction Methods

Pitaya fruits commonly known as dragon fruits are categorized under the cactus species in the genus of Hylocereus and Stenocereus. There are three varieties of dragon fruits: Hylocereus polyrhizus (which has red-magneta skin and red pulp), *Hylocereus undathus* (which has red-magenta skin and white pulp), and *Hylocereus megalanthus* (which has yellow skin and white pulp). The peels from the dragon fruit account for one-third of the fruit's weight (Chua et al., 2020). Peels are always thrown away as waste, particularly in the agriculture, food and beverage industries. If dragon fruit peels can be reused, that would be beneficial to the environment and reduce the costs for further processing of the peel waste.

Dragon fruit peels are a rich source of bioactive compounds that have a variety of health-promoting properties. These compounds include polyphenols, carotenoids, flavonoids, and DF (Ben-Othman et al., 2020). Peels from dragon fruit have been shown to contain a variety of phenolic compounds, such as flavonoids, anthocyanins, hydroxycinnamic acids, gallic acid, and protocatechuic acid (Chen et al., 2021). There are several health benefits associated with gallic acid, such as anti-inflammatory and anti-cancer properties. The vibrant colour of dragon fruit skins are attributed to anthocyanins, which also have strong anti-inflammatory and antioxidant effects.

Peels from dragon fruit contain the flavonoids quercetin, kaempferol, myricetin, and isorhamnetin (Saenjum et al., 2021). Quercetin and kaempferol are associated with cancer prevention and cardiovascular health while myricetin is linked to neuroprotective, anti-inflammatory, and anti-cancer properties. Peels from dragon fruit are rich in carotenoids such as betacarotene, lycopene, and betalains (Taliana, 2020). Some types of dragon fruit may contain betalains, a unique class of pigments. Betalains are responsible for the vibrant purple and scarlet colour of these fruits. The potential health benefits of betalains include their ability to reduce inflammation and oxidative stress as well as their anti-inflammatory and antioxidant properties. Pectin, a complex polysaccharide consisting of galacturonic acid units joined by α-1,4-glycosidic connections, is present in substantial amounts in dragon fruit peels (Chen et al., 2022). Pectin derived from dragon fruit peels is typically classified as low methoxyl pectin (LMP) which can effectively decrease the interfacial tension, resulting in excellent emulsification properties when used individually or in conjunction with proteins (Verkempinck et al., 2018). LMP is commonly employed in the food industry as a stabiliser, thickening agent, and gelling agent in the absence of sugar and acid. This illustrates the potential of dragon fruit peels as a cost-effective and sustainable source of bioactive compounds.

Nevertheless, information regarding the emulsification properties of pectin derived from dragon fruit peels is limited. Thus, pectin derived from dragon fruit peels was extracted using the hot acid method by Chen et al. (2022), and comparative test was performed to assess the extracted pectin's emulsification properties. This involved stirring dried dragon fruit peel powder in water, adjusted to pH 2 using citric acid, for 2 h at 90 °C. The crude extract was then centrifuged and subjected to ethanolic precipitation for 12 h to obtain the crude pectin. Lastly, the crude pectin was purified using dialysis and freeze-dried. The pure dragon fruit peel pectin's molecular weight, galacturonic acid content, and degree of esterification were determined to be 1,181.75 kDa, 87.02%, and 37.26%, respectively. The extracted pectin is an ultra-LMP, as evidenced by the low molecular weight and degree of esterification value. The extracted dragon fruit peel pectin also demonstrated better emulsifying properties at low concentrations than the two commercial pectins (APA140 and LMP). This is attributed to its lower molecular weight and higher rhamnogalacturonan I (i.e., RG-I consists of repetitive disaccharide units with various glycan chains) content, indicating its potential for use in food applications. The stability of the betacyanin extract from dragon fruit peels was examined by Priatni and Pradita (2015). Water, methanol, acidified water, and acidified ethanol were used in a maceration process that took place in a refrigerator for 24 h to obtain the extract. The amount of betacyanin content that was extracted using methanol at pH 5 (515.20 μg/100 g) was more than the amount that was extracted using water at pH 5 (491.16 g/100 g). The temperature and solvent type had an impact on the stability of betacyanin extract and the yield of betacyanin in the methanol extract dropped by approximately 10.44% after 5 h at room temperature, while the yield of betacyanin in the water extract dropped by approximately 22.58%. The accelerated reaction method's prediction of betacyanin's stability revealed that it was susceptible to thermal degradation. The primary structure of betacyanin, betanin, will be broken down by isomerization and decarboxylation during heat processing, which will eventually cause the colour to turn lighter brown and gradually lose its red hue. Phongtongpasuk et al. (2016) employed a dragon fruit peel water extract as an environmentally friendly approach to synthesize biogenic silver nanoparticles. The phytochemical compounds present in the plant extract serve as a capping and reducing agent to stabilize silver nanoparticles and contribute to the reduction of Ag+ to Ag0. The production of biosynthesized silver nanoparticles with good dispersion and stability in the suspension was validated by UV-Vis spectroscopy, Transmission Electron Microscopy (TEM), energy dispersive X-ray analysis (EDX), X-ray Diffraction (XRD), and zeta potential analysis. When tested against pathogenic bacteria, all biosynthesized nanoparticles showed strong antibacterial activity. The biological synthetic approach is more favorable than the physicochemical method for synthesizing silver nanoparticles since it is straightforward, affordable, environmentally benign, and simple to scale up for mass production. The physicochemical method involves chemical toxicity and is costly. Due to its biocompatibility, the silver nanoparticle produced using a biological synthetic approach could be applied safely in numerous therapeutic applications.

MAE outperforms the conventional approach by decreasing the extraction time and solvent usage while simultaneously increasing the extraction rate. A four-variable, three-level Box-Behnken design combined with response surface methodology was applied by Thirugnanasambandham et al. (2014) to improve the MAE process. They evaluated the impact of different MAE extraction conditions on the yield of dragon fruit peel pectin. The solvent applied was water with pH adjusted using 0.05 M nitric acid. The ideal extraction conditions, which yielded the optimum amount of pectin (7.5%), were 400 W of power, 45 °C, 20 min of extraction time, and a solid-liquid ratio of 24 g/ml. Using the fuzzy assessment method, Tongkham et al. (2017) assessed the MAE conditions to identify the ideal conditions for achieving a better yield and less pectin degradation. The MAE approach yielded a higher pectin production than the CE method at 85 °C for 1 h. The maximum pectin yield of 23.11% was achieved with 600 W microwave power and a 10 min heating duration. However, pectin degradation occurred during the MAE extraction, resulting in the viscosity decreasing. According to the best overall performance index derived from FAM, 450 W of microwave power and 5 min of extraction time were ideal for producing a large amount of pectin with minimal degradation.

UAE is an effective method to extract pectin from dragon fruit peel and dragon fruit peel pectin has higher antioxidant activity than commercial citrus pectin. Chua et al. (2020) investigated the effects of pre-treatment (drying) temperature and ultrasonic-assisted extraction parameters on the extraction of pectin from dragon fruit peels. The optimum oven-drying temperature for pre-treatment was found to be 45 °C with the lowest ash content (8.34%) and the maximum pectin yield (6.27%). Then, the ultrasound-assisted pectin extraction was optimized for time, temperature, and solid-to-liquid ratio using a three-factor, three-level Box-Behnken design. A maximum pectin yield of 23.09% was achieved under the optimum conditions of 32 min at 85 °C and a solid-to-solvent ratio of 1:30 g/ml using 0.1 M citric acid at pH 2.0 as the solvent. Pectin's degree of esterification both before and after optimization (36.36% and 33.27%, respectively) indicated that it had a low methoxyl content. Pectin's ash level decreased by 23.65%, indicating that higher-purity pectin was extracted after optimization. Fourier Transform Infrared (FTIR) spectroscopy analysis shows that the extracted pectin from dragon fruit peels has a high potential to be an alternative source of pectin in food applications because it is rich in polygalacturonic acid and has functional groups that are like those of commercial grade pectin. The physio-chemical properties and antioxidant activity of pectin from white dragon fruit peel were evaluated by Nguyen and Pirak (2019) at different extraction conditions using conventional and UAE. It was found that the UAE increased the extracted pectin's antioxidant activity. The maximum antioxidant activity in dragon fruit peel extracted with the UAE at 45 °C for 30 min was as high as 51.58 mg GAE/100 g by DPPH, and 39.81 mg GAE/100 g by ABTS. Further to this, the UAE can extract a higher amount of pectin yield of 9.38% than the CE treatments.

4.6.2 Pink Guava By-Products and Extraction Methods

Pink guava (*Psidium guajava*) is a type of pink-fleshed tropical food known for its sweet and slightly tangy flavour. It originated from Central America and it is now widely cultivated in tropical and subtropical regions around the world. It is well known to possess lots of nutrients, and it is an excellent source of vitamin C, with even higher content compared to other citrus fruits like oranges. Two types of guava fruit are being cultivated on a large scale, namely, white- and pink-fleshed guava. Guava is a highly productive and profitable crop due to its extensive adaptability and high yield (Hassan et al., 2012; Kumar et al., 2022). From the literature, it was reported that white-flesh guava (142.6 mg/100 g) is a better source of vitamin C than pink-flesh guava (72.2 mg/100 g) (Chang et al., 2020). Besides, pink guava also contains vitamin A, vitamin B, potassium, DF, phenolic compounds, and carotenoids, particularly lycopene which gives the flesh its pink colour. Pink-fleshed guava is known to possess higher antioxidant properties compared to white-fleshed guava (Chang et al., 2020).

In the processing and manufacturing industry this fruit is mostly processed into puree, juice or concentrate, jellies, soft drinks, as well as baby food. Processing guava fruit into various products generates a significant number of wastes and by-products, including skin, seeds and pomace. The worldwide production of guava is approximately 500,000 metric tons (Kumar et al., 2022). Approximately 80 kg of waste is generated per metric ton of guava fruit being processed. Depending on the guava variety, about 4–30% of the fruit's weight is converted into waste and by-product residues. This solid agro-industrial residue consists of a mixture of peel, pulp, and seeds, which are rich in phenolic compounds and possess significant antioxidant activity. Guava residue typically contains 15–50% seeds by weight and around 15% peel (Rojas-Garbanzo et al., 2019; Kumar et al., 2022). In the production of guava pulp, approximately 30% of the weight of the fruit becomes residue/waste, with seeds comprising about 6% of this residue (da Silva et al., 2020; Kumar et al., 2022). Whereas guava peels, which are considered a waste product of guava processing, account for roughly 20% of the fruit's fresh weight (Rojas-Garbanzo et al., 2019). Guava waste, including the peel and seeds, can be utilized to prevent oxidative stress and related disorder in pharmaceuticals and other industries, as they are rich in bioactive compounds such as phenolic compounds, IDFs, carotenoids, saponins, phytosterol, and fatty acid (Kumar et al., 2022). From the literature, guava seeds contain unsaturated fatty acids such as oleic acid and linoleic acid, IDF and other bioactive compounds like ascorbic acid and carotenoids (Uchôa-thomaz et al., 2014). Guava seeds are also rich in numerous amino acids such as tryptophan and methionine (Nicanor et al., 2001).

Guava pomace, a major by-product generated in the guava juice and puree industry, is a rich source of bioactive compounds. To produce pink guava puree, the processes involved are crushing, pulping, refining and sieving, and finally decanting to eliminate the guava stone cells (Nagarajan et al., 2019). In Malaysia, around 25% of the pink guava fruits were discarded as by-products during puree processing (Kong et al, 2010). These by-products from the decanter contain a significant number of bioactive compounds and essential nutrients. Kong and Ismail (2011) found that the lycopene content in the pink guava decanter by-products (5.51 mg/100 g wwb) is significantly higher compared to those of whole fruit (3.78 mg/100 g wwb). Lycopene is a naturally occurring bioactive compound classified under carotenoid pigment. It was found in high concentrations of tomatoes, watermelons, and pink grapefruits. Lycopene is known for its antioxidant properties, which means it helps neutralize harmful free radicals in the body, reducing the oxidative stress of our cells, thus preventing cellular and DNA damage as well as downregulating the inflammatory mediators (Vasconcelos et al., 2017; Kong et al., 2010). Pink guava by-products are also rich in other phytochemicals such as phenolic compounds, DF and low molecular weight carbohydrates (Chang et al., 2020). Thus, these pink guava by-products are potential sources to be valorised and extracted to be utilized in various ways. The phenolic compounds of the guava pomace extract (3.40 mg gallic acid equivalent (GAE)/g) were myricetin, quercetin, epicatechin, isovaleric and gallic acid found to be the major phenolic compounds. The guava pomace extracts also exhibited

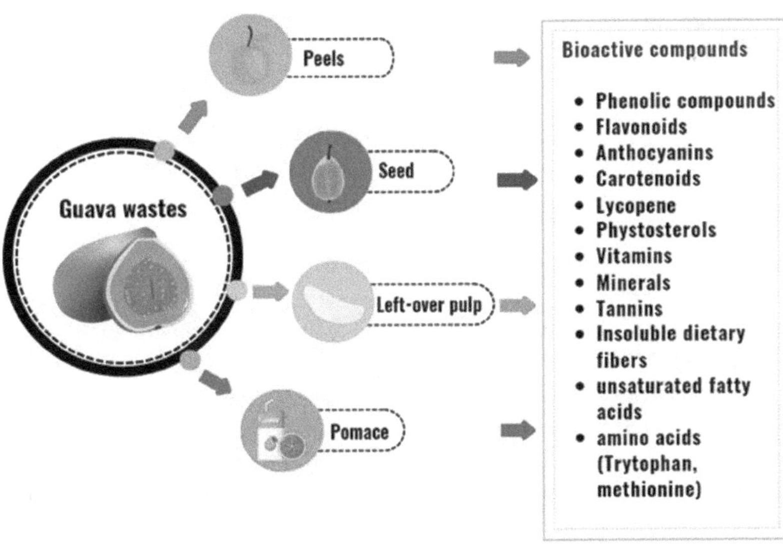

FIGURE 4.5 Bioactive compounds present in guava wastes.

anti-inflammatory and antinociceptive activities (Denny et al., 2013). The bioactive compounds present in the guava waste are presented in Figure 4.5.

Lycopene can be extracted from pink guava waste using solvent extraction. The solvents are organic solvents like acetone: hexane (2:5), acetone with 0.5% butylated hydroxytoluene (BHT): 95% ethanol: hexane (1:1:2) (Kong and Ismail, 2011) methanol: ethyl ether (1:1) or solely hexane. Killadi et al. used a sequential extraction method where acetone was used for extraction at the beginning, followed by petroleum ether (Killadi et al., 2018). As lycopene is very light sensitive, extra precautions were taken, and the extraction was carried out in dark conditions to avoid degradation by light. The extraction of bioactive compounds from guava waste and by-products typically involves using large quantities of organic solvents, which poses environmental hazards and health risks to operators. Several studies have employed traditional extraction methods, such as maceration and Soxhlet extraction, to extract bioactive compounds from guava wastes. These conventional methods are often time-consuming, require many organic solvents, and are challenging to scale up commercial applications. In contrast, emerging green extraction methods like pressurized liquid extraction, UAE, and supercritical fluid extraction can minimise the use of toxic solvents, reduce extraction time, and operate under mild conditions. For example, the use of SFE with carbon dioxide (SC-CO_2) and 10% ethanol as a co-solvent to extract phytochemicals from pink guava decanter by-products (Kong et al., 2010). This advanced extraction method stands out as a cleaner extraction process as organic solvents are often deemed toxic to the human body and the environment. It was reported that SC-CO_2 can produce an overall yield of 3.15 mg/100 g dried sample. The only issues are lower lycopene content of 13.66 mg/g extract and antioxidant activity compared to the solvent extraction method. da Silva et al. (2020) extracted the lycopene and β-carotenoids from guava pulp and guava waste using both the maceration and UAE methods. Their finding indicated that UAE had superior extraction capacity compared to the conventional maceration method, achieving a maximum lycopene yield of 135 mg/100 g from guava pulp and 76.64 mg/100 g guava waste, with reduced extraction time and lower solvent consumption.

Pectin can be extracted from guava waste too. Bhat and Singh (2014) optimized the acid extraction of pectin from guava peel as shown in Figure 4.6. The efficiency of two citric acid and hydrochloric acid were tested at varying times of 30–60 min, temperature of 65–85 °C and pH conditions from 2–3. The optimal extraction conditions were achieved using HCl at 85 °C for 60 min at pH 2, resulting in a pectin yield of 16.8%, with methoxyl content of 4.25%, anhydro-uronic acid content of 67.4%, ash content of 1.8% and moisture content of 3.2%. Alternatively, the environmentally friendly deep eutectic solvent (DES) can be utilized to extract pectin from guava waste as a substitute for conventional solvents. Yusof et al. (2020) found that the pectin extracted using 10% choline chloride/ethylene glycol (ChCl/EG) demonstrated higher gel-forming ability compared to pectin extracted using 10% citric acid. The pectin extracted using 10% ChCl/EG also showed high methoxyl content (15.6%), higher degree of esterification (98.47%) and lower ash content (2.02%) than that extracted using 10% citric acid.

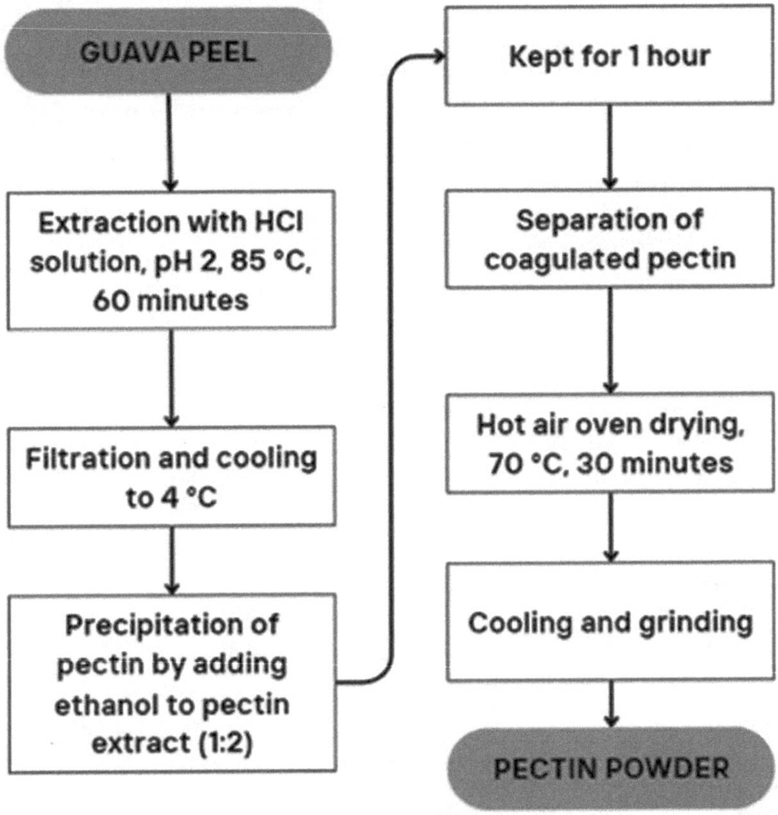

FIGURE 4.6 Production of pectin from guava peel.

4.7 CONCLUSION

The extraction and utilisation of bioactive compounds present in expired food and FW were discussed in this chapter. The compounds identified include polyphenols, DF, proteins, lipids, vitamins, and minerals that can offer opportunities for creating novel and health-enhancing products. The extraction methods are chemical, microbial, enzymatic, and green extraction methods. The efficiency of valorisation technologies that can handle the diverse composition of expired food or FW needs to be optimized. Optimization of processes including energy-efficient extraction, and design of innovative systems for waste-to-product conversion. In addition, there is limited large scale or commercial research scale carried out on the valorisation of fruit and vegetable by-products, the process is still in its infancy and faces challenges like practical implementation from lab-scale extraction to pilot-scale and later on industrial-scale processing. Besides, the recovery of the solvent and the treatment of the left over filtrate also need to be studied.

EXERCISES

1. What are the three major stages in valorisation of food and vegetable waste into bioactive compounds?
2. What are the techniques usually employed for pre-treatment stages and how does pre-treatment stage enhance the efficiency of the bioactive compound extraction from food and vegetable waste?
3. What is the most important parameter or key consideration during the extraction process?
4. What physical methods are involved in extracting fibers from expired food or food waste?
5. What are the most common sources of polyphenols found in expired food or food waste?
6. Are specific expired food or food waste streams affluent in fibers, proteins, polyphenols, or aromas?
7. What are the potential applications of extracted protein in various industries?
8. How does the valorisation of expired food and food waste through extraction align with sustainable practices?

REFERENCES

Adesogan, A. T. et al. (2020) 'Animal source foods: Sustainability problem or malnutrition and sustainability solution? Perspective matters', *Global Food Security*, 25, p. 100325. doi: 10.1016/j.gfs.2019.100325.

Aiello, G. et al. (2022) 'Quality assessment of the protein ingredients recovered by ultrasound-assisted extraction from the press cakes of coconut and almond beverage preparation', *Foods*, 11(22), p. 3693. doi: 10.3390/foods11223693.

Al-Dhabi, N. A., Ponmurugan, K., and Maran Jeganathan, P. (2017) 'Development and validation of ultrasound-assisted solid-liquid extraction of phenolic compounds from waste spent coffee grounds', *Ultrasonics Sonochemistry*, 34, pp. 206–213. doi: 10.1016/j.ultsonch.2016.05.005.

Aliakbarian, B., Paini, M., Adami, R., Perego, P., and Reverchon, E. (2017) 'Use of supercritical assisted atomization to produce nanoparticles from olive pomace extract', *Innovative Food Science & Emerging Technologies*, 40, pp. 2–9.

Aliakbarian, B., Sampaio, F. C., de Faria, J. T., Pitangui, C. G., Lovaglio, F., Casazza, A. A., Converti, A., and Perego, P. (2018) 'Optimization of spray drying microencapsulation of olive pomace polyphenols using response surface methodology and artificial neural network', *Lwt*, 93, pp. 220–228.

Alvi, T., Asif, Z., and Iqbal Khan, M. K. (2022) 'Clean label extraction of bioactive compounds from food waste through microwave-assisted extraction technique-a review', *Food Bioscience*, 46, p. 101580. doi: 10.1016/j.fbio.2022.101580.

Ali, H. I. et al. (2022) 'Total dietary fiber analysis in dates and other dry fruits without starch and protein hydrolyzing enzymes', *Journal of Food Composition and Analysis*, 108(September 2021), pp. 10–13. doi: 10.1016/j.jfca.2022.104415.

Alves, D. R. S. et al. (2019) 'Palatability of protein hydrolysates from industrial byproducts for Nile Tilapia juveniles', *Animals*, 9(6), p. 311. doi: 10.3390/ani9060311.

Ambigaipalan, P., and Shahidi, F. (2017) 'Bioactive peptides from shrimp shell processing discards: Antioxidant and biological activities', *Journal of Functional Foods*, 34, pp. 7–17. doi: 10.1016/j.jff.2017.04.013.

Ateeq, H. et al. (2023) 'Valorization and structural characterization of onion peel powder for the development of functional bread', *International Journal of Food Properties*, 26(1), pp. 2553–2562. doi: 10.1080/10942912.2023.2250573.

Ayar, A. et al. (2018) 'Probiotic properties of ice creams produced with dietary fibres from by-products of the food industry', *International Journal of Dairy Technology*, 71(1), pp. 174–182. doi: 10.1111/1471-0307.12387.

Baker, P. W. et al. (2023) 'Valorisation of waste olive pomace: Laboratory and pilot scale processing to extract dietary fibre', *Cleaner and Circular Bioeconomy*, 5(May), p. 100045. doi: 10.1016/j.clcb.2023.100045.

Balasubramani, V. et al. (2024) 'Extraction of lignocellulosic fiber and cellulose microfibrils from agro waste-palmyra fruit peduncle: Water retting, chlorine-free chemical treatments, physio-chemical, morphological, and thermal characterization', *International Journal of Biological Macromolecules*, 259(P2), p. 129273. doi: 10.1016/j.ijbiomac.2024.129273.

Ballardo, C. et al. (2016) 'Valorization of soy waste through SSF for the production of compost enriched with Bacillus thuringiensis with biopesticide properties', *Journal of Environmental Management*, 169, pp. 126–131. doi: 10.1016/j.jenvman.2015.12.029.

Barrios, C. et al. (2022) 'A techno-economic perspective on a microwave extraction process for efficient protein recovery from agri-food wastes', *Industrial Crops and Products*, 186, p. 115166. doi: 10.1016/j.indcrop.2022.115166.

Battista, F. et al. (2020) 'Valorization of residual orange peels: Limonene recovery, volatile fatty acids, and biogas production', *ACS Sustainable Chemistry and Engineering*, 8(17), pp. 6834–6843. doi: 10.1021/acssuschemeng.0c01735.

Behere, M., Patil, S. S., and Rathod, V. K. (2021) 'Rapid extraction of watermelon seed proteins using microwave and its functional properties', *Preparative Biochemistry & Biotechnology*, 51(3), pp. 252–259. doi: 10.1080/10826068.2020.1808792.

Bezerra, T. K. A. et al. (2019) 'Identification of angiotensin I-converting enzyme-inhibitory and anticoagulant peptides from enzymatic hydrolysates of chicken combs and wattles', *Journal of Medicinal Food*, 22(12), pp. 1294–1300. doi: 10.1089/jmf.2019.0066.

Ben-Othman, S., Jõudu, I., and Bhat, R. (2020) 'Bioactives from agri-food wastes: Present insights and future challenges', *Molecules*. doi: 10.3390/molecules25030510.

Bhat, S. A., and Singh, E. R. (2014) 'Extraction and characterization of pectin from guava fruit peel', *International Journal of Research in Engineering and Advanced Technology*, 2(3), pp. 447–454.

Cabral, E. M. et al. (2023) 'Recovery of protein from industrial hemp waste (Cannabis sativa, L.) using high-pressure processing and ultrasound technologies', *Foods*, 12(15), p. 2883. doi: 10.3390/foods12152883.

Cabrera, L. A. et al. (2023) 'Order matters: Methods for extracting cellulose from rice straw by coupling alkaline, ozone and enzymatic treatments', *Carbohydrate Polymers*, 328 (December 2023), p. 121746. doi: 10.1016/j.carbpol.2023.121746.

Caldas, T. W., Mazza, K. E., Teles, A. S., Mattos, G. N., Brígida, A. I. S., Conte-Junior, C. A., … & Tonon, R. V. (2018). 'Phenolic compounds recovery from grape skin using conventional and non-conventional extraction methods', *Industrial Crops and Products*, 111, pp. 86–91.

Chang, Y. P., Woo, K. K., and Gnanaraj, C. (2020) 'Pink guava'. In C. M. Galanakis (Ed.), *Valorization of fruit processing by-products* (pp. 227–252). Academic Press.

Cao, J. et al. (2019) 'Extraction and physicochemical characterization of pepsin soluble collagens from golden pompano (*Trachinotus blochii*) skin and bone', *Journal of Aquatic Food Product Technology*, 28(8), pp. 837–847. doi: 10.1080/10498850.2019.1652216.

Chen, H., Liu, Y., Zhang, J., Jiang, Y., and Li, D. (2022) 'Pectin extracted from dragon fruit peel: An exploration as a natural emulsifier', *International Journal of Biological Macromolecules*, 221, pp. 976–985. doi: 10.1016/j.ijbiomac.2022.09.069.

Chen, L. et al. (2023) 'Dietary fiber extraction from citrus peel pomace: Yield optimization and evaluation of its functionality, rheological behavior, and microstructure properties', *Journal of Food Science*, 88(8), pp. 3507–3523. doi: 10.1111/1750-3841.16702.

Chen, Z., Zhong, B., Barrow, C. J., Dunshea, F. R., and Suleria, H. A. R. (2021) 'Identification of phenolic compounds in Australian grown dragon fruits by LC-ESI-QTOF-MS/MS and determination of their antioxidant potential', *Arabian Journal of Chemistry*, 14, p. 103151. doi: 10.1016/j.arabjc.2021.103151

Chua, B. L., Tang, S. F., Ali, A., and Chow, Y. H. (2020) 'Optimisation of pectin production from dragon fruit peels waste: Drying, extraction and characterisation studies', *SN Applied Sciences*, 2. doi: 10.1007/s42452-020-2415-y.

Corrales, M. L., Toepfl, S., Butz, P., Knorr, D., and Tauscher, B. (2008) 'Extraction of anthocyanins from grape by-products assisted by ultrasonics, high hydrostatic pressure or pulsed electric fields: A comparison', *Innovative Food Science and Emerging Technologies*, 9, pp. 85–91.

Chockchaisawasdee, S. et al. (2023) 'Development of a gluten free bread enriched with faba bean husk as a fibre supplement', *Lwt*, 173(August 2022), p. 114362. doi: 10.1016/j.lwt.2022.114362.

Cudjoe, D., Chen, W., and Zhu, B. (2022) 'Valorization of food waste into hydrogen: Energy potential, economic feasibility and environmental impact analysis', *Fuel*, 324(PA), p. 124476. doi: 10.1016/j.fuel.2022.124476.

Dao, D. T. A. et al. (2022) 'Recovery of amino acids and peptides from pig bone soup using thermal pre-treatment and enzymatic hydrolysis', *Journal of Food Processing and Preservation*, 46(7). doi: 10.1111/jfpp.16700.

da Silva Lima, R., Nunes, I. L., and Block, J. M. (2020) 'Ultrasound-assisted extraction for the recovery of carotenoids from guava's pulp and waste powders', *Plant Foods for Human Nutrition*, 75, pp. 63–69.

Denny, C., Melo, P. S., Franchin, M., Massarioli, A. P., Bergamaschi, K. B., de Alencar, S. M., and Rosalen, P. L. (2013) 'Guava pomace: A new source of anti-inflammatory and analgesic bioactives', *BMC Complementary and Alternative Medicine*, 13(1), pp. 1–7.

Deb, S., Kumar, Y., and Saxena, D. C. (2022) 'Functional, thermal and structural properties of fractionated protein from waste banana peel', *Food Chemistry:* X, 13, p. 100205. doi: 10.1016/j.fochx.2022.100205.

Dikkala, P. K. et al. (2023) 'Valorization of peanut skin: Development of functional skin-on peanut butter and quality characteristics', *Waste and Biomass Valorization*. doi: 10.1007/s12649-023-02367-2.

Dong, Z. Y. et al. (2019) 'Effects of ultrasonic pretreatment on the structure and functionality of chicken bone protein prepared by enzymatic method', *Food Chemistry*, 299, p. 125103. doi: 10.1016/j.foodchem.2019.125103.

Du, G. et al. (2022) 'Optimized strategy valorizing unautoclaved cottonseed hull as ruminant alternative feeds via solid-state fermentation: Detoxifying polyphenols, restraining hazardous microflora and antibiotic-resistance gene hosts', *Environmental Technology and Innovation*, 28(99), p. 102937. doi: 10.1016/j.eti.2022.102937.

Fagbemi, K. O., Aina, D. A., and Olajuyigbe, O. O. (2021) 'Soxhlet extraction versus hydrodistillation using the clevenger apparatus: A comparative study on the extraction of a volatile compound from Tamarindus indica seeds', *The Scientific World Journal*, 2021, p. 5961586. doi: 10.1155/2021/5961586

Ferrentino, G., Morozova, K., Mosibo, O. K., Ramezani, M., Scampicchio, M. (2018) 'Biorecovery of antioxidants from apple pomace by supercritical fluid extraction', *Journal of Cleaner Production*, 186, pp. 253–261. doi: 10.1016/j.jclepro.2018.03.165.

Fidriyanto, R. et al. (2023) 'Multivariate analysis of structural and functional properties of fibres from apple pomace using different extraction methods', *Food Production, Processing and Nutrition*, 5(1). doi: 10.1186/s43014-022-00119-8.

Franco-Arnedo, G., Buelvas-Puello, L. M., Miranda-Lasprilla, D., Martínez-Correa, H. A., Parada-Alfonso, F. (2020) 'Obtaining antioxidant extracts from tangerine (C. reticulata var. Arrayana) peels by modified supercritical CO_2 and their use as protective agent against the lipid oxidation of a mayonnaise', *The Journal of Supercritical Fluids*, 165, p. 104957. https://doi.org/https://doi.org/10.1016/j.supflu.2020.104957

Ganguly, P. et al. (2020) 'Valorization of food waste: Extraction of cellulose, lignin and their application in energy use and water treatment', *Fuel*, 280(January), p. 118581. doi: 10.1016/j.fuel.2020.118581.

Gao, J. et al. (2023) 'The foaming properties of sweet potato protein hydrolysates produced by Alcalase and Ficin', *Journal of the Science of Food and Agriculture*, 103(8), pp. 4157–4163. doi: 10.1002/jsfa.12420.

Garcia-Amezquita, L. E. et al. (2018) 'Influence of drying method on the composition, physicochemical properties, and prebiotic potential of dietary fibre concentrates from fruit peels', *Journal of Food Quality*. doi: 10.1155/2018/9105237.

Golmohammadi, M. et al. (2018) 'Optimization of essential oil extraction from orange peels using steam explosion', *Heliyon*, 4(11). doi: 10.1016/j.heliyon.2018.e00893.

González-Serrano, D. J. et al. (2022) 'Bioactive peptide fractions from collagen hydrolysate of common carp fish byproduct: Antioxidant and functional properties', *Antioxidants*, 11(3), p. 509. doi: 10.3390/antiox11030509.

Görgüç, A., Özer, P., and Yılmaz, F. M. (2020) 'Microwave-assisted enzymatic extraction of plant protein with antioxidant compounds from the food waste sesame bran: Comparative optimization study and identification of metabolomics using LC/Q-TOF/MS', *Journal of Food Processing and Preservation*, 44(1). doi: 10.1111/jfpp.14304.

Guandalini, B. B. V., Rodrigues, N. P., and Marczak, L. D. F. (2019) 'Sequential extraction of phenolics and pectin from mango peel assisted by ultrasound', *Food Research International*, 119, pp. 455–461. doi: 10.1016/j.foodres.2018.12.011.

Hamza, H. et al. (2021) 'Characterization and valorization of dietary fibre concentrate from parthenocarpic date fruits', *European Journal of Biology and Biotechnology*, 2(3), pp. 11–18. doi: 10.24018/ejbio.2021.2.3.192.

Hassan, I., Khurshid, W., and Iqbal, K. (2012) 'Factors responsible for decline in guava (Psidium guajava) yield', *Journal of Agriculture Research*, 50(1), pp. 129–134.

Helstad, A. et al. (2022) 'Protein extraction from cold-pressed hempseed press cake: From laboratory to pilot scale', *Journal of Food Science*, 87(1), pp. 312–325. doi: 10.1111/1750-3841.16005.

Hosseini, S. S., Khodaiyan, F., Kazemi, M., and Najari, Z. (2019) 'Optimization and characterization of pectin extracted from sour orange peel by ultrasound assisted method', *International Journal of Biological Macromolecules*, 125, pp. 621–629. doi: 10.1016/j.ijbiomac.2018.12.096.

Huang, C.-Y. et al. (2016) 'Isolation and characterization of fish scale collagen from tilapia (Oreochromis sp.) by a novel extrusion–hydro-extraction process', *Food Chemistry*, 190, pp. 997–1006. doi: 10.1016/j.foodchem.2015.06.066.

Huang, G., Ren, Z., and Jiang, J. (2011) 'Separation of iron-binding peptides from shrimp processing by-products hydrolysates', *Food and Bioprocess Technology*, 4(8), pp. 1527–1532. doi: 10.1007/s11947-010-0416-3.

Huynh, N. T. et al. (2014) 'Enzyme-assisted extraction enhancing the phenolic release from cauliflower (Brassica oleracea L. var. botrytis) outer leaves', *Journal of Agricultural and Food Chemistry*, 62(30), pp. 7468–7476. doi: 10.1021/jf502543c.

Jaeger, A. et al. (2020) 'Brewer's spent yeast (BSY), an underutilized brewing by-product', *Fermentation*, 6(4), p. 123. doi: 10.3390/fermentation6040123.

Jha, P. et al. (2019) 'Valorisation of orange peel: Supplement in fermentation media for ethanol production and source of limonene', *Environmental Sustainability*, 2(1), pp. 33–41. doi: 10.1007/s42398-019-00048-2.

Jiang, B. et al. (2019) 'Reutilization of food waste: One-step extration, purification and characterization of ovalbumin from salted egg white by aqueous two-phase flotation', *Foods*, 8(8), p. 286. doi: 10.3390/foods8080286.

Jiang, Y. et al. (2022) 'Recovery of emulsifying and gelling protein from waste chicken exudate by using a sustainable pH-shifting treatment', *Food Chemistry*, 387, p. 132886. doi: 10.1016/j.foodchem.2022.132886.

Kadi, A., Boudries, H., Bachir-bey, M., Teffane, M., Taibi, A., and Boulekbache-Makhlouf, L. (2022) 'Optimization of microwave-assisted extraction of carotenoids from citrus clementina peels', *Current Bioactive Compounds*, 18(6), pp. 63–73(11). doi: 10.2174/15734072186662112301 52122.

Kamal, H. et al. (2023) 'Food waste valorization potential: Fiber, sugar, and color profiles of 18 date seed varieties (Phoenix dactylifera, L.)', *Journal of the Saudi Society of Agricultural Sciences*, 22(2), pp. 133–138. doi: 10.1016/j.jssas.2022.11.001.

Kaur, B., Panesar, P. S., and Thakur, A. (2021) 'Extraction and evaluation of structural and physicochemical properties of dietary fiber concentrate from mango peels by using green approach', *Biomass Conversion and Biorefinery*, (0123456789). doi: 10.1007/s13399-021-01740-2.

Kaur, R., Panesar, P. S., and Riar, C. S. (2023) 'Green extraction of dietary fiber concentrate from pearl millet bran and evaluation of its microstructural and functional properties', *Biomass Conversion and Biorefinery*, (0123456789). doi: 10.1007/s13399-023-04627-6.

Kapoor, S., Gandhi, N., Tyagi, S. K., Kaur, A., and Mahajan, B. V. C. (2020) 'Extraction and characterization of guava seed oil: A novel industrial byproduct', *LWT*, 132, p. 109882. doi: 10.1016/j.lwt.2020.109882.

Khan, M. K. et al. (2010) 'Ultrasound-assisted extraction of polyphenols (flavanone glycosides) from orange (Citrus sinensis L.) peel', *Food Chemistry*, 119(2), pp. 851–858. doi: 10.1016/j.foodchem.2009.08.046.

Khanpit, V. V., Mandavgane, S. A., and Tajane, S. P. (2023) 'Waste to wealth-recovery of total dietary fibers from waste peel: A generalized model for predicting operating parameters', *Biomass Conversion and Biorefinery*, 13(10), pp. 9155–9164. doi: 10.1007/s13399-021-01791-5.

Khanpit, V. V., Tajane, S. P., and Mandavgane, S. A. (2021) 'Dietary fibers from fruit and vegetable waste: Methods of extraction and processes of value addition', *Biomass Conversion and Biorefinery*, (0123456789). doi: 10.1007/s13399-021-01980-2.

Killadi, B., Chaurasia, R., and Shukla, D. K. (2018) 'Maturity indices and quality attributes during growth and development in guava cultivars "shweta" and "lalitâ"', *Journal of Pharmacognosy and Phytochemistry*, 7(2), pp. 2102–2106.

Kim, J. M., Liceaga, A. M., and Yoon, K. Y. (2019) 'Purification and identification of an antioxidant peptide from perilla seed (*Perilla frutescens*) meal protein hydrolysate', *Food Science & Nutrition*, 7(5), pp. 1645–1655. doi: 10.1002/fsn3.998.

Kohli, D. et al. (2019) 'Waste utilization of amla pomace and germinated finger millets for value addition of biscuits', *Current Research in Nutrition and Food Science*, 7(1), pp. 272–279. doi: 10.12944/CRNFSJ.7.1.27.

Kong, K. W., and Ismail, A. (2011). 'Lycopene content and lipophilic antioxidant capacity of by-products from *Psidium guajava* fruits produced during puree production industry', *Food and Bioproducts Processing*, 89(1), pp. 53–61.

Kong, K. W., Rajab, N. F., Prasad, K. N., Ismail, A., Markom, M., and Tan, C. P. (2010) 'Lycopene-rich fractions derived from pink guava by-product and their potential activity towards hydrogen peroxide-induced cellular and DNA damage', *Food Chemistry*, 123(4), pp. 1142–1148.

Krenz, L. M. M. et al. (2023) 'Valorization of faba bean (Vicia faba) by-products', *Biomass Conversion and Biorefinery*. doi: 10.1007/s13399-023-03779-9.

Krishnan, R., Arshinder, K., and Agarwal, R. (2022) 'Robust optimization of sustainable food supply chain network considering food waste valorization and supply uncertainty', *Computers and Industrial Engineering*, 171(July), p. 108499. doi: 10.1016/j.cie.2022.108499.

Kumar, D. et al. (2023) 'Antioxidant potential and amino acid profile of ultrafiltration derived peptide fractions of spent hen meat protein hydrolysate', *Journal of Food Science and Technology*, 60(3), pp. 1195–1201. doi: 10.1007/s13197-022-05437-w.

Kumar, V. et al. (2021) 'Optimization of Enzymatic Extraction of ACE Inhibitory Peptide from Rohu (Labeo rohita) Fish Waste using RSM', *Indian Journal of Animal Research*, (Of). doi: 10.18805/IJAR.B-4542.

Kumar, K., Ahmed, N., Jan, S., Thakur, P., Chauhan, D., and Kaur, J. (2022) 'Guava wastes and by-products: Chemistry, processing, and utilization'. In K. Muzaffar, S. A. Sofi, and S. A. Mir (Eds.), *Handbook of fruit wastes and by-products* (pp. 99–112). CRC Press.

Kuprina, E. E. et al. (2023) 'Obtaining and study of peptide compositions based on hydrolysates of collagen-containing fish raw materials', *Fermentation*, 9(5). doi: 10.3390/fermentation9050458.

Lalnunthari, C., Devi, L. M., and Badwaik, L. S. (2020) 'Extraction of protein and pectin from pumpkin industry by-products and their utilization for developing edible film', *Journal of Food Science and Technology*, 57(5), pp. 1807–1816. doi: 10.1007/s13197-019-04214-6.

Lau, K. Q. et al. (2022) 'Incorporation of sugarcane bagasse in the development of high dietary fibre noodles', *International Journal of Food Science and Technology*, 57(7), pp. 4586–4593. doi: 10.1111/ijfs.15796.

Lasunon, P., and Sengkhamparn, N. (2022) 'Effect of ultrasound-assisted, microwave-assisted and ultrasound-microwave-assisted extraction on pectin extraction from industrial tomato waste', *Molecules*. doi: 10.3390/molecules27041157.

Li, Y. et al. (2024) 'High energy media mill modified pea dietary fiber: Physicochemical property and its mechanism in stabilizing pea protein beverage', *Food Hydrocolloids*, 147(PA), p. 109392. doi: 10.1016/j.foodhyd.2023.109392.

Liew, S. Q. et al. (2016) 'Sequential ultrasound-microwave assisted acid extraction (UMAE) of pectin from pomelo peels', *International Journal of Biological Macromolecules*, 93, pp. 426–435. doi: 10.1016/j.ijbiomac.2016.08.065.

Lin, C. B. E. E., Kai, N. G. Y. E. E., and Ali, A. (2018) 'Ultrasound assisted extraction of pectin from dragon fruit peels', *Journal of Engineering Science and Technology*, 13, pp. 65–81.

Liu, X. et al. (2023) 'Analysis of the fate of valuable bioactive polyphenols during commercial winemaking and their partitioning into wastes for valorization', *Cleaner Waste Systems*, 5(April), p. 100104. doi: 10.1016/j.clwas.2023.100104.

Liu, X., Hu, K. K. Y., and Haritos, V. S. (2024) 'Enzymatic production of cello-oligosaccharides with potential human prebiotic activity and release of polyphenols from grape marc', *Food Chemistry*, 435(March 2023), p. 137562. doi: 10.1016/j.foodchem.2023.137562.

López-Pedrouso, M. et al. (2023) 'Valorisation of pork by-products to obtain antioxidant and antihypertensive peptides', *Food Chemistry*, 423, p. 136351. doi: 10.1016/j.foodchem.2023.136351.

Ma, Q. et al. (2022) 'The effects of enzymatic modification on the functional ingredient - Dietary fiber extracted from potato residue', *Lwt*, 153(June 2021), p. 112511. doi: 10.1016/j.lwt.2021.112511.

Mansour, M. S. M., Abdel-Shafy, H. I., and Mehaya, F. M. S. (2018) 'Valorization of food solid waste by recovery of polyphenols using hybrid molecular imprinted membrane', *Journal of Environmental Chemical Engineering*, 6(4), pp. 4160–4170. doi: 10.1016/j.jece.2018.06.019.

Maran, J. P., and Priya, B. (2014) 'Ultrasound-assisted extraction of polysaccharide from Nephelium lappaceum L. Fruit peel', *International Journal of Biological Macromolecules*, 70, pp. 530–536. doi: 10.1016/j.ijbiomac.2014.07.032.

Maran, J. P., Priya, B., Al-Dhabi, N. A., Ponmurugan, K., Moorthy, I. G., and Sivarajasekar, N. (2017) 'Ultrasound assisted citric acid mediated pectin extraction from industrial waste of Musa balbisiana', *Ultrasonics Sonochemistry*, 35, pp. 204–209.

Martínez-Abad, A., Ramos, M., Hamzaoui, M., Kohnen, S., Jiménez, A., and Garrigós, M. C. (2020) 'Optimisation of sequential microwave-assisted extraction of essential oil and pigment from lemon peels waste', *Foods*. doi: 10.3390/foods9101493.

Marzorati, S., Schievano, A., Idà, A., and Verotta, L. (2020) 'Carotenoids, chlorophylls and phycocyanin from Spirulina: Supercritical CO_2 and water extraction methods for added value products cascade', *Green Chemistry*, 22, pp. 187–196. doi: 10.1039/C9GC03292D.

Melgosa, R. et al. (2021) 'Subcritical water extraction and hydrolysis of cod (Gadus morhua) frames to produce bioactive protein extracts', *Foods*, 10(6), p. 1222. doi: 10.3390/foods10061222.

Mellinas, A. C., Jiménez, A., and Garrigós, M. C. (2020) 'Optimization of microwave-assisted extraction of cocoa bean shell waste and evaluation of its antioxidant, physicochemical and functional properties', *LWT*, 127, p. 109361. doi: 10.1016/j.lwt.2020.109361.

Minjares-Fuentes, R. et al. (2014) 'Ultrasound-assisted extraction of pectins from grape pomace using citric acid: A response surface methodology approach', *Carbohydrate Polymers*, 106(1), pp. 179–189. doi: 10.1016/j.carbpol.2014.02.013.

Moayedi, A. et al. (2018) 'Peptidomic analysis of antioxidant and ACE-inhibitory peptides obtained from tomato waste proteins fermented using Bacillus subtilis', *Food Chemistry*, 250, pp. 180–187. doi: 10.1016/j.foodchem.2018.01.033.

Moorthy, I. G., Maran, J. P., Ilakya, S., Anitha, S. L., Sabarima, S. P., and Priya, B. (2017) 'Ultrasound assisted extraction of pectin from waste Artocarpus heterophyllus fruit peel', *Ultrasonics Sonochemistry*, 34, pp. 525–530.

Moorthy, I. G., Maran, J. P., Surya, S. M., Naganyashree, S., and Shivamathi, C. S. (2015) 'Response surface optimization of ultrasound assisted extraction of pectin from pomegranate peel', *International Journal of Biological Macromolecules*, 72, pp. 1323–1328. doi: 10.1016/j.ijbiomac.2014.10.037.

Moradi, E., and Fathi, M. (2023) 'Production of cellulose nanocrystals from tomato pomace as a food waste and their application for stabilizing of Pickering emulsions', *Bioactive Carbohydrates and Dietary Fibre*, 30(October 2022), p. 100378. doi: 10.1016/j.bcdf.2023.100378.

Mosa, Z. M., and Kkalil, A. F. (2015) 'Comparative study between the effects of mango and orange peels preparations on the total dietary fiber', *IOSR Journal of Environmental Science Ver.* II, 9(11), pp. 2319–2399. doi: 10.9790/2402-09112129136.

Nagarajan, J., Krishnamurthy, N. P., Ramanan, R. N., Raghunandan, M. E., Galanakis, C. M., and Ooi, C. W. (2019) 'A facile water-induced complexation of lycopene and pectin from pink guava byproduct: Extraction, characterization and kinetic studies', *Food Chemistry*, 296, pp. 47–55.

Nepote, V., Grosso, N. R., and Guzmán, C. A. (2005) 'Optimization of extraction of phenolic antioxidants from peanut skins', *Journal of the Science of Food and Agriculture*, 85, pp. 33–38. doi: 10.1002/jsfa.1933.

Nguyen, B. M. N., and Pirak, T. (2019) 'Physicochemical properties and antioxidant activities of white dragon fruit peel pectin extracted with conventional and ultrasound-assisted extraction', *Cogent Food & Agriculture*, 5, p. 1633076. doi: 10.1080/23311932.2019.1633076.

Nguyen, N. D. T. et al. (2022) 'Enzymatic treatment of spent green tea leaves and their use in high-fibre cookie production', *Food Technology and Biotechnology*, 60(3), pp. 396–405. doi: 10.17113/ftb.60.03.22.7474.

Nicanor, A. B., Ortíz Moreno, A., Martinez Ayala, A. L., and Dávila Ortíz, G. (2001) 'Guava seed protein isolate: Functional and nutritional characterization', *Journal of Food Biochemistry*, 25(1), pp. 77–90.

Nieto-Calvache, J. E., de Escalada Pla, M., and Gerschenson, L. N. (2019) 'Dietary fibre concentrates produced from papaya by-products for agroindustrial waste valorisation', *International Journal of Food Science and Technology*, 54(4), pp. 1074–1080. doi: 10.1111/ijfs.13962.

Nieto-Veloza, A. et al. (2021) 'Utilization of tofu processing wastewater as a source of the bioactive peptide lunasin', *Food Chemistry*, 362, p. 130220. doi: 10.1016/j.foodchem.2021.130220.

Obafaye, R. O., and Omoba, O. S. (2018) 'Orange peel flour: A potential source of antioxidant and dietary fiber in pearl-millet biscuit', *Journal of Food Biochemistry*, 42(4), pp. 1–8. doi: 10.1111/jfbc.12523.

Oliveira, C. F. de, Giordani, D., Lutckemier, R., Gurak, P. D., Cladera-Oliver, F., Marczak, L. D. F. (2016) 'Extraction of pectin from passion fruit peel assisted by ultrasound', *LWT - Food Science and Technology*, 71, pp. 110–115. doi: 10.1016/j.lwt.2016.03.027.

Ostolski, M., Adamczak, M., Brzozowski, B., and Wiczkowski, W. (2021) 'Antioxidant activity and chemical characteristics of supercritical CO_2 and water extracts from willow and poplar', *Molecules*. doi: 10.3390/molecules26030545.

Ozturk, B., Winterburn, J., and Gonzalez-Miquel, M. (2019) 'Orange peel waste valorisation through limonene extraction using bio-based solvents', *Biochemical Engineering Journal*, 151(June), p. 107298. doi: 10.1016/j.bej.2019.107298.

Özyurt, V. H., Tetik, I., and Ötleş, S. (2021) 'Influence of process conditions on ultrasound-assisted protein extraction from cold pressed tomato seed waste', *Journal of Food Processing and Preservation*, 45(12). doi: 10.1111/jfpp.16079.

Pal, C. B. T., and Jadeja, G. C. (2020) 'Microwave-assisted extraction for recovery of polyphenolic antioxidants from ripe mango (Mangifera indica L.) peel using lactic acid/sodium acetate deep eutectic mixtures', *Food Science and Technology International*, 26(1), pp. 78–92. doi: 10.1177/1082013219870010.

Palma, M., and Barroso, C. G. (2002) 'Ultrasound-assisted extraction and determination of tartaric and malic acids from grapes and winemaking by-products', *Analytica Chimica Acta*, 458, pp. 119–130. doi: 10.1016/S0003-2670(01)01527-6.

Pan, Z., Qu, W., Ma, H., Atungulu, G. G., and McHugh, T. H. (2012) 'Continuous and pulsed ultrasound-assisted extractions of antioxidants from pomegranate peel', *Ultrasonics Sonochemistry*, 19, pp. 365–372. doi: 10.1016/j.ultsonch.2011.05.015.

Pandey, K. S., and Rizvi, S. I. (2009) 'Plant polyphenols as dietary antioxidants in healyth and disease', *Oxidative Medicine and Cellular Longevity*, 2(5), pp. 270–278. doi: 10.4161/oxim.2.5.9498.

Paroha, S. (2020) 'Sugarcane bagasse as dietary fibre', *Indian Journal of Pure & Applied Biosciences*, 8(6), pp. 590–597. doi: 10.18782/2582-2845.8442.

Patsalou, M. et al. (2020) 'A biorefinery for conversion of citrus peel waste into essential oils, pectin, fertilizer and succinic acid via different fermentation strategies', *Waste Management*, 113, pp. 469–477. doi: 10.1016/j.wasman.2020.06.020.

Pavlić, B. et al. (2023) 'Sustainable raw materials for efficient valorization and recovery of bioactive compounds', *Industrial Crops and Products*, 193(September 2022). doi: 10.1016/j.indcrop.2022.116167.

Paulo, F., and Santos, L. (2021) 'Deriving valorization of phenolic compounds from olive oil by-products for food applications through microencapsulation approaches: A comprehensive review', *Critical Reviews in Food Science and Nutrition*, 61(6), pp. 920–945.

Petcharat, T. et al. (2021) 'Ultrasound-assisted extraction of collagen from clown featherback (*Chitala ornata*) skin: Yield and molecular characteristics', *Journal of the Science of Food and Agriculture*, 101(2), pp. 648–658. doi: 10.1002/jsfa.10677.

Petkowicz, C. L. O., and Williams, P. A. (2020) 'Pectins from food waste: Characterization and functional properties of a pectin extracted from broccoli stalk', *Food Hydrocolloids*, 107(April), p. 105930. doi: 10.1016/j.foodhyd.2020.105930.

Petrotos, K. B., Karkanta, F. K., Gkoutsidis, P. E., Giavasis, I., Papatheodorou, K. N., and Ntontos, A. C. (2012) 'Production of novel bioactive yogurt enriched with olive fruit polyphenols', *World Academy of Science, Engineering and Technology*, 64, pp. 867–872.

Phongtongpasuk, S., Poadang, S., and Yongvanich, N. (2016) 'Environmental-friendly method for synthesis of silver nanoparticles from dragon fruit peel extract and their antibacterial activities', *Energy Procedia*, 89, pp. 239–247. doi: 10.1016/j.egypro.2016.05.031.

Plakantonaki, S. et al. (2023) 'Dietary fiber from plant-based food wastes: A comprehensive approach to cereal, fruit, and vegetable waste valorization', *Processes*, 11(5), pp. 1–21. doi: 10.3390/pr11051580.

Plazzotta, S., Sillani, S., and Manzocco, L. (2018) 'Exploitation of lettuce waste flour to increase bread functionality: Effect on physical, nutritional, sensory properties and on consumer response', *International Journal of Food Science and Technology*, 53(10), pp. 2290–2297. doi: 10.1111/ijfs.13820.

Prakash Maran, J., Sivakumar, V., Thirugnanasambandham, K., and Sridhar, R. (2013) 'Optimization of microwave assisted extraction of pectin from orange peel', *Carbohydrate Polymers*, 97, pp. 703–709. doi: 10.1016/j.carbpol.2013.05.052.

Prakash Maran, J. et al. (2017) 'Ultrasound assisted extraction of bioactive compounds from Nephelium lappaceum L. fruit peel using central composite face centered response surface design', *Arabian Journal of Chemistry*, 10, pp. S1145–S1157. doi: 10.1016/j.arabjc.2013.02.007.

Priatni, S., and Pradita, A. (2015) 'Stability study of betacyanin extract from red dragon fruit (Hylocereus polyrhizus) peels', *Procedia Chemistry*, 16, pp. 438–444. doi: 10.1016/j.proche.2015.12.076.

Radenkovs, V. et al. (2018) 'Non-waste technology through the enzymatic hydrolysis of agro-industrial by-products', *Trends in Food Science and Technology*, 77(May), pp. 64–76. doi: 10.1016/j.tifs.2018.05.013.

Raunhan, R. et al. (2023) 'Valorization of food waste derived anaerobic digestate into polyhydroxyalkanoate (PHA) using Thauera mechernichensis TL1', *Waste Management*, 171(April), pp. 248–258. doi: 10.1016/j.wasman.2023.08.039.

Ravikumar, H. et al. (2021) 'Pre-treatment temperature and multi-response surface optimisation of ultrasound-assisted extraction of antioxidants from red dates', *Journal of Experimental Biology and Agricultural Sciences*, 9(2320), pp. 148–160.

Ravikumar, H., Lin, C. B., and Hui, M. S. (2022) 'Bioassay-guided fractionation of antioxidants from crataegus -monogyna (common hawthorn) and its role in treating cardiovascular disease using H9c2 rat myoblast cardiac cell line', *Journal of Positive School Psychology*, 6(10), pp. 3205–3218.

Ramić, M., Vidović, S., Zeković, Z., Vladić, J., Cvejin, A., and Pavlić, B. (2015) 'Modeling and optimization of ultrasound-assisted extraction of polyphenolic compounds from Aronia melanocarpa by-products from filter-tea factory', *Ultrasonics Sonochemistry*, 23, pp. 360–368.

Renuka, V. et al. (2019) 'Characterization of gelatin obtained from unicorn leatherjacket (Aluterus monoceros) and reef cod (Epinephelus diacanthus) skins', *LWT*, 116, p. 108586. doi: 10.1016/j.lwt.2019.108586.

Rifna, E. J., Misra, N. N., and Dwivedi, M. (2023) 'Recent advances in extraction technologies for recovery of bioactive compounds derived from fruit and vegetable waste peels: A review', *Critical Reviews in Food Science and Nutrition*, 63, pp. 719–752. doi: 10.1080/10408398.2021.1952923.

Rivadeneira, J. P., Wu, T., Ybanez, Q., Dorado, A. A., Migo, V. P., Nayve, F. R. P., and Castillo-Israel, K. A. T. (2020) 'Microwave-assisted extraction of pectin from "Saba" banana peel waste: Optimization, characterization, and rheology study', *International Journal of Food Science*, 2020, p. 8879425. doi: 10.1155/2020/8879425.

Rodrigues, L. A., Matias, A. A., and Paiva, A. (2021) 'Recovery of antioxidant protein hydrolysates from shellfish waste streams using subcritical water extraction', *Food and Bioproducts Processing*, 130, pp. 154–163. doi: 10.1016/j.fbp.2021.09.011.

Rojas-Garbanzo, C., Winter, J., Montero, M. L., Zimmermann, B. F., and Schieber, A. (2019) 'Characterization of phytochemicals in Costa Rican guava (Psidium friedrichsthalianum-Nied.) fruit and stability of main compounds during juice processing-(U) HPLC-DAD-ESI-TQ-DMSn', *Journal of Food Composition and Analysis*, 75, pp. 26–42.

Rosas Ulloa, P. et al. (2023) 'Protein isolate from orange (Citrus sinensis L.) seeds: Effect of high-intensity ultrasound on its physicochemical and functional properties', *Food and Bioprocess Technology*, 16(3), pp. 589–602. doi: 10.1007/s11947-022-02956-4.

Saba, B. et al. (2023) 'Characterization and potential valorization of industrial food processing wastes', *Science of the Total Environment*, 868(November 2022), p. 161550. doi: 10.1016/j.scitotenv.2023.161550.

Saenjum, C., Pattananandecha, T., and Nakagawa, K. (2021) 'Antioxidative and anti-inflammatory phytochemicals and related stable paramagnetic species in different parts of dragon fruit', *Molecules*. doi: 10.3390/molecules26123565.

Safdar, M. N., Kausar, T., Jabbar, S., Mumtaz, A., Ahad, K., and Saddozai, A. A. (2017) 'Extraction and quantification of polyphenols from kinnow (Citrus reticulate L.) peel using ultrasound and maceration techniques', *Journal of Food and Drug Analysis*, 25, pp. 488–500. doi: 10.1016/j.jfda.2016.07.010.

Scarabattoli, L. et al. (2023) 'Use of carbohydrases to promote protein extraction from rice bran and soybean meal: A comparative study', *LWT*, 184, p. 115060. doi: 10.1016/j.lwt.2023.115060.

Sharayei, P. et al. (2019) 'Ultrasound assisted extraction of bioactive compounds from pomegranate (Punica granatum L.) peel', *Lwt*, 101(December 2017), pp. 342–350. doi: 10.1016/j.lwt.2018.11.031.

Shaw, D., Tripathi, A. D., Paul, V., Agarwal, A., Mishra, P. K., and Kumar, M. (2023) 'Valorization of essential oils from citrus peel powder using hydro-distillation', *Sustainable Chemistry and Pharmacy*, 32, p. 101036. doi: 10.1016/j.scp.2023.101036.

Shehata, M. G., Abd El Aziz, N. M., Youssef, M. M., and El-Sohaimy, S. A. (2021) 'Optimization conditions of ultrasound-assisted extraction of phenolic compounds from orange peels using response surface methodology', *Journal of Food Processing and Preservation*, 45, p. e15870. doi: 10.1111/jfpp.15870.

Shi, P. et al. (2023) 'Valorization of spent barley grains: Isolation of protein and fibers for starch-free noodles and its effect on glycemic response in healthy individuals', *Frontiers in Sustainable Food Systems*, 7. doi: 10.3389/fsufs.2023.1146614.

Silva, G. C., Rodrigues, R. A. F., and Bottoli, C. B. G. (2021) 'Passion fruit seed extract enriched in piceatannol obtained by microwave-assisted extraction', *Sustainable Chemistry and Pharmacy*, 22, p. 100472. doi: 10.1016/j.scp.2021.100472.

Singanusong, R. et al. (2015) 'Low power ultrasound-assisted extraction of phenolic compounds from mandarin (Citrus reticulata Blanco cv. Sainampueng) and lime (Citrus aurantifolia) peels and the antioxidant', *Food Analytical Methods*, 8(5), pp. 1112–1123. doi: 10.1007/s12161-014-9992-6.

Skenderidis, P., Leontopoulos, S., Petrotos, K., and Giavasis, I. (2020) 'Optimization of vacuum microwave-assisted extraction of pomegranate fruits peels by the evaluation of extracts' phenolic content and antioxidant activity', *Foods*. doi: 10.3390/foods9111655.

Song, H. et al. (2023) 'Recovery of proteins from soybean oil-body wastewater at various pH levels and their structural and functional characterization', *LWT*, 174, p. 114455. doi: 10.1016/j.lwt.2023.114455.

Su, D. L. et al. (2019) 'Efficient extraction and characterization of pectin from orange peel by a combined surfactant and microwave assisted process', *Food Chemistry*, 286(January), pp. 1–7. doi: 10.1016/j.foodchem.2019.01.200.

Suárez, M., Romero, M. P., Ramo, T., Macià, A., and Motilva, M. J. (2009) 'Methods for preparing phenolic extracts from olive cake for potential application as food antioxidants', *Journal of Agricultural and Food Chemistry*, 57(4), pp. 1463–1472.

Taliana, L. (2020) 'Facial skin health: Antioxidant facial scrub from red dragon fruit extract', *Journal of Asian Multicultural Research for Medical and Health Science Study*, 1, pp. 1–5.

Tagliani, C. et al. (2019) 'Blueberry pomace, valorization of an industry by-product source of fibre with antioxidant capacity', *Food Science and Technology (Brazil)*, 39(3), pp. 644–651. doi: 10.1590/fst.00318.

Tan, Y., and Chang, S. K. C. (2018) 'Isolation and characterization of collagen extracted from channel catfish (Ictalurus punctatus) skin', *Food Chemistry*, 242, pp. 147–155. doi: 10.1016/j.foodchem.2017.09.013.

Tarna, R. et al. (2021) 'Recovery of residual brewer's yeast by electroactivation', *Food and Nutrition Sciences*, 12(12), pp. 1177–1190. doi: 10.4236/fns.2021.1212086.

Teterycz, D., and Sobota, A. (2023) 'Use of high-protein and high-dietary-fibre vegetable processing waste from bell pepper and tomato for pasta fortification', *Foods*, 12(13). doi: 10.3390/foods12132567.

Thirugnanasambandham, K., Sivakumar, V., Prakash Maran, J., and Maran, J. P. (2014) 'Process optimization and analysis of microwave assisted extraction of pectin from dragon fruit peel', *Carbohydrate Polymers*, 112, pp. 622–626. doi: 10.1016/j.carbpol.2014.06.044.

Thu Dao, T. A., Webb, H. K., and Malherbe, F. (2021) 'Optimization of pectin extraction from fruit peels by response surface method: Conventional versus microwave-assisted heating', *Food Hydrocolloids*, 113, p. 106475. doi: 10.1016/j.foodhyd.2020.106475.

Tkaczewska, J., Bukowski, M., and Mak, P. (2018) 'Identification of antioxidant peptides in enzymatic hydrolysates of carp (Cyprinus carpio) skin gelatin', *Molecules*, 24(1), p. 97. doi: 10.3390/molecules24010097.

Tongkham, N., Juntasalay, B., Lasunon, P., and Sengkhamparn, N. (2017) 'Dragon fruit peel pectin: Microwave-assisted extraction and fuzzy assessment', *Agriculture and Natural Resources*, 51, pp. 262–267.

Tsouko, E. et al. (2020) 'Integrated biorefinery development for the extraction of value-added components and bacterial cellulose production from orange peel waste streams', *Renewable Energy*, 160, pp. 944–954. doi: 10.1016/j.renene.2020.05.108.

Ummels, M. et al. (2023) 'Evaluating brewers' spent grain protein isolate postprandial amino acid uptake kinetics: A randomized, cross-over, double-blind controlled study', *Nutrients*, 15(14). doi: 10.3390/nu15143196.

Uchôa-Thomaz, A. M. A., Sousa, E. C., Carioca, J. O. B., de Morais, S. M., de Lima, A., Martins, C. G., Alexandrino, C. D., Ferreira, P. A. T., Rodrigues, A. L. M., Rodrigues, S. P., Thomaz, J. C. A., Silva, J. N., and Rodrigues, L. L. (2014) 'Chemical composition, fatty acid profile and bioactive compounds of guava seeds (Psidium guajava L.)', *Food Science and Technology*, 34(3), pp. 485–492.

Vasconcelos, A. G., das GN Amorim, A., Dos Santos, R. C., Souza, J. M. T., de Souza, L. K. M., de SL Araújo, T., Nicolau, L. A. D., de Lima Carvalho, L., de Aquino, P. E. A., da Silva Martins, C., and Ropke, C. D. (2017) 'Lycopene rich extract from red guava (Psidium guajava L.) displays anti-inflammatory and antioxidant profile by reducing suggestive hallmarks of acute inflammatory response in mice', *Food Research International*, 99, pp. 959–968.

Veggi, P. C., Martinez, J., and Meireles, M. A. A. (2013) 'Fundamentals of microwave extraction BT - Microwave-assisted extraction for bioactive compounds: Theory and practice'. In Chemat, F., and Cravotto, G. (Eds.), Springer US, pp. 15–52. doi: 10.1007/978-1-4614-4830-3_2.

Verkempinck, S. H. E., Kyomugasho, C., Salvia-Trujillo, L., Denis, S., Bourgeois, M., Van Loey, A. M., Hendrickx, M. E. and Grauwet, T. (2018). 'Emulsion stabilizing properties of citrus pectin and its interactions with conventional emulsifiers in oil-in-water emulsions'. *Food Hydrocolloids*, 85, pp. 144–157.

Viji, P. et al. (2019) 'Functional and antioxidant properties of gelatin hydrolysates prepared from skin and scale of sole fish', *Journal of Aquatic Food Product Technology*, 28(10), pp. 976–986. doi: 10.1080/10498850.2019.1672845.

Vu, H. T., Scarlett, C. J., and Vuong, Q. V. (2019) 'Maximising recovery of phenolic compounds and antioxidant properties from banana peel using microwave assisted extraction and water', *Journal of Food Science and Technology*, 56, pp. 1360–1370. doi: 10.1007/s13197-019-03610-2.

Wang, Q. et al. (2018) 'HPLC–DAD–ESI–MS2 analysis of phytochemicals from Sichuan red orange peel using ultrasound-assisted extraction', *Food Bioscience*, 25(July), pp. 15–20. doi: 10.1016/j.fbio.2018.07.001.

Yang, C. et al. (2023) 'Valorization of food waste digestate to ash and biochar composites for high performance adsorption of methylene blue', *Journal of Cleaner Production*, 397(February), p. 136612. doi: 10.1016/j.jclepro.2023.136612.

Yao, Y. et al. (2020) 'Insights into the improvement of the enzymatic hydrolysis of bovine bone protein using lipase pretreatment', *Food Chemistry*, 302. doi: 10.1016/j.foodchem.2019.125199.

Yusof, R., Zaini, S.Z., Azman, M.A. (2020) 'Characterization of pectin extracted from guava peels using deep eutectic solvent and citric acid'. In Alias, N. Z., and Yusof, R. (Eds.), *Charting the Sustainable Future of ASEAN in Science and Technology*, Springer, Singapore, pp. 421–433

Zhang, L. et al. (2019) 'Identification and active evaluation of antioxidant peptides from protein hydrolysates of skipjack tuna (Katsuwonus pelamis) head', *Antioxidants*, 8(8), p. 318. doi: 10.3390/antiox8080318.

Zhang, Y. et al. (2020) 'Properties of dietary fiber from citrus obtained through alkaline hydrogen peroxide treatment and homogenization treatment', *Food Chemistry*, 311, p. 125873. doi: 10.1016/j.foodchem.2019.125873.

Zhou, J. et al. (2023) 'Preparation, purification and characterization of antibacterial and ACE inhibitory peptides from head protein hydrolysate of kuruma shrimp, Marsupenaeus japonicus', *Molecules*, 28(2), p. 894. doi: 10.3390/molecules28020894.

Zhu, L. et al. (2022) 'Process optimization, amino acid composition, and antioxidant activities of protein and polypeptide extracted from waste beer yeast', *Molecules*, 27(20). doi: 10.3390/molecules27206825.

5 Valorise food waste into biochar

Aan Mohammad Nusrat Aman, and Anurita Selvarajoo

5.1 INTRODUCTION

Biochar is the carbon skeleton product of an organic material when kept under high temperatures and in the absence of oxygen. It has been under the spotlight of the research community due to its advantageous chemical and physical properties like its porosity, high carbon content, water absorption capabilities, high surface area, thermal conductivity, high chemical stability, low density, abundant functional groups, and small particle size. Several materials have been investigated as parent materials for biochar production like agriculture waste, food waste, wood waste, rice waste, marine biowaste, manure, and animal carcasses. Food waste has unique handling challenges compared to other waste, as it is a complex mixture of various organic materials that require specialized management approaches. The food waste category can be further broken down into different classifications such as food waste from urban household areas, urban restaurant areas, seafood-based restaurants, cafeterias, and fruit and vegetable-based waste, as shown in Figure 5.1. Various thermochemical conversion methods can be selected: pyrolysis, torrefaction, gasification, and hydrothermal combustion, each distinguished by the amount of biochar, bio-oil, and syngas produced.

There are environmental, economic, and social benefits in converting food waste into biochar, a valuable carbonaceous material. Multiple research projects reported that the sustainable management of food waste via the conversion into biochar is an evolving paradigm shift from traditional landfilling and incineration. Various types of food waste-derived biochar have been engineered for specific applications, including use in electrochemical devices.The engineered biochar refers to the pre- and post-modification of the material physically and chemically to increase its efficiency. Su et al. (2022) have provided a comprehensive review of pyrolysis and co-pyrolysis investigation on food waste, additionally reviewing materials that would aid the pyrolysis processes like CaO and tyre waste. The authors also compare the pyrolysis mechanism breakdown of food waste with the usual lignocellulose biomass. Moreover, the key distinction between them is that the food waste made of organic content is mostly biodegradable, with a mixture of compounds like carbohydrates, proteins, lipids, and other nutrients, whereas the lignocellulose biomass is made of cellulose and lignin (Su et al. 2022).

Wood is the most used biomass for biochar conversion due to its low moisture content, compared to that of food waste. Therefore, food waste requires a large amount of energy to convert into biochar through pyrolysis, so in most cases, pretreatment is required in terms of pre-drying. However, in other methods like hydrothermal

 DOI: 10.1201/9781032718507-5

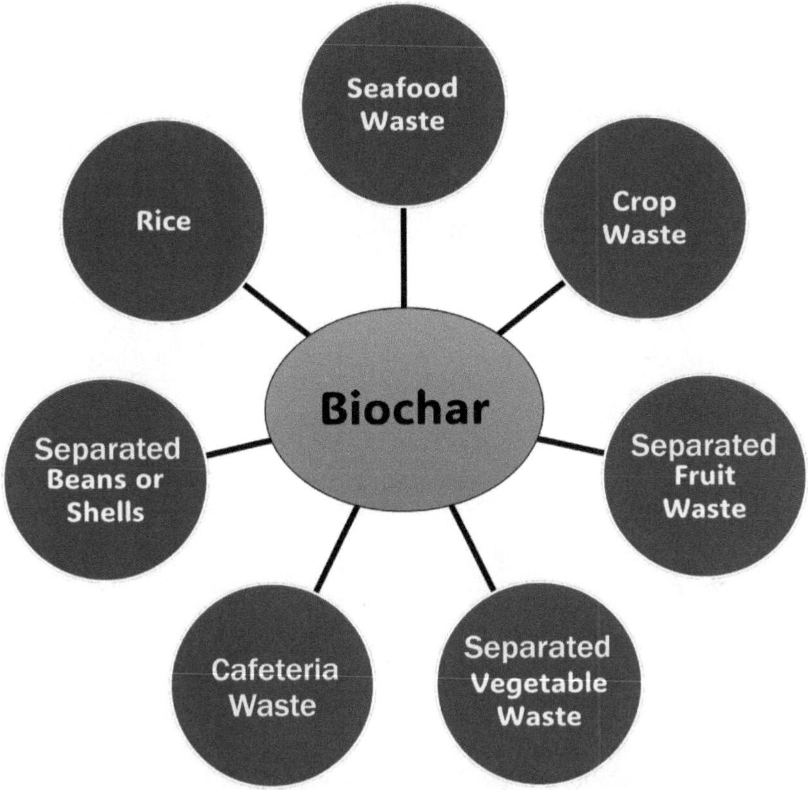

FIGURE 5.1 Types of food waste used for biochar.

carbonisation (HTC), pre-drying is not required; therefore, it would be more commonly used for food waste materials (Shen 2020; Yuan et al. 2021). Table 5.1 shows the different operating parameters of the thermochemical conversion methods and their usual end products. Moreover, the benefits of using food waste biochar from the common lignocellulosic biomass are low ash content, low temperature for carbonization, and activation process, suitable for hydrothermal carbonisation to produce hydrochar uniform structures compared to lignocellulosic biomass.

5.1.1 Valuable impact of food waste biochar

The production of biochar has been widely known to be one of the methods for environmental management when utilising an abundant amount of waste while also sequestering carbon. It would also decrease the economic aspect by utilising waste from different industries. The materials used to produce biochar are primarily agricultural and forest residues. By drawing attention to the highly performant biochar that can be derived from these readily available feedstocks, researchers are promoting the large-scale deployment of biochar production and utilisation. However, the commercial adaptation of these procedures is still lacking as research is at the early stages of exploring their

TABLE 5.1
Operating parameters of different thermochemical conversion methods.

Method	Temperature (°C)	Heating rate (°C/min)	Residence time	Products	References
Pyrolysis	300–850	-	1–3 h	Biochar	Kapoor and Shah (2023)
	200–800	2-20	10–155 min		Nusrat Aman et al. (2023)
	350–700 (slow)	-	hours		Spokas et al. (2012)
	300–800 (fast)		Less than 1 s		
Hydrothermal carbonisation	120–260	-	1–16 h	Hydrochar	Kapoor and Shah (2023)
	180–300	-	1–16 h		Funke and Ziegler (2010)
Torrefaction	200–300		30–120 min	Biochar	Ismail et al. (2021)
Gasification	More than 800	-	10–20 s	Syngas	Kapoor and Shah (2023)
	750–900		10–20 s		Klinghoffer, Castaldi, and Nzihou (2015)

cost-effectiveness. These aspects are usually evaluated with life cycle assessment (LCA) and life cycle cost analysis or similar approaches. Biochar price is the most crucial factor in determining the feasibility of a biochar production plant. A few examples of the cost of thermochemical conversion methods are shown in Table 5.2.

In Korea, two methods of biofuel production pyrolysis and transesterification in terms of cost and environmental impact from spent coffee grounds were compared. In the study, transesterification is a process of using cooking oils in the presence of spent coffee grounds to produce biodiesel. Gu et al. found that the pyrolysis process is the more economically viable option for producing biofuel from spent coffee grounds, compared to the transesterification process. Although the transesterification pathway results in lower CO_2 emissions, the economic advantages of the pyrolysis process outweigh these environmental benefits when considering this specific feedstock. Another study conducted a comparative techno-economic analysis (TAE) on microwave and conventional pyrolysis (Haeldermans et al. 2020). Both technologies were viable for large-scale biochar production, with conventional pyrolysis being the more established and economical option. Similarly, a study was conducted in India of TAE and LCA on both laboratory and commercial scale of microwave pyrolysis food waste and low-density polyethylene (Neha, Prasanna Kumar Ramesh, and Remya 2022). They reported that the drying and pyrolysis unit consumed 91% of the total energy and contributed the most to the environmental impact. This is mainly highlighted in most investigations into the processing of food waste. The commercial pyrolysis plant would show a lower environmental impact and energy efficiency which include 0.54 kWh/kg and projected emissions of 38.92 kg CO_2 eq, 48.00×10^{-3} kg SO_2 eq, 7.70×10^{-3} kg PO_4^{3-} eq, and 3.50×10^{-3} kg C_2H_6 for processing 100 kg of food waste.

TABLE 5.2
Thermochemical conversion methods costing.

Biomass	Country	Cost	Application	References
Spent coffee grounds	Korea	Transesterification: \$30.76–\$43.71 per GJ Pyrolysis: \$8.82–\$12.35 per GJ	Biofuel	Gu et al. (2023)
Coffee husks	Belgium	Conventional pyrolysis: €436/tonne to €863/tonne	Biochar, Electricity	Haeldermans et al. (2020)
Palm date fronds	Qatar	Microwave pyrolysis: €564/tonne to €979/tonne		
Olive stones	Italy			
Combined food waste	India	Lab-scale reactor (capacity - 1 kg/d) was 10,00,000 INR/kg and 8,248 INR/kg Commercial plant (capacity - 100 kg/d), the estimated fixed and monthly operating cost was 53,12,585 INR/kg and 24,420 INR/kg, respectively	Bio-oil Biochar	Neha, Prasanna Kumar Ramesh, and Remya (2022)
Multiple feedstocks	USA	Minimum product selling price: \$774–\$1256 per tonne	Biochar	Cheng, Luo, and Colosi (2020)
Kitchen food waste and rice starw	China	A positive net present value after 2nd Year. The project had an internal rate of return of 0.62. The profitability index was 3.85 over 10 years.	Biofuel	Nyambura et al. (2023)

The LCA and the economic impacts of slow pyrolysis with different feedstocks were investigated by Cheng, Luo, and Colosi (2020) using machine learning. It was concluded that the best overall energy and climate change performance is achieved with pyrolysis of lignocellulosic biomass at high temperatures, while the best minimum product selling price (MPSP) is with pyrolysis of sludge at low temperatures. Nyambura et al. (2023) conducted a TAE study in China of co-pyrolyzing kitchen food waste and rice straw. The study reported that the simulated heat and power generation plant had a boiler efficiency of 96.30%, steam cycle thermal efficiency of 36.46%, electrical efficiency of 41.61%, and overall system efficiency of 35.11%. It was estimated to produce 744,576 kWh/year of electricity, 1,727,649 L/year of hot water, and 162,045 kg/year of biochar (Nyambura et al. 2023). This determined that the technology is profitable and an effective waste management solution for food waste.

Thermochemical conversion methods gasification/esterification and HTC are the best methods in food waste management (Batool et al. 2024). Another LCA study was conducted on food waste containing noodles and lettuce waste from Jinan China that undergoes pyrolysis at 550°C (Sun et al. 2022). The pyrolysis temperature and the food waste types were compared, and it was found that the environmental impact of activated carbon from noodles was minor than that of lettuce. In addition, using lower temperatures for pyrolysis at 350°C would be preferable as less energy consumption is used.

Similarly, another study in China compared the environmental and energy impact of two different food waste disposal methods of anaerobic digestion and hyperthermal process: solid-liquid mixed and liquid phase (Fei et al. 2021). The solid-liquid mix has a lesser negative environmental impact than the liquid phase and recommended that more studies be made on different types of food waste. Even though, the anaerobic digestion methods are different, it is still important to study the different impacts of thermochemical conversion processes to promote awareness in the industry. In LCA with cost analysis using a catalyst study, the hydrothermal process has the potential to reduce Green House Gas (GHG) emissions by 2.6% and with a minimum selling price of $1.06 per gasoline gallon equivalent (LeClerc et al. 2022). Further to this, biochar properties can be maximised using pre- and post-treatment. The proposed methods include acidic or alkaline treatment, impregnation, steam activation, microwave treatment, ozone treatment, plasma treatment, etc. Pre-treatment using KOH can increase the specific surface area to enhance energy storage (Atinafu et al. 2023). Different pyrolysis temperatures were tested to determine the most appropriate production parameters for the intended application. It exceeded the original form of the material energy storage from 243.9 to 346.9. The conversion of various food wastes into biochar is in compliance with the standards set by the International Biochar Initiative (IBI) and the European Biochar Certificate (EBC) for use in agricultural applications (Raček et al. 2024).

5.2 VERSATILE APPLICATIONS OF FOOD WASTE BIOCHAR

Since biochar has various advantageous properties, research has been conducted to test its capabilities in a wide range of applications. These applications include its use as a fuel source, supplementary cementitious material (Gupta, Kua, and Koh 2018), adsorbents (Qing et al. 2024), and catalyst, improving soil conditions (Raček et al. 2024), water storage, energy storage (Atinafu et al. 2023; Packiyalakshmi, Chandrasekhar, and Kalaiselvi 2019), and CO_2 capture. There have been numerous studies that have proven that food waste biochar from materials such as potato peel, apple waste, orange, banana, and watermelon peel showed a good potential for an electrochemical application such as supercapacitor (Khalafallah et al. 2021), hydrogen evolution reaction (HER) (Atchudan et al. 2022), and oxygen reduction reaction for Zn-air battery (Li et al. 2021). Figure 5.2 illustrates different production techniques and applications where food waste biochar has been researched.

The pyrolysis temperature was set at 600°C with an additive like wood waste and zeolite catalyst to obtain a more prominent resultant biochar and understand different yields and chemical properties. This food waste was then dried for 14 days, and it consisted of fruits, vegetables, scraps, meat, fish, pasta, noodles, etc. collected from restaurants and canteens in the Czech Republic. Different pyrolysis temperatures of food waste biochar production 350°C, 450°C, and 500°C was used for heavy metal adsorption. The food waste collected in Changsha China processed at 450°C has been concluded to be most effective in the adsorption of chromium and copper at capacities of Q_e 40.283 and 21.456 mg/g. To make food waste a more prominent resource for utilisation in water purification, more studies need to be done at a large scale as these tests were carried out with high concentrations and actual wastewater would be more complex to analyse (Qing et al. 2024). There have been investigations

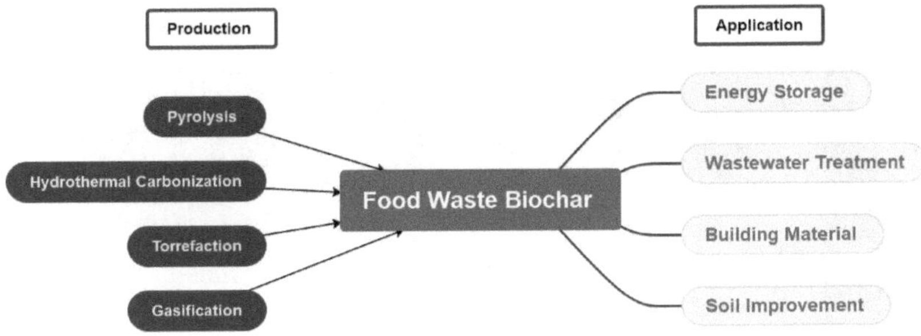

FIGURE 5.2 Food waste biochar production techniques and applications.

on using biochar in cement composites, and some studies have explored food waste biochar. A study used food biochar as an aggregate. The different pyrolysis temperatures used were 550°C, 650°C, and 750°C to understand the different biochar characteristics that can change the actual cement-based composites' mechanical properties (Chen et al. 2024). It has been concluded that the higher pyrolysis temperature of 750°C had the highest 28-day strength and lowest density 57.8 MPa and 1.77 g/cm^3 resulted in a favorable lightweight with high strength.

5.3 THERMOCHEMICAL CONVERSION METHODS

5.3.1 Pyrolysis

Pyrolysis is often the selected thermochemical conversion when the biochar product is prioritized. The factors that are most commonly interchangeable are the set temperature, the heating rate, and the residence time and less commonly the pressure and the gas flow rate used. Different types of gases are used to conduct a pyrolysis experiment to expel the oxygen and often inert gases like nitrogen or carbon dioxide are chosen. There are two types of pyrolysis, fast and slow, and each differs at the heating rate and set temperature. The usual temperature ranges are 390–980°C for slow and 380–780°C for fast (Arni 2018). From the two different types when maximum biochar yield is the desired product, slow pyrolysis is mostly preferred. There have been multiple studies that altered the biochar characteristics based on the required application by changing the production parameters. Moreover, to ease the experimental research by saving time and resources, these studies have been conducted with the help of machine learning techniques like response surface methodology (Nusrat Aman et al. 2023; Anthony et al. 2023; Yadav and Jagadevan 2021). Figure 5.3 illustrates a schematic diagram of a typical pyrolysis set-up using a Carbolite Long Tubular furnace.

As food waste has a high moisture content, using the materials for pyrolysis with the addition of pretreatment would require high energy. Therefore, it has been researched that if aerobic digestion is first conducted on the materials and then followed by pyrolysis, a high demand of energy is not required by destroying the food cell walls (Lü et al. 2015). An investigation was conducted with pyrolysis on two different materials: food waste and food waste solid digestate (Zhao et al. 2022). The authors experimented with different pyrolysis temperatures ranging from 500

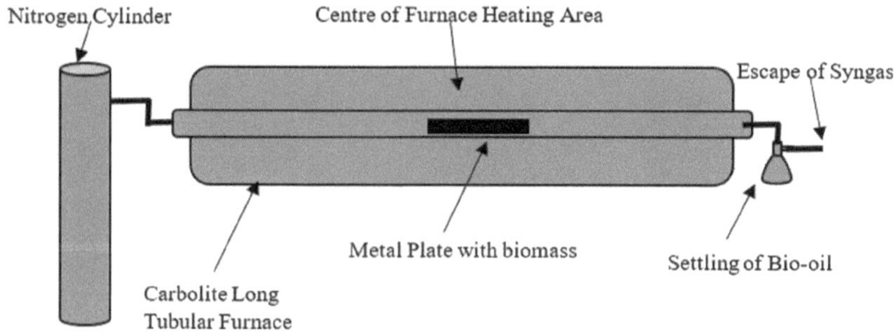

FIGURE 5.3 Schematic diagram for pyrolysis.

to 800°C and studied the pyrolytic products from each temperature. They have concluded that the pyrolytic gas quality from food waste was higher than food waste solid digest; however, the opposite was found for the pyrolytic oil. A case study experimented on chitosan (biopolymer from shellfish) to be added into a cement-based composite with biochar. The pyrolysis temperatures used were 440°C and 800°C, and the mechanical aspects of the composites were tested (Nisticò et al. 2020). The results proved that biochar would not aid in all mechanical aspects as it decreased the flexural strength but increased the toughness of the composites.

5.3.2 Hydrothermal Carbonization

Hydrothermal carbonisation (HTC) is a thermochemical conversion process that can convert wet biomass and organic waste into a solid product known as hydrochar. This process operates at temperatures typically ranging from 180 to 250°C under low pressure, making it favorable for hydrochar production. Unlike other conversion methods, HTC does not require preheating, which helps in saving energy. In a critical review of HTC parameters, including temperature, residence time, feedwater pH, substrate concentration, and heating rate, it has been investigated how each parameter individually affects the HTC process (Wang et al. 2018). This review provides valuable insights into optimizing the HTC process for efficient hydrochar production. A study focusing on the LCA of food waste conversion into energy using both anaerobic digestion and hydrothermal carbonisation found that HTC has a lower environmental impact, with a reduction of 141.46 kg CO_2 compared to anaerobic digestion, which resulted in a reduction of 38.87 kg CO_2 (Zhao et al. 2024). The effects of different hydrothermal carbonisation temperatures and process water cycling on the physicochemical properties of hydrochar fuel are extensively explored. The study reports that increasing the number of process water cycles enhances the hydrochar yield and energy yield while also enhancing its aromatic nature (Wu et al. 2023). Another study used a catalyst hydroxyapatite to help enhance the production of biocrude through hydrothermal liquefaction (LeClerc et al. 2022). Overall, these references highlight the advantages of hydrothermal carbonisation in the conversion of wet biomass and food waste. The HTC process can be conducted at lower temperatures, requires less energy, and produces hydrochar with desirable physicochemical properties, making it a promising method for sustainable waste-to-energy conversion.

5.3.3 Torrefaction

In the process of torrefaction, low temperatures are used ranging from 200 to 300°C with residence time lower than 3 h. A study by Ismail et al. (2021) investigated olive stone biochar produced through torrefaction using a machine learning approach. Specifically, an Artificial Neural Network (ANN) was employed to analyse the effects of torrefaction parameters temperature, particle size, and residence time on the yields of solid, liquid, and gas yields. Similarly, another study conducted on ground coffee residue with different torrefaction parameters such as temperature (200–300°C), residence time (30–60 min), and sweep gas flow rate (50–250 ml/min) did not use machine learning (Pathomrotsakun et al. 2020). They have concluded that the ground coffee food waste is liable to produce solid fuel to replace coal. A study in South Korea investigated the fuel characteristics of food waste subjected to torrefaction ranging from 150 to 600°C and residence time of 0 to 50 min (Poudel, Ohm, and Oh 2015). The study reported that optimal torrefaction conditions for FW were identified as between 290 and 330°C, where the energy yield remained above 80%.

Another research conducted in South Korea investigated the effects of torrefaction (250–275°C) on the product distribution and quality of bio-oil obtained from the fast pyrolysis of mixed food waste under nitrogen and carbon dioxide atmospheres (Ly et al. 2022). The study highlighted that torrefaction pretreatment and the use of carbon dioxide as the pyrolysis atmosphere can effectively improve the quality and reduce the environmental impact of bio-oils produced from mixed food waste pyrolysis, making this process a potentially feasible approach for energy production and environmental pollution control. A case study conducted in India evaluated the changes in physical and chemical properties of sun-dried food waste during torrefaction in a fixed bed reactor at temperatures of 230, 260, and 290°C, and residence times of 15, 30, 45, 60, 75, and 90 min (Singh and Yadav 2021). The study reported that torrefaction increased the carbon content from 46.0% to 58.5% and the higher heating value from 16.1 to 21.9 MJ/kg. Moreover, the study concluded that torrefaction can effectively improve the physicochemical properties of kitchen food waste, making it a more suitable feedstock for energy applications.

5.3.4 Gasification

Gasification is the thermochemical conversion that reaches high temperatures up to 1200°C with short residence time. This method is mainly used when the products syngas and oil are favored more than biochar. A comparative study was conducted on both pyrolysis and gasification on food waste (Ahmed and Gupta 2010). Both were conducted at two distinct temperatures 800 and 900°C. It was reported that gasification was found to be more beneficial than pyrolysis in terms of syngas flow rate, hydrogen flow rate, output power, total syngas, hydrogen and energy yield, and apparent thermal efficiency. However, gasification took longer due to the slower char gasification reactions. Another study compared the co-gasification of sewage sludge and food wastes (You et al. 2016). The study reported that food wastes are more favorable than sewage sludge for co-gasification in terms of residue generation and energy output. The study suggested two decentralized gasification-based waste treatment schemes were proposed for Singapore to dispose of sewage sludge and food wastes with cost-benefit analysis to improve the economics of the waste treatment schemes. A study conducted in South Korea investigated the use of biochar produced from a mixture of food waste and wood waste via gasification as an adsorbent for CO_2 capture (Igalavithana et al. 2020). The materials were mixed with different ratios and tested on their effective absorption capabilities. The study concluded that 20% of food waste and 80% of wood waste were the most effective. However, when the mix with 40% food waste was modified with potassium hydroxide, it increased the mix's micropore development and therefore its CO_2 absorption efficiency. Table 5.3 shows different applications and analyses of food waste biochar from different methods.

5.4 CASE STUDIES

5.4.1 Case study 1

Company: Planters International Berhad

Country: Malaysia

Address: PLANTERS INTERNATIONAL BERHAD, Level 23, Menara Exchange 106, Lingkaran TRX, Tun Razak Exchange, 55188 Kuala Lumpur, Malaysia

Link: https://www.planters.life/

Material Used for Biochar Production: Bamboo

Biochar Utilisation: Planters are expanding their operations to promote the development and adoption of green energy services.

Engineering Solutions: The Planters employ a comprehensive approach mainly for forest management, encompassing a wide range of activities designed to revitalize and conserve forest ecosystems. This includes planting, replanting, transplanting, and focused conservation efforts to heal damaged land and promote the long-term health of these vital natural resources. To closely monitor tree growth and detect wildlife activity, The Planters leverage advanced technologies such as drones, satellite imagery, and thermal imaging. This data-driven approach allows the company to make informed, real-time decisions to optimize forest management strategies. Planters have used the Active Forest Asset Management (AFAM) system which is an advanced technology platform modernizing forest management. Utilising innovative technologies like satellite connectivity, ground sensors, GPS tracking, RFID systems, climate sensors, geotagging, drones, and cutting-edge remote sensing tools like LiDAR, AFAM is at the forefront of modern forest management strategies. In addition to their comprehensive forest management initiatives, The Planters are also committed to promoting green energy services such as bamboo pellets and bamboo biochar production. As biochar is a multipurpose material with an understanding of its characteristics and properties (Table 5.4), its utilisation can be linked to different applications. Planters have suggested that a sustainable method to utilise this bamboo is to produce pellets to allow emission and carbon-neutral energy sources when converted to biochar. By maximising its expertise and innovative technological capabilities, the company aims to drive the transition toward renewable and environmentally friendly energy sources. Through strategic partnerships and targeted investments, The Planters are exploring diverse green energy applications, from bioenergy to other sustainable energy technologies. This multifaceted approach allows the company to

TABLE 5.3
Different applications and analyses of food waste biochar from different methods.

App	Temp (°C)	Analysis	Study objective	References
Cement additive	500	C: 70.90%, O: 8.42 %, Ca: 0.55%, Mg: 0.14% Silica: 0.16 % Water Absorption: 3 g/g BET: 9.70 m^2/g, Micropore: 0.00288cm^3/g Carbon: 47.71% O:C :0.09	Investigating the potential use of food waste, wood waste, and rice waste as a supplementary cementitious additive.	Gupta, Kua, and Koh (2018)
Wastewater	350	pH:8.3 C:55.93%, H:4.65%, O:8.69%, N:6.79%, S:0.39% Ash:23.55% BET:0.3386 m^2/g, Micropore: 0.000945cm^3/g	Investigation most efficient biochar production temperature for Cu(II) and CR(II) adsorption.	Qing et al. (2024)
	450	pH:11.0 C:49.20%, H: 2.73%, O:9.44%, N:6.63%, S:0.37% Ash:31.63% BET:0.7995 m^2/g, Micropore: 0.002509 cm^3/g		
	500	pH:11.8 C:48.3%, H:1.9%, O:8.95%, N:6.34%, S:0.27% Ash:34.24% BET:0.4776 m^2/g, Micropore: 0.001734 cm^3/g		
Energy	500	C:71.30%, H:2.10%, O:6.44%, N:2.64%, S:0.12% VM:13.30% FC:69.30% Ash:17.4% HHV:24.80 MJ/kg	Assess energy potential of food waste by digestion and then pyrolysis.	Opatokun, Strezov, and Kan (2015)
	800	C:84.30%, H:0.51%, O:2.50%, N:3.07% HHV:31.44 MJ/kg	Investigated different food waste for energy.	Grycová, Koutník, and Pryszcz (2016)

(Continued)

TABLE 5.3 *(Continued)*

App	Temp (°C)	Analysis	Study objective	References
Concrete	550	C:27.3%, H:2.3%, N:2.6%, S:0.9%, O:12.0%, VM:39.2% FC:5.9% Ash:54.9% Biochar yield:42.1% Water absorption:45.1% SA:14.1 m^2/g Pore volume: 0.047 cm^3/g Avg. pore diameter: 14.5 nm	Design high strength lightweight concrete using biochar as aggregate.	Chen et al. (2024)
	650	C:27.5%, H:1.3%, N:1.7%, S:0.9%, O:10.9%, VM:36.2% FC:6.1% Ash:57.7% Biochar yield: 40.4% Water absorption: 46.9 % SA: 14.6 m^2/g Pore volume: 0.051 cm^3/g Avg. pore diameter: 14.7 nm		
	750	C:28.3%, H:1.1%, N:1.6%, S:1.7%, O:5.5%, VM:31% FC:7.2% Ash:61.8% Biochar yield:38.9 % Water absorption: 49.5% SA: 13 m^2/g Pore volume: 0.051 cm^3/g Avg. pore diameter: 18.2 nm		

Fuel	220	C: 70.87%, H:8.12%, O:16.65%, N:4.36 % FC:36.11% VM:62.77% Ash:1.12 % Fuel ratio: 0.5753 Hydrochar yield: 47.25% HHVs:30.66 MJ/kg Energy density: 1.51 Energy yield:71.33 % Carbon retention: 70.10%	Produce fuel through HTC with different temperatures.	Wu et al. (2024)
	220	FC: 35.8% VM:63.3% Ash: 0.9% C:68.4%, H:8.0%, O:19.5%, N:4.1% HHV:29.9 MJ/kg	Study different types of char and the dynamic parameters were determined using the Flynn-Wall-Ozawa (FWO) and Friedman methods.	Wu et al. (2023)
	300	C: 69.9%, H: 8.4%, N: 3.8%, O:17.9% Oil yield: 14% HHV: 32.1 MJ/kg Energy recovery: 18 %	Studying the fuel yield produced from hydrothermal liquefaction with aid of a catalyst.	LeClerc et al. (2022)
	300	C:74.08 %, H: 5.60 %, N:3.97%, S:0.19%, O:11.42% Ash: 4.74%, VM: 42.20%, FC: 53.06 % HHV: (MJ/kg)	Investigate the effect of different torrefaction parameters on the chemical, thermal, and surface functional groups of ground coffee residue.	Pathomrotsakun et al. (2020)

App: Application, Temp: Temperature, Ref: References

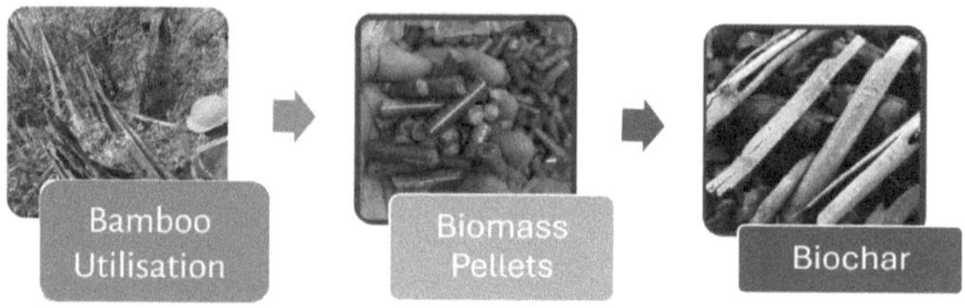

FIGURE 5.4 Bamboo utilisation.

contribute to the global shift toward a more sustainable energy landscape, further aligning its business model with the principles of a circular bioeconomy.

5.4.1.1 Expected benefits aligning with Sustainable Development Goals

The Planters aim to unite environmental sustainability with advanced technologies, driving the transition to a circular bioeconomy. By bringing together cutting-edge bioenergy and biotechnology solutions, the company seeks to fulfill several key Sustainable Development Goals (SDGs). They are (1) No Poverty; (2) Zero Hunger; (3) Good Health and Wellbeing; (4) Quality Education; (6) Clean Water and Education; (8) Decent Work and Economic Growth; (9) Industry, Innovation and Infrastructure; (13) Climate Action, (15) Life on Land; and (17) Partnerships for the Goals. Through their comprehensive forest management services, The Planters will provide a full lifecycle of forest-related offerings. This includes conservation efforts, nutritional support, and large-scale planting initiatives. The company's approach is designed to promote the long-term viability and health of forest ecosystems. Furthermore, The Planters intends to obtain carbon and sustainable forestry certifications. This will not only validate the environmental integrity of their operations but also enable the company to further contribute to global climate change mitigation efforts by aligning their business model with the principles of a circular bioeconomy.

5.4.1.2 Properties and characteristics of biochar

TABLE 5.4
Bamboo biochar properties.

BET surface area (m^2/g):	**17.50**
Pore size (nm):	1.78
Pore volume (cm^3/g):	0.01
Proximate analysis	
Inherent moisture (%):	5.3
Volatile matter (%):	13.6
Fixed carbon (%):	71.4
Ash content (%):	9.7
Ultimate analysis:	
Carbon (%):	78.8
Hydrogen (%):	2.5
Nitrogen (%):	0.7
Oxygen (%):	7.6
Sulfur (%):	0.3

5.4.2 Case study 2

Company: Henan Mingjie Environmental Equipment Co.

Country: China

Address: 6th Floor, 101# Zijingshan Road, Guancheng District, Zhengzhou City -450000, Henan Province, China.

Link: https://www.mingjiegroup.com

Material Used for Biochar Production: Woody waste, agriculture waste, tyre waste, aluminized paper, and medical waste

Biochar Utilisation: Renewable energy solid fuel (heating and cooking) and soil improvement

Engineering Solutions: Henan Mingjie provide a holistic approach to solid waste recycling, transforming discarded materials into environmentally friendly products that can be utilise as fuel. Beyond merely providing recycled materials, the company offers advanced technological solutions to facilitate this process. Henan Mingjie's offerings include state-of-the-art equipment such as pyrolysis plants, carbonisation plants, and oil distillation plants, enabling their clients to seamlessly integrate these technologies into their operations. This holistic strategy empowers businesses to reduce waste, minimise environmental impact, and generate valuable fuel sources from what would otherwise be discarded. Henan Mingjie's commitment to innovation and end-to-end solutions positions them as a leader in the circular economy, driving sustainable practices and resource optimization across industries. Figures 5.5 and 5.6 show the 3D images of the equipment the company offers to clients through the purchasing process. Figure 5.7 shows an example of the resultant wood biochar from the carbonisation plant. The large-scale equipment offered by the company adheres to specific types of materials used to maximise the efficiency of the conversion process. An example of this equipment is shown in Figure 5.8. A few examples are pyrolysis plants for medical waste and waste tires. Moreover, for the carbonisation plants for woody waste and agriculture waste, the company offers a clear and concise technical parameter to their clients as shown in Table 5.5. Figure 5.4 illustrates a pyrolysis machine that the company has delivered to a client. Table 5.2 shows the analysis conducted on biochar that has been produced from pyrolysis machines provided by the company.

Expected Benefits aligning with SDG: It can be reported that the company's aim is accomplished by utilising solid waste, converting it into fuel, and exporting equipment to be used globally. The company has exported in more than 30 countries like Canada, Ireland, Czech Republic, Egypt, Brazil, Philippines, Korea, Jordan, Thailand, Malaysia, and England. The following SDGs are fulfilled: (3) Good Health and Wellbeing; (4) Quality Education; (8) Decent Work and Economic Growth; (9) Industry, Innovation and Infrastructure; (13) Climate Action; (15) Life on Land; (17) Partnerships for the Goals.

5.5 CONCLUSIONS

Utilisation of food waste through the production of biochar offers multiple applications and benefits. LCA reports indicate that using food waste biochar for carbon

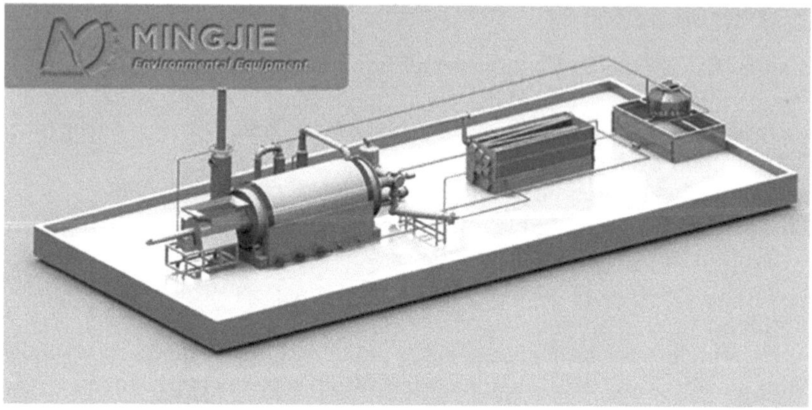

FIGURE 5.5 3D model of pyrolysis plant.

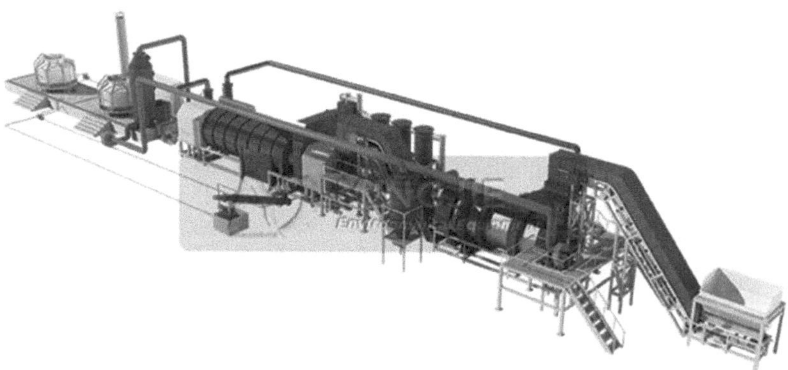

FIGURE 5.6 3D model of carbonisation plant.

FIGURE 5.7 Woody biochar.

FIGURE 5.8 Pyrolysis machine.

storage is advantageous in terms of reducing global emissions compared to landfilling. However, there are challenges to overcome in the process. Higher temperatures are required for the efficient conversion of food waste into biochar. Despite these challenges, treating food waste as a valuable resource and transforming it into high-value carbon material can help close material loops and reduce the demand for new resources. The implementation of large-scale thermochemical conversion processes for food waste biochar is still under optimization, particularly in terms of energy requirements and parameter adjustments. Extensive research is needed to ensure the quality and suitability of waste for various applications while considering the environmental impact of the conversion process. Economic feasibility studies are necessary for practical large-scale implementation, and regulatory and policy frameworks should be established to support the utilisation of food waste biochar.

5.6 EXERCISE

1. What are the differences between the mentioned thermochemical conversion methods?
2. Why hasn't the industry adopted using food waste biochar on a larger scale?
3. What is the estimated production temperature of the bamboo biochar used in case study 1?
4. Discuss the importance of pretreatment for converting food waste into biochar through different thermochemical processes.
5. What are some of the valuable applications and impacts of using food waste-derived biochar?

TABLE 5.5
Technical parameters of the carbonisation plant offered to clients.

	MJT-500	MJT-1000	MJT-2000	MJT-3000	MJT-5000
Reactor structure	Dual cylinders	Dual cylinders	Dual cylinders	Dual cylinders	Dual cylinders
Reactor material	304 Stainless steel	304 Stainless steel	304 Stainless steel	304 Stainless steel	304 Stainless steel
Input capacity	300–500 KG/H	800–1000 KG/H	1500–2000 KG/H	2500–3000 KG/H	4500–5000 KG/H
Feedstock size	<20mm	<20mm	<20mm	<20mm	<20mm
Feedstock moisture	<15%	<15%	<15%	<15%	<15%
Charcoal output ratio	28–35%	28–35%	28–35%	28–35%	28–35%
Working method	Fully continuous	Fully continuous	Fully continuous	Fully continuous	Fully continuous
Fuel consumption	35–50 m^3 Natural gas for initial heating	40–60 m^3 Natural gas for initial heating	60–80 m^3 Natural gas for initial heating	80–100 m^3 Natural gas for initial heating	120–140 m^3 Natural gas for initial heating
Land required	L35*W12*H6m	L35*W12*H7m	L45*W15*H7m	L45*W15*H7m	L50*W15*H7m
Final products	Biochar/charcoal, tar and wood vinegar, Syngas				

TABLE 5.6
Biochar analysis report.

Items	Rice husk	Coffee shell	Coconut shell	Corn straw
Total moisture (%)	6.09	11.78	9.1	0.78
Mad (%)	3.58	1.21	4.19	0.15
Ash content (%)	39.08	2.75	0.18	50.96
Volatiles (%)	19.12	30.96	6.95	10.96
Fixed carbon (%)	38.22	65.08	88.68	37.99
Total sulfur (%)	0.12	0.02	0.09	0.02
High calorific value Kcal/kg	3860	7050	7277	N/A
Low calorific value Kcal/kg	3559	6053	6724	N/A

Mad: Moisture air-dried

ACKNOWLEDGMENT

The author would like to acknowledge Henan Mingjie Environmental Equipment Co., Ltd, for sharing the pyrolysis machine Figure 5.8.

REFERENCES

Ahmed, I. I., and A. K. Gupta. 2010. "Pyrolysis and Gasification of Food Waste: Syngas Characteristics and Char Gasification Kinetics." *Applied Energy* 87 (1). Elsevier: 101–8. doi:10.1016/J.APENERGY.2009.08.032.

Anthony, Onokwai O., Ibikunle A. Rotimi, Ajisegiri S. A. Emmanuel, Onokpite Ejiroghene, Ezekiel B. Omoniyi, Anyaegbuna E. Benjamin, Aliyu J. Samuel, and Igbebuike U. Kingsley. 2023. "Optimization of Process Parameters for Intermediate Pyrolysis of Sugarcane Bagasse for Biochar Production Using Response Surface Methodology." *2023 International Conference on Science, Engineering and Business for Sustainable Development Goals (SEB-SDG)*. Institute of Electrical and Electronics Engineers Inc. doi:10.1109/SEB-SDG57117.2023.10124642.

Arni, Saleh Al. 2018. "Comparison of Slow and Fast Pyrolysis for Converting Biomass into Fuel." *Renewable Energy* 124 (August). Elsevier Ltd: 197–201. doi:10.1016/J.RENENE.2017.04.060.

Atchudan, Raji, Suguna Perumal, Thomas Nesakumar Jebakumar Immanuel Edison, S. Aldawood, Rajangam Vinodh, Ashok K. Sundramoorthy, Gajanan Ghodake, and Yong Rok Lee. 2022. "Facile Synthesis of Novel Molybdenum Disulfide Decorated Banana Peel Porous Carbon Electrode for Hydrogen Evolution Reaction." *Chemosphere* 307 (November). Pergamon: 135712. doi:10.1016/J.CHEMOSPHERE.2022.135712.

Atinafu, Dimberu G., Beom Yeol Yun, Ji Yong Choi, Xiangzhou Yuan, Yong Sik Ok, and Sumin Kim. 2023. "Introduction of Sustainable Food Waste-Derived Biochar for Phase Change Material Assembly to Enhance Energy Storage Capacity and Enable Circular Economy." *Journal of Energy Storage* 72 (November). Elsevier: 108338. doi:10.1016/J.EST.2023.108338.

Batool, Fatima, Tonni Agustiono Kurniawan, Ayesha Mohyuddin, Mohd Hafiz Dzarfan Othman, Faissal Aziz, Hussein E. Al-Hazmi, Hui Hwang Goh, and Abdelkader Anouzla. 2024. "Environmental Impacts of Food Waste Management Technologies: A Critical Review of Life Cycle Assessment (LCA) Studies." *Trends in Food Science & Technology* 143 (January). Elsevier: 104287. doi:10.1016/J.TIFS.2023.104287.

Chen, Liang, Xiaohong Zhu, Yu Zheng, Lei Wang, Chi Sun Poon, and Daniel C. W. Tsang. 2024. "Development of High-Strength Lightweight Concrete by Utilizing Food Waste Digestate Based Biochar Aggregate." *Construction and Building Materials* 411 (January). Elsevier: 134142. doi:10.1016/j.conbuildmat.2023.134142.

Cheng, Fangwei, Hongxi Luo, and Lisa M. Colosi. 2020. "Slow Pyrolysis as a Platform for Negative Emissions Technology: An Integration of Machine Learning Models, Life Cycle Assessment, and Economic Analysis." *Energy Conversion and Management* 223 (November). Pergamon: 113258. doi:10.1016/J.ENCONMAN.2020.113258.

Fei, Xionghui, Wenbao Jia, Ting Chen, and Yongsheng Ling. 2021. "Life-Cycle Assessment of Two Food Waste Disposal Processes Based on Anaerobic Digestion in China." *Journal of Cleaner Production* 293 (April). Elsevier: 126113. doi:10.1016/J.JCLEPRO.2021.126113.

Funke, Axel, and Felix Ziegler. 2010. "Hydrothermal Carbonization of Biomass: A Summary and Discussion of Chemical Mechanisms for Process Engineering." *Biofuels, Bioproducts and Biorefining* 4 (2). John Wiley & Sons, Ltd: 160–77. doi:10.1002/BBB.198.

Grycová, Barbora, Ivan Koutník, and Adrian Pryszcz. 2016. "Pyrolysis Process for the Treatment of Food Waste." *Bioresource Technology* 218 (October): 1203–7. doi:10.1016/J.BIORTECH.2016.07.064.

Gu, Jiwon, Aejin Lee, Changgwon Choe, and Hankwon Lim. 2023. "Comparative Study of Biofuel Production Based on Spent Coffee Grounds Transesterification and Pyrolysis: Process Simulation, Techno-Economic, and Life Cycle Assessment." *Journal of Cleaner Production* 428 (November). Elsevier: 139308. doi:10.1016/J.JCLEPRO.2023.139308.

Gupta, Souradeep, Harn Wei Kua, and Hui Jun Koh. 2018. "Application of Biochar from Food and Wood Waste as Green Admixture for Cement Mortar." *Science of the Total Environment* 619–20 (April). Elsevier: 419–35. doi:10.1016/j.scitotenv.2017.11.044.

Haeldermans, T., L. Campion, T. Kuppens, K. Vanreppelen, A. Cuypers, and S. Schreurs. 2020. "A Comparative Techno-Economic Assessment of Biochar Production from Different Residue Streams Using Conventional and Microwave Pyrolysis." *Bioresource Technology* 318 (December). Elsevier: 124083. doi:10.1016/J.BIORTECH.2020.124083.

Igalavithana, Avanthi Deshani, Seung Wan Choi, Pavani Dulanja Dissanayake, Jin Shang, Chi Hwa Wang, Xiao Yang, Sumin Kim, Daniel C.W. Tsang, Ki Bong Lee, and Yong Sik Ok. 2020. "Gasification Biochar from Biowaste (Food Waste and Wood Waste) for Effective CO_2 Adsorption." *Journal of Hazardous Materials* 391 (June). Elsevier: 121147. doi:10.1016/J.JHAZMAT.2019.121147.

Ismail, Hamza Y., Sary Fayyad, Mohammad N. Ahmad, James J. Leahy, Mu. Naushad, Gavin M. Walker, Ahmad B. Albadarin, and Witold Kwapinski. 2021. "Modelling of Yields in Torrefaction of Olive Stones Using Artificial Intelligence Coupled with Kriging Interpolation." *Journal of Cleaner Production* 326 (December). Elsevier: 129020. doi:10.1016/j.jclepro.2021.129020.

Kapoor, Riti Thapar, and P. Maulin Shah, eds. 2023. *Designer Biochar Assisted Bioremediation of Industrial Effluents*. Taylor & Francis Group.

Khalafallah, Diab, Xinyao Quan, Chong Ouyang, Mingjia Zhi, and Zhanglian Hong. 2021. "Heteroatoms Doped Porous Carbon Derived from Waste Potato Peel for Supercapacitors." *Renewable Energy* 170 (June). Pergamon: 60–71. doi:10.1016/J.RENENE.2021.01.077.

Klinghoffer, Naomi B., Marco J. Castaldi, and Ange Nzihou. 2015. "Influence of Char Composition and Inorganics on Catalytic Activity of Char from Biomass Gasification." *Fuel* 157 (October). Elsevier: 37–47. doi:10.1016/J.FUEL.2015.04.036.

LeClerc, Heather O., Geoffrey A. Tompsett, Alex D. Paulsen, Amy M. McKenna, Sydney F. Niles, Christopher M. Reddy, Robert K. Nelson, Feng Cheng, Andrew R. Teixeira, and Michael T. Timko. 2022. "Hydroxyapatite Catalyzed Hydrothermal Liquefaction Transforms Food Waste from an Environmental Liability to Renewable Fuel." *IScience* 25 (9). Elsevier: 104916. doi:10.1016/J.ISCI.2022.104916.

Li, Lu, Zhiheng Wu, Jin Zhang, Yige Zhao, and Guosheng Shao. 2021. "Watermelon Peel-Derived Nitrogen-Doped Porous Carbon as a Superior Oxygen Reduction Electrocatalyst for Zinc-Air Batteries." *ChemElectroChem* 8 (24). John Wiley & Sons, Ltd: 4790–96. doi:10.1002/CELC.202101339.

Lü, Fan, Qi Zhou, Duo Wu, Tianfeng Wang, Liming Shao, and Pinjing He. 2015. "Dewaterability of Anaerobic Digestate from Food Waste: Relationship with Extracellular Polymeric Substances." *Chemical Engineering Journal* 262 (February). Elsevier: 932–38. doi:10.1016/J.CEJ.2014.10.051.

Ly, Hoang Vu, Byeongwan Kwon, Jinsoo Kim, Changho Oh, Hyun Tae Hwang, Jung Suk Lee, and Seung Soo Kim. 2022. "Effects of Torrefaction on Product Distribution and Quality of Bio-Oil from Food Waste Pyrolysis in N_2 and CO_2." *Waste Management* 141 (March). Pergamon: 16–26. doi:10.1016/J.WASMAN.2022.01.013.

Neha, Shukla, Kondragunta Prasanna Kumar Ramesh, and Neelancherry Remya. 2022. "Techno-Economic Analysis and Life Cycle Assessment of Microwave Co-Pyrolysis of Food Waste and Low-Density Polyethylene." *Sustainable Energy Technologies and Assessments* 52 (August). Elsevier: 102356. doi:10.1016/J.SETA.2022.102356.

Nisticò, Roberto, Luca Lavagna, Daniele Versaci, Pavlo Ivanchenko, and Paola Benzi. 2020. "Chitosan and Its Char as Fillers in Cement-Base Composites: A Case Study." *Boletín de La Sociedad Española de Cerámica y Vidrio* 59 (5). Elsevier: 186–92. doi:10.1016/j.bsecv.2019.10.002.

Nusrat Aman, Aan Mohammad, Anurita Selvarajoo, Teck Leong Lau, and Wei Hsin Chen. 2023. "Optimization via Response Surface Methodology of Palm Kernel Shell Biochar for Supplementary Cementitious Replacement." *Chemosphere* 313 (February). Pergamon: 137477. doi:10.1016/J.CHEMOSPHERE.2022.137477.

Nyambura, Samuel Mbugua, Chao Li, Jialiang Xu, Jufei Wang, Hua Li, Zhaotong Zhang, Gbenontin V. Bertrand, et al. 2023. "Techno-Economic Assessment of Co-Pyrolyzed Kitchen Waste and Rice Straw for Co-Generation of Heat and Power." *Sustainable Production and Consumption* 42 (November). Elsevier: 434–45. doi:10.1016/J.SPC.2023.06.015.

Opatokun, Suraj Adebayo, Vladimir Strezov, and Tao Kan. 2015. "Product Based Evaluation of Pyrolysis of Food Waste and Its Digestate." *Energy* 92 (December). Pergamon: 349–54. doi:10.1016/J.ENERGY.2015.02.098.

Packiyalakshmi, Parameswaran, Bongu Chandrasekhar, and Nallathamby Kalaiselvi. 2019. "Domestic Food Waste Derived Porous Carbon for Energy Storage Applications." *ChemistrySelect* 4 (27). John Wiley & Sons, Ltd: 8007–14. doi:10.1002/SLCT.201900818.

Pathomrotsakun, Jiaranai, Kamonwat Nakason, Wasawat Kraithong, Pongtanawat Khemthong, Bunyarit Panyapinyopol, and Prasert Pavasant. 2020. "Fuel Properties of Biochar from Torrefaction of Ground Coffee Residue: Effect of Process Temperature, Time, and Sweeping Gas." *Biomass Conversion and Biorefinery* 10 (3). Springer: 743–53. doi:10.1007/s13399-020-00632-1

Poudel, Jeeban, Tae In Ohm, and Sea Cheon Oh. 2015. "A Study on Torrefaction of Food Waste." *Fuel* 140 (January). Elsevier: 275–81. doi:10.1016/J.FUEL.2014.09.120.

Qing, Mengxia, Wenbin Liu, Liang Liu, Sibiao Huang, Zihang He, Yanshan Yin, and Jun Xiang. 2024. "Effective Fixation of Cu(II) and Cr(III) in Solution by Food Waste Biochar – Innovative and Valuable Treatment Method for Municipal Solid Waste." *Fuel* 361 (April). Elsevier: 130679. doi:10.1016/J.FUEL.2023.130679.

Raček, Jakub, Tomáš Chorazy, Marco Carnevale Miino, Martina Vršanská, Martin Brtnický, Ludmila Mravcová, Jiří Kučerík, and Petr Hlavínek. 2024. "Biochar Production from the Pyrolysis of Food Waste: Characterization and Implications for Its Use." *Sustainable Chemistry and Pharmacy* 37 (February). Elsevier: 101387. doi:10.1016/J.SCP.2023.101387.

Shen, Yafei. 2020. "A Review on Hydrothermal Carbonization of Biomass and Plastic Wastes to Energy Products." *Biomass and Bioenergy* 134 (March). Pergamon: 105479. doi:10.1016/J.BIOMBIOE.2020.105479.

Singh, Dharminder, and Sanjeev Yadav. 2021. "Evaluation of the Physico-Chemical Development of Kitchen Food Wastes through Torrefaction - a Biodiversity Case Study." *Biomass Conversion and Biorefinery* 11 (4). Springer Science and Business Media Deutschland GmbH: 1353–62. doi: 10.1007/s13399-019-00526-x

Spokas, Kurt A., Keri B. Cantrell, Jeffrey M. Novak, David W. Archer, James A. Ippolito, Harold P. Collins, Akwasi A. Boateng, Isabel M. Lima, Marshall C. Lamb, Andrew J. McAloon, Rodrick D. Lentz, and Kristine A. Nichols. 2012 (Jul–Aug). Biochar: A Synthesis of its Agronomic Impact beyond Carbon Sequestration. *Journal of Environmental Quality* 41 (4). 973–89. doi: 10.2134/jeq2011.0069. PMID: 22751040.

Su, Guangcan, Hwai Chyuan Ong, I. M. Rizwanul Fattah, Yong Sik Ok, Jer Huan Jang, and Chin Tsan Wang. 2022. "State-of-the-Art of the Pyrolysis and Co-Pyrolysis of Food Waste: Progress and Challenges." *Science of the Total Environment* 809 (February). Elsevier: 151170. doi:10.1016/J.SCITOTENV.2021.151170.

Sun, Yifan, Chao Li, Mengjiao Fan, Lijun Zhang, Shu Zhang, Guangzhi Hu, and Xun Hu. 2022. "Distinct Nature of Biochar and Activated Carbon from Pyrolysis-Activation of Vegetable (Lettuce) and Staple Food (Noodles) in Food Waste." *Resources, Conservation & Recycling Advances* 15 (November). Elsevier: 200118. doi:10.1016/J.RCRADV.2022.200118.

Wang, Tengfei, Yunbo Zhai, Yun Zhu, Caiting Li, and Guangming Zeng. 2018. "A Review of the Hydrothermal Carbonization of Biomass Waste for Hydrochar Formation: Process Conditions, Fundamentals, and Physicochemical Properties." *Renewable and Sustainable Energy Reviews* 90 (July). Pergamon: 223–47. doi:10.1016/J.RSER.2018.03.071.

Wu, Shuang, Qing Wang, Da Cui, Xinmin Wang, Dongyang Wu, Jingru Bai, Faxing Xu, Zhenye Wang, and Jinghui Zhang. 2023. "Analysis of Fuel Properties of Hydrochar Derived from Food Waste and Biomass: Evaluating Varied Mixing Techniques Pre/Post-Hydrothermal Carbonization." *Journal of Cleaner Production* 430 (December). Elsevier: 139660. doi:10.1016/J.JCLEPRO.2023.139660.

Wu, Shuang, Qing Wang, Dongyang Wu, Da Cui, Chunlei Wu, Jingru Bai, Faxing Xu, Bin Liu, Zhiming Shan, and Jinghui Zhang. 2024. "Influence of Temperature and Process Water Circulation on Hydrothermal Carbonization of Food Waste for Sustainable Fuel Production." *Journal of the Energy Institute* 112 (February). Elsevier: 101459. doi:10.1016/J.JOEI.2023.101459.

Yadav, Krishna, and Sheeja Jagadevan. 2021. "Effect of Pyrolysis of Rice Husk–Derived Biochar on the Fuel Characteristics and Adsorption of Fluoride from Aqueous Solution." *Bioenergy Research* 14 (3). Springer: 964–77. doi: 10.1007/s12155-020-10189-6

You, Siming, Wei Wang, Yanjun Dai, Yen Wah Tong, and Chi Hwa Wang. 2016. "Comparison of the Co-Gasification of Sewage Sludge and Food Wastes and Cost-Benefit Analysis of Gasification- and Incineration-Based Waste Treatment Schemes." *Bioresource Technology* 218 (October). Elsevier: 595–605. doi:10.1016/J.BIORTECH.2016.07.017.

Yuan, Xiangzhou, Pavani Dulanja Dissanayake, Bin Gao, Wu Jun Liu, Ki Bong Lee, and Yong Sik Ok. 2021. "Review on Upgrading Organic Waste to Value-Added Carbon Materials for Energy and Environmental Applications." *Journal of Environmental Management* 296 (October). Academic Press Inc.: 113128. doi:10.1016/J.JENVMAN.2021.113128.

Zhao, Juan, Zhi Wang, Jian Li, Beibei Yan, and Guanyi Chen. 2022. "Pyrolysis of Food Waste and Food Waste Solid Digestate: A Comparative Investigation." *Bioresource Technology* 354 (June). Bioresour Technol. doi:10.1016/J.BIORTECH.2022.127191.

Zhao, Zhenghui, Shulong Qi, Ruikun Wang, Huijun Li, Gaoke Song, Hangjin Li, and Qianqian Yin. 2024. "Life Cycle Assessment of Food Waste Energy and Resource Conversion Scheme via the Integrated Process of Anaerobic Digestion and Hydrothermal Carbonization." *International Journal of Hydrogen Energy* 52 (January). Pergamon: 122–32. doi:10.1016/J.IJHYDENE.2023.08.203.

6 Bioconversion of food waste into value-added products via black soldier fly larvae (BSFL)

Teck Ann Yeow, Malvin Ma, Yuana Elly Agustin, Said Ismail Amer Aref Yaser, Jocelyn Jean Yi Lim, Man-Kei Wong, Yew Woh Hui, Norazira Abdu Rahman, Swee Tiam Tan, and Jian Ping Tan

6.1 THE POTENTIAL OF BSFL FOR FW MANAGEMENT

The growing population across the world has led to the substantial generation of food waste (FW) which has become a pressing environmental issue that needs to be tackled. According to the European Commission, FW has been classified into the following three categories: (1) food products lost during the production phase such as the agricultural residues, (2) unavoidable waste during the consumption phase such as the fruit peels and cores, and (3) avoidable waste which refers to FW that could have been consumed such as the leftover meals from the restaurants. Overall, it is estimated that 28% of the global fertile land (~ 1.4 billion hectares) has been used to produce food products that end up as FW. Additionally, activities associated with FW production have resulted in the emission of 3.3 billion tonnes of CO_2 into the atmosphere every year (Paritosh et al. 2017). Therefore, the accumulation of FW not only signifies inefficient resource utilisation but also poses serious environmental threats.

The main challenge of managing FW lies in its poor segregation from the municipal solid waste (MSW). In many countries, inadequate funding and ineffective implementation of household FW recycling programs have resulted in failure to segregate FW from MSW. These mixed wastes are often landfilled without proper segregation which leads to the emission of undesirable greenhouse gases due to decaying inside the soil. Hence, addressing the environmental impact of FW necessitates the implementation of affordable management methods. This is particularly crucial in developing nations where financial constraints often hinder effective solutions.

Notably, the bioconversion of FW by black soldier fly larvae (BSFL) has demonstrated significant advantages in managing FW due to its ability in digesting a

DOI: 10.1201/9781032718507-6

wide variety of organic waste. Research showed that BSFL could effectively consume polluted organic waste without having adverse growth effect (Purschke et al., 2017), and this is particularly helpful when FW are not effectively segregated from MSW containing high level of mycotoxins. On top of that, the reared BSFL would not only reduce the volume of waste upon composting but also result in the generation of value-added products such as nutritious insect meals and fertilisers (the applications are explored in the following sections). Most importantly, BSFL have minimal risk of transmitting harmful pathogens unlike other insect species. Due to the aforementioned benefits, BSFL are often considered symbiotic and beneficial. In fact, in Europe, BSFL are legally allowed to be farmed as a commercial species under the European law (EG no. 1069/2009) for systematic production (Menino and Murta, 2021). Hence, in this chapter, the basic know-how of rearing BSFL would be revealed, followed by its state-of-the-art applications.

6.2 BLACK SOLDIER FLY (BSF)

Insect-based composting for FW has gained increasing recognition owing to its low installation costs and the lower greenhouse gases (GHG) emissions compared to alternative treatments. While many insect species are known for their capabilities in degrading a wide range of organic waste, most of them are traditionally captured from the wild and can hardly be farmed. The larvae of black soldier fly (*Hermetia illucens*) and yellow mealworms (*Tenebrio molitor*) are among the species that are industrially bred for commercial purposes. These insects upon rearing become a good source of proteins and lipids. By comparing these two species, BSFL demonstrates superior commercial value compared to yellow mealworms (Figure 6.1). A fully reared BSFL would consist approximately 2 times more protein, 7.5 times more vitamins, and 2.5 times more minerals – rendering higher value to be processed as insect meals on top of composting organic waste (Thrastardottir, Olafsdottir, and Thorarinsdottir 2021). Given the incentives, FW treatment through BSFL composting has gained increasing attention and has been studied by scientists across the world in recent years.

Biologically, BSF belongs to the order Diptera in the family of Stratiomyidae. BSF is commonly found to inhabit in tropical regions across the world. Many Asian countries especially in South East Asia have suitable natural conditions to rear BSF. With average lifespan of 50 days, BSF would go through four main stages in their lifecycle, i.e. the egg, larval, pupal, and adult phases as illustrated in Figure 6.2. The lifecycle starts when the female BSF are drawn by the scent of organic waste and lay their eggs on the dry surfaces. On average, a female BSF could lay up to 1000 eggs per oviposition. Subsequently, the hatching process would normally take 4 days to a week under conducive conditions (26–27°C and relative humidity of 60–70%) (Barragan-Fonseca, Dicke, and van Loon 2017). Upon hatching, larvae would emerge, and this would be the only feeding stage throughout the entire lifecycle. Overall, BSFL can be categorised into six different stages (known as instars, from 1st to 5th instars, and prepupae) according to their sizes and maturity. BSFL at different instars consist of various compositions and behaviours, especially in food consumption.

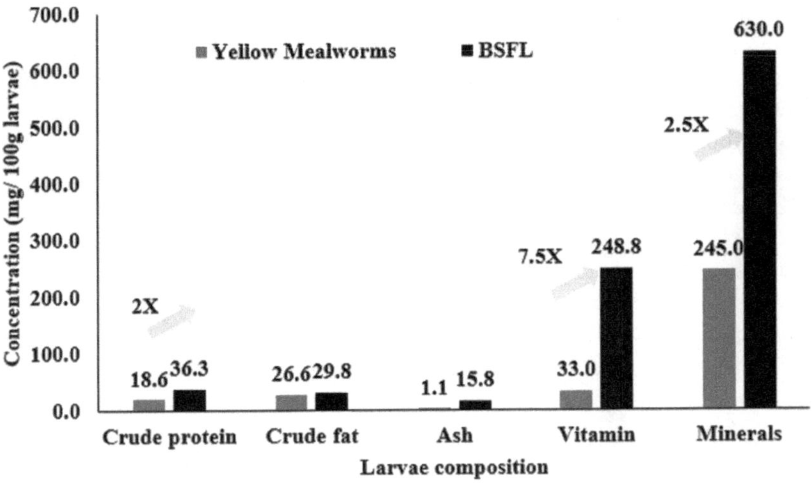

FIGURE 6.1 Comparative analysis of larvae compositions.

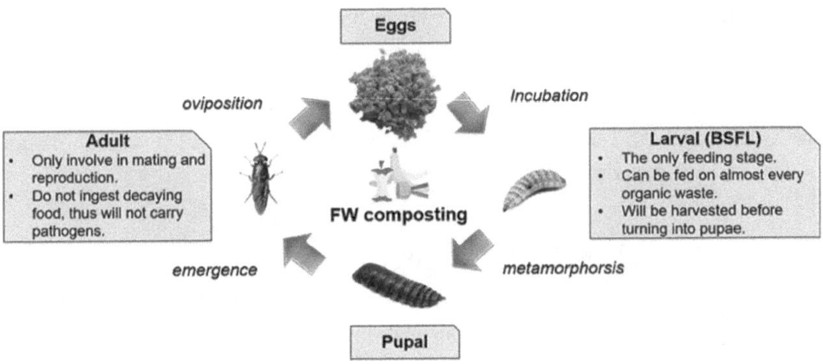

FIGURE 6.2 The complete lifecycle of BSF.

BSFL could break-down a wide range of organic waste with their strong mouth parts and diverse digestive enzymes from both their saliva glands and guts (Cho et al. 2020). In a study, BSFL guts were the main organs in secreting digestive enzymes consisting of active protease and lipase that helped digest FW. Amylase on the other hand was produced from their saliva glands and was observed to have lower enzymatic activities (Kim et al. 2011). Studies found that BSFL could effectively degrade highly indigestible waste such as vegetable residues and even plant debris (Kim et al. 2021). By supplying an optimal amount of substrate (2:1 w:w of larvae-to-substrate ratio) consisting of moisture content of 52–70%, BSFL could pupate within 2 weeks (Pastor et al. 2015). However, the pupation rates are affected by other factors beyond just the amount of substrate supplied. Other factors such as the size of the rearing space, the total number of larvae reared, and the feeding rate could influence the pupation rate. The larvae stage is often perceived as the most valuable stage as the prepupae could consist of up to 40% and 30% of crude proteins and fats, respectively (Cummins et al. 2017). Several studies demonstrated the potential of processing the harvested BSFL into animal feeds and biofuels. However, unlike the larvae, the composition of BSF pupae mainly consists of chitin which may not be as easily extractable as the proteins and fats. Some industries such as cosmetics, dietetics, and medicines could benefit from BSFL chitin.

The emergence of adult flies from the pupal stage would normally take up to a week. In contrast to houseflies, the adult BSF would not ingest any food but only rely on the fats preserved during their larvae feeding stage to sustain their metabolic activities. Nevertheless, BSF still requires water for moisturisation. The typical lifespan of adult BSF lasts approximately 10 days during mating and reproduction (Ferrarezi and Bailey 2016). With their role solely for mating, BSF would not intentionally come into contact with the decaying organic waste. Hence, it is believed that they hardly carry the pathogens from these wastes and spread as disease vectors. On top of that, composting organic waste with BSFL has been found to suppress the growth of houseflies (Furman, Young, and Catts 1959). This is primarily attributed to the rapid consumption of organic waste by BSFL while keeping the compost moisturised – creating an unfavourable breeding environment for houseflies. Additionally, the natural antibodies found in BSFL could eliminate pathogens in the organic waste such as *Escherichia coli* and *Salmonella enterica* and hence reducing the pathogenic population in the compost favouring their use as fertiliser (Erickson et al. 2004). Overall, cultivating BSFL for FW management offers numerous advantages beyond waste reduction; it also helps in reducing the toxicity of the compost prior to fertilisation and has the potential to generate additional revenue through the production of valuable insect meals.

6.2.1 Case study: A practical approach to rearing BSF

In this case study, we examined the BSF rearing implemented at an insect farm facility located within Xiamen University Malaysia (XMUM) capable of processing approximately 10 kg of organic waste per day. Firstly, the fresh eggs would be placed on top of the substrate with a separator (normally a sieve) as shown in Figure 6.3 (a) within a tray. The hatched larvae was attracted by the scene of organic waste underneath them and wiggled through the sieve separating from the unhatched eggs. The starter feed would normally be those with strong odour when decaying and easily digestible such as okara. This is aimed at expediting the growth of newly hatched larvae to a more manageable size. After approximately 4 days of rearing, the larvae reach a larger size as shown in Figure 6.3 (b); this would enhance their ability to effectively degrade less palatable substrates such as vegetable residues. Notably, while BSFL can be exposed to light during the rearing process, they are sensitive to this abiotic stress and it would impede their pupation (Lailatul and Nahid 2021). Hence, the trays would need to be covered to provide conducive dark conditions for the growth of BSFL. Subsequently, the larvae could be transferred into other trays containing low-value organic wastes as shown in Figure 6.3 (c). Depending on the quality of substrate, the larvae shall pupate at different rates. Typically, 20 days would be sufficient for the pupation of larvae into prepupae given sufficient supply of substrate. About 70% of the prepupae would be harvested for subsequent processes

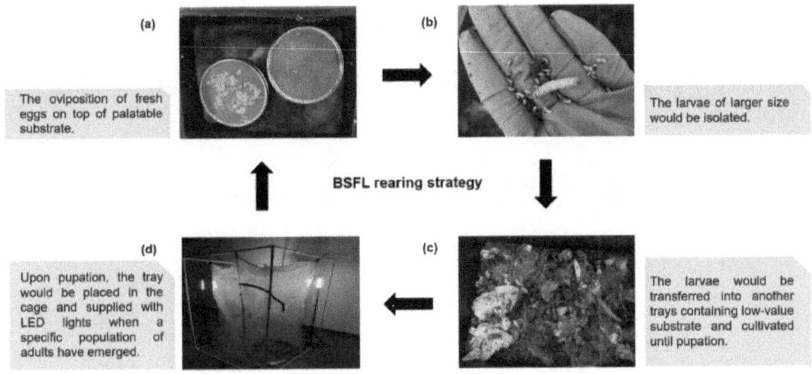

FIGURE 6.3 The rearing strategy for BSFL implemented at Xiamen University Malaysia.

(analysis or to be processed into other value-added products) while, the remaining would be moved to a rearing net as illustrated in Figure 6.3 (d).

When a specific population of adults emerges, LED-emitting lights at the wavelength 440 nm (blue) and 550 nm (green) are used to stimulate the BSFs' photoreceptors and induce mating behaviours. Naturally, both male and female BSF would 'rest' in their own territories. Mating takes place when the male BSF approaches the 'resting' female in the territory on the ground where both males and females are facing opposite directions. Notably, similar to most species in nature, lekking behaviour is also observed during the mating process of BSF. A study showed that lekking took place when a male BSF flew past 0.5 m to 1.5 m above the 'resting' territory of another male BSF (Tomberlin and Sheppard 2001). Both males would grapple within this region until they left the vicinity, where one would return to the 'resting' territory while the other would leave. During oviposition, the female BSF would lay eggs in the crevices of the dry cardboard placed on top of the waste. The eggs are periodically harvested from the cardboards for subsequent oviposition.

The conditions of the farm facility should be controlled at 30°C and with 70% relative humidity for optimal BSFL cultivation. On top of that, the growth historic data of BSFL must be logged to analyse their growth characteristic with the parameters summarised in Table 6.1. Besides quantifying the growth of BSFL, other parameters such as the biodegradability of the substrate fed can be considered observing. These parameters would allow for deeper insights into the rearing process and the palatability of the substrate. Additionally, these parameters would enable the farmers to identify and address bottlenecks effectively, particularly when BSFL experiences slow growth.

6.3 FOOD WASTE AS SUBSTRATE FOR BLACK SOLDIER FLY LARVAE

BSFL is saprophagous where it degrades a wide variety of organic waste ranging from FW to highly recalcitrant agricultural waste. Since the larvae stage is the only feeding stage throughout the BSF lifecycle, BSFL would consume large amounts of substrate in order to accumulate sufficient nutrients such as proteins, lipids, and

TABLE 6.1
Parameters of BSFL growth characteristics and their corresponding computations

Parameter (unit)	Equations	
Larvae survival rate (%)	$\frac{N_{L,\text{final}}}{N_{L,\text{initial}}} \times 100$	(6.1)
Development time (days)	$A_{L,\text{final}} - A_{L,\text{initial}}$	(6.2)
Substrate conversion rate (mg/day)	$\frac{m_{L,\text{final}} - m_{L,\text{initial}}}{t}$	(6.3)
Substrate overall degradation (%)	$\frac{m_{S,\text{initial}} - m_{S,\text{final}}}{m_{S,\text{initial}}} \times 100\%$	(6.4)
Waste reduction index (g/day)	$\frac{m_{S,\text{initial}} - m_{S,\text{final}}}{t}$	(6.5)

*N_L: Number of larvae; A_L: Age of larvae; m_L: Mass of larvae; m_S: Mass of substrate; t: duration in days. Subscript "initial" refers to the data logged during the initial cultivation phase of BSFL while "final" refers to the final phase of cultivation, normally at their prepupae stage.

minerals prior to pupation (Raksasat et al. 2020). Based on a study, by exploiting 3 kg of BSFL, approximately 10 kg of FW could be degraded in just 12 days demonstrating tremendous composting ability (Salomone et al. 2017). The final quality of BSFL depends on the physical properties and chemical compositions of the substrates. Proteins are the most essential component required by BSFL from the substrate as they may account for as much as 40% body weight of a fully reared BSFL (Barragan-Fonseca, Dicke, and van Loon 2018). Additionally, proteins in the substrate were found to have a more significant influence on the growth of BSFL as compared to lipids and carbohydrates (Beniers and Graham 2019). Therefore, conditioning substrate with the right proportion of proteins and other components could help to achieve optimal BSFL quality upon harvesting.

Choosing the right substrate for BSFL is challenging as it requires considerations in economics, availability, and sustainability of the source. Highly nutritious FW such as okara, by-products of soybean processing, is often valorised into animal feed in the farm. Thus, exploiting okara as substrate for BSFL indirectly competes with the farming sector for substrates and may incur undesirable add-on cost of purchase. Utilising expensive substrates to cultivate BSFL are not economically sustainable unless farmers have identified avenues to profit from the insect meals and compost (Raksasat et al. 2020). Further to this, agricultural wastes resulting from food processing are abundantly available and have the potential to be leveraged as feed for BSFL. However, most of these wastes are nutritionally imbalanced and mostly consist of lignin and structural carbohydrates, natural forms of biopolymers. These biopolymers in the substrate would hinder bioconversion by BSFL and lead to slow growth and underweighted insect meals. Though feeding BSFL with large amounts of substrates may help to overcome the poor quality of substrates, this method has not been validated. Conventionally, apart from feeding BSFL with only (1) single substrate, (2) blending of multiple substrates, and (3) the incorporation of microbial fermentation of substrates as pre-treatment have also been examined by several studies. In the following sections, studies related to the respective feeding method were reviewed.

6.3.1 Single substrate

The idea of upcycling fruit and vegetable (FV) and restaurant wastes (RW) as substrates for BSFL have been validated by several researchers. Spranghers et al. (2017) found that the duration required for BSFL reared on FV and RW to reach maturity were 15 and 19 days, respectively (Spranghers et al. 2017). The slower growth observed in BSFL reared on RW was mainly due to the poor conditions of the substrates covered by grease. BSFL had difficulties converting grease into their body weights and hence required longer time to develop into pupae when being reared on RW. Additionally, the ash content of BSFL reared on FV waste was 2.6 times higher than on RW. This was suggested because FV waste consisted of a higher amount of minerals which was more easily assimilated into BSFL body mass. Though FV waste are a promising source of substrate for BSFL, the low protein content in FV remained a hurdle in hastening the growth of BSFL (Leong et al. 2016). Therefore, to exploit FV waste while maintaining rapid eclosion of larvae into pupae, it is essential to condition the protein content of the substrate by blending with substrates rich in protein. On the other hand, the use of RW as substrate for BSFL could be improved by removing the excessive grease prior to feeding. Overall, feeding only a single substrate to BSFL could hardly achieve optimal growth as most organic wastes are deprived of essential nutrients and composed of large amounts of recalcitrant biopolymers. Thereby, blending with proteinic substrates could possibly help to fortify the overall nutritional profile of the lignocellulosic feed prior to feeding.

6.3.2 Blending of substrates

Blending of substrates could help to overcome the nutritional shortcoming of a substrate – by reducing the C/N ratio and improving the buffer capacities of the feed. Optimising the ratio of substrates not only would help to achieve optimal BSFL growth but also help to degrade highly recalcitrant organic wastes. For instance, dairy manure, a source of abundant and low-value organic waste due to farming, is often used as compost. However, this approach has brought about various environmental issues such as the spread of diseases, unpleasant odours and may possibly cause water pollution. Several studies have then been carried out to mitigate these manures. Among the studies conducted, Ur Rehman et al. (2010) showed that feeding 1000 BSFL with 1 kg of dairy manure would require as long as 24 days for the first prepupal to emerge, indicating low bioconversion efficacy (Rehman et al. 2017). This may be possibly due to the high concentration of lignin, cellulose, and hemicellulose in dairy manure which were not easily digestible by BSFL (D. Wang et al. 2017). However, scientists found a way to increase the palatability of dairy manure – that was through blending with okara. The slightly acidic okara (~pH 5.7) could help to offset the high pH value of dairy manure (~pH 8.4) and to achieve an optimal buffer capacity that was more palatable for BSFL (Shuhong Li et al. 2016). Following the same study, by blending dairy manure and okara in the ratio of 1:4 portion, it was found that about 32%, 75%, and 70% of lignin, hemicellulose, and cellulose were reduced upon feeding BSFL for 21 days (Rehman et al. 2017). This showed higher palatability of substrates as only 25%, 45%, and 52% of the aforementioned compounds were reduced in 4:1 ratio of dairy manure: okara throughout the same feeding duration.

On the other hand, waste coconut endosperm (WCE) is also an abundantly available FW derived from coconut milk processing and has received attention by researchers to be valorised as feed for BSFL. Though WCE is rich in soluble nutrients, it consists of a high level of biopolymers (~ 30%) and only small portions of proteins (~ 5%) and fats (~ 9%) (Raksasat et al. 2020). Similarly, okara has been exploited as the blending substrate to fortify the recalcitrance of WCE. Further to this, BSFL fed with a blended ratio of WCE-to-okara (3:2) yielded the highest final weight of larvae (67.5 mg/larval) upon rearing (Lim et al. 2019). Oppositely, feeding BSFL with only WCE as control attained BSFL weighted only 32.5 mg/ larval – almost two times lighter than the optimally reared. The authors suggested that this may be due to the insufficient protein content in WCE that led to immature larvae development and resulted in smaller prepupae formation in the control. Looking into the protein and lipid compositions of harvested insect meals, BSFL reared with 3:2 WCE-to-okara yielded 2.5 and 1.3 times more protein and lipids, respectively than the control. However, further increasing the portion of okara beyond the optimal ratio would instead reduce the final weight of BSFL. This may be due to the degradation of excessive protein in okara which acidified the substrates. Also, the ammonia gas released upon okara degradation might have a negative effect on the digestive tract of the larvae, resulting in poor BSFL development (Tschirner and Simon 2015). In essence, blending substrates improves the overall nutritional balance of the feed. However, this remains a passive strategy as it heavily relies on the inherent chemical and physical properties of the substrates. In the subsequent section, we will introduce microbial fermentation of substrates as a proactive approach to amplify their palatability for BSFL.

6.3.3 Microbial fermentation

Microbial fermentation could be executed to modify the composition of low-value FW. There are currently two modes of fermentation being studied, that is through (1) *in situ* or (2) *ex situ* fermentation. The *in situ* fermentation implies the inoculation of microorganisms into the substrates simultaneously during the feeding of BSFL. On the other hand, *ex situ* fermentation refers to the execution of microbial fermentation prior to BSFL feeding. Implementing microbial fermentation on the substrates have proven success in increasing the palatability of lignocellulosic feed for BSFL as the microorganisms would help to break down complex biopolymers via metabolic mechanisms into end-products that are more readily accessible by BSFL (Wong et al. 2020). Apart from improving the palatability of substrates, inoculating microorganisms, especially the exogenous species could colonise the feed – thus eliminating the parasites and pathogens found in the substrates, consequently creating a more conducive growth environment for the larvae.

6.3.3.1 *In situ* fermentation

The incorporation of single cell yeast, *Saccharomyces cerevisiae, has* been exploited to ferment WCE while feeding BSFL (Wong et al. 2020). The study showed that incorporating 11.5% of yeast in the feed helped to attain a waste-to-biomass conversion of 42.5 mg/larval/day as compared to 3.25 mg/larval/day when only 2.5% of yeast was inoculated. The presence of yeast boosted the breakdown of carbohydrate compounds in WCE and led to better digestibility of the substrates. The degradation of these compounds into monosaccharides improved the nutrient assimilations into BSFL bodies. Similarly, the commercial bacteria consortia product (Rid-X) consisting of an effective microbial consortium was applied for *in situ* fermentation of blended rice straw and restaurant waste (Zheng, Hou, et al. 2012). Upon rearing with BSFL, it was found that both cellulose and hemicellulose in the fermented substrates were reduced by 65% and 55%, as opposed to the control which only observed reduction of 27% and 32%, respectively. These studies showed that the incorporation of microorganisms helps in BSFL ingestion mainly through their strong ability in utilising highly recalcitrant compounds in the substrates.

In the other study where okara was being fermented *in situ* with *Lactobacillus buchneri*, it was found that BSFL reared on the fermented substrates required shorter time for pupation as compared to control (Somroo et al. 2019). Additionally, the crude protein and lipid contents of BSFL reared on the fermented substrates were spurred to 30% and 55%, as compared to the control which only observed 26% and 53%, respectively. The better growth of BSFL in the fermented treatment was due to the improved digestive enzyme production.

6.3.3.2 *Ex situ* fermentation

Mohd-Noor etal. (2017) studied the effect of self-fermentation duration on WCE and its palatability for BSFL prior to feeding (Mohd-Noor et al. 2017). By self-fermenting WCE for 4 weeks prior to feeding, the reared BSFL gained 56% more weight than those reared with fresh WCE. On top of that, the harvested BSFL reared on 4-weeks old WCE observed crude protein and lipid contents of 15% and 58%, as compared to control of 12% and 21%, respectively. The improved growth of BSFL may be due to the breakdown of polysaccharides and fibres into bioproducts which were essential for BSFL growth during fermentation. However, longer fermentation duration had otherwise deteriorated the growth of BSFL. The longer duration was believed to have depleted nutrients in substrates due to consumption and colonisation by the overgrown microorganisms. The nutrients in substrates would no longer be easily assessed by BSFL as they would be colonised by microorganisms with rigid cell walls (Leong et al. 2016). Therefore, it is essential to optimise the duration for *ex situ* fermentation to avoid unnecessary nutrient depletion due to competitions from the overgrown microorganisms.

6.4 COMMERCIAL APPLICATIONS OF BLACK SOLDIER FLY LARVAE

The ability of BSFL in rapidly decomposing organic waste yielding valuable insect meals rich in proteins and fats forms suggest a promising waste-to-wealth scheme. The low technological and capital requirements to initiate the production cycle of BSFL position itself as one of the most financially feasible ventures in comparison to other insect species. In reality, the industrialised production of BSFL is no longer a theoretical concept as it has been successfully implemented by various companies worldwide. These companies employ diverse substrates for BSFL depending on feed availability in the different regions. For instance, several companies in Malaysia have attempted to utilise residual palm waste and FW to rear BSFL. The resulting BSFL can be processed into different value-added products to meet local demands. The upcoming section will introduce the concept of BSFL production and its associated products.

Upon rearing, the prepupae would migrate to dryer spots separated from their substrates. With this behaviour, by building a pathway from their rearing container that leads to the collecting containers would allow for effective self-harvesting of BSFL (Dolezal, Helbling, and Hutter 2008; Mutafela and Richard 2015). Subsequently, the harvested BSFL would need to be inactivated. Boiling and freezing using liquid nitrogen are among the available methods to kill BSFL (Larouche et al. 2019). Following that, the inactivated BSFL could be defatted via hammer-milling or mechanical pressing (Y. S. Wang and Shelomi 2017). The oils collected upon defatting have shown high potential to be processed into biodiesel through transesterification. Alternatively, depending on the local demand, raw inactivated BSFL could also be sold as animal feed without further defatting while the compost could be used to fertilise the crops and subsequently the agricultural residuals can be upcycled and fed back to the larvae. This could potentially realise a circular and self-sustaining loop of material flow. The general idea of the overall process flow has been illustrated in Figure 6.4 while the details of each process would be elaborated in the subsequent sections.

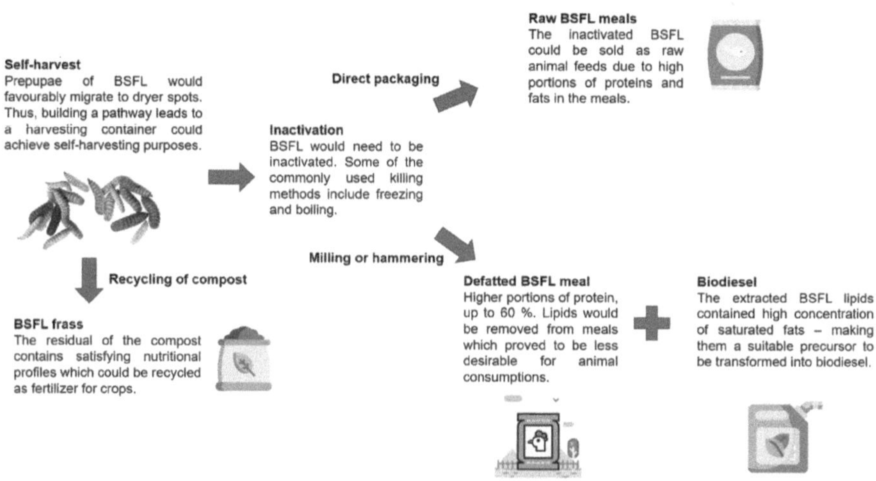

FIGURE 6.4 An overview of the industrial production cycle of BSFL.

6.4.1 Applications of Black Soldier Fly Larvae as Animal Feed

Raw harvested BSFL contains high portions of crude proteins and fats making them good candidates for partial or full replacement for animal feeds in the livestock and pet industries. On top of that, studies found that the level of essential minerals and vitamins in BSFL is superior to most insect meals (Finke 2015). Particularly, BSFL accumulates at least one order higher level of calcium as compared to most of the insect meals (up to 9.3% of body weight while other insects normally contain less than 1% of calcium) (Spranghers et al. 2017; Makkar et al. 2014), and most importantly, BSFL does not accumulate sodium and sulphur – making this a critical edge over other insects nutritionally. Besides that, the BSFL meals are rich in C12:0 medium-chain fatty acids which serves as prebiotics for the gut microbiota in the livestock (Mohana Devi and Kim 2014) and combats gastrointestinal disease-causing bacteria (Skrivanova et al. 2007). As such, feeding animals with BSFL meals has been increasingly considered as an alternative for the banned in-feed, artificial antibiotics (Y. S. Wang and Shelomi 2017).

Raw BSFL meal has already been considered an alternative animal-grade 'meat' to fish meal in feeding carnivorous fishes (Kroeckel et al. 2012). Studies proved that supplementing BSFL meals with fish offals to rainbow trout (*Oncorhynchus mykiss*) yielded no significant difference in their growth, especially in vision and fillet development (Sealey et al. 2011). Similar results have also been proven on the feeding of African catfish (*Clarias gariepinus*) where total substitution of BSFL meal in the diet originally containing 25% fish meals showed no significant difference in terms of their nutrient utilisation indices and growth rate (Okey Aniebo, Erondu, and Owen 2009). Apart from its use as feed for carnivorous fishes, poultry animals have also been found to benefit from BSFL. For instance, feeding raw BSFL meals to broiler quails (*Coturnix japonica*) showed no significant difference in productive performance and breast meat weight (Cullere et al. 2016). Additionally, analysis found that the slaughtered meat of broiler quails reared with BSFL meals had improved amino acid profiles and resulted in no significant difference in the sensory and flavour aspects, oxidative status, and cholesterol compositions (Cullere et al. 2018). Similar responses were observed in laying hens, demonstrating minimal impacts on the health of the hens and eggs when being fed with 50% or total replacement of BSFL meals for soybean cake in their diets (Maurer et al. 2016). Notably, BSFL meals had higher apparent attractiveness for the poultry, a study reported that the laying hens preferably sought for BSFL meals instead of the original feeds after being fed with the insect meals. These findings suggested that supplementing BSFL meals for poultry could become feasible as it is highly palatable and capable of improving the protein characteristics in the meat.

Nevertheless, the major disadvantage of exploiting BSFL meals as animal feed lies in its fatty acid profile which contains low levels of polyunsaturated fats. The proportion of polyunsaturated fats in BSFL are significantly inferior as compared to other commercialised insect species such as the larvae of mealworms and adults of crickets (Makkar et al. 2014; Starčević et al. 2017). The consumption of BSFL meals by animals may otherwise result in the reduced level of polyunsaturated fats but increased levels of undesirable saturated and monounsaturated fats (Senlin Li et al. 2016; Schiavone,

Cullere, et al. 2016). For instance, supplementing BSFL meals to broiler quails showed increased levels of saturated and monounsaturated fatty acids in the slaughtered meats (Cullere et al. 2018) while similar results were observed in broiler chickens (*Gallus domesticus*) (Schiavone, De Marco, et al. 2017). We propose two suggestions to overcome the concerns about saturated fats, among them, (1) defatting the insect meals could help to lower the fat proportions while increasing the crude protein content as high as 60% (Spranghers et al. 2017), and (2) changing the substrate to modify their fatty acid compositions upon rearing. The first suggestion is considered a more practical approach as it has been proven that broiler chickens fed with defatted BSFL meals had reduced proportions of saturated fats in their meats (Schiavone, De Marco, et al. 2017). The latter suggestion may limit the choice of substrates for BSFL, diminishing its advantage in valorising a wide range of low-value organic wastes. A study in which BSFL reared with vegetable waste and biogas fermentation waste found no significant difference in the fatty acid profiles in the insect meals (Sprangers et al. 2017). This may suggest that alternating low-value organic waste, such as FW, may not help lower saturated fats in the insect meals. Overall, there is still room for improvement in terms of the choice of feed for BSFL in yielding a preferable fatty acid profile.

6.4.2 Black Soldier Fly Larvae as Source of Proteins for Human Consumption

Entomophagy, a term referring to the human consumption of insects, has been practiced to date. BSFL with its strong composting ability on a wide range of organic waste has shown promising prospects of being processed into proteins for human consumption. The greatest advantage of consuming BSFL proteins as opposed to animal proteins lies in its lower environmental impacts. For instance, rearing BSFL produces fewer GHG and ammonia emissions as compared to the conventional livestock, occupying less land to generate higher amounts of protein with the same amount of water needed as compared to livestock or even plant proteins (Mekonnen and Hoekstra 2012). To date, BSFL is still not being regarded as one of the edible insect species due to its perception of being unhygienically reared on dirty wastes. The sole consumption record of BSFL originates from the indigenous Kadazan-Dusun tribes residing in Sabah, Malaysia, where they consume raw BSFL with their special fermented beverage known as *tapai* (Chung 2010). Based on comments retrieved, those who consumed BSFL reported that "BSFL released pungent smells containing slight fishy aroma", though the taste may be better consisting of chocolate-like flavours. The unfavourable smell may be overcome through cooking (Y. S. Wang and Shelomi 2017).

The biggest obstacle that prevents BSFL from being commercialised as human foods comes from the psychological fear of consuming insects itself. Several creative ideas have been tested to disguise the insect perception of BSFL meals such as being marketed in powders (Lensvelt and Steenbekkers 2014) and blending with snacks or beers (Chung 2010). Additionally, packaging BSFL in the form of protein bars or meat products mixed with livestock or plant proteins may be an idea that is more tolerated by new consumers. More ideas should be attempted so that in the future, BSFL meals could be commercialised as an alternative protein that effectively mitigates the negative effects caused by livestock farming and serves the stomachs of people.

6.4.3 Defatting Black Soldier Fly Larvae for Biodiesel Production

The production of biodiesel often interrupts the food supply chain due to the use of food crops such as soybean and palm oil as raw material – raising concerns in food-fuel competition. On the other hand, tapping into the use of waste for biodiesel production faces significant technical challenges. Exploiting BSFL in converting waste into lipids stored in their body could be a promising waste-to-biodiesel pathway. Containing as much as 25% lipid in the body, the extracted BSFL lipids which consist of high levels of saturated fatty acids demonstrate the high potential of being processed into value-added biodiesel.

BSFL lipids constitute low levels of polyunsaturated fatty acids but high proportions of saturated fatty acids (about 67% of total fatty acid). The proportionality is higher than those in soybean (11%) and palm oil (37%) (Hasnol et al. 2020) – a highly desirable trait that shows high oxidation stability against air exposure (Ramos et al. 2009). Furthermore, lauric acid, palmitic acid, and myristic acid were the majority of fatty acid methyl esters analysed in the BSFL fats (Figure 6.5), showing comparable profiles with the conventional biodiesel (Q. Li, Zheng, Cai, et al. 2011). Coupling with its high oxidation stability, BSFL lipids-transformed biodiesel was considered as a high-quality biodiesel. A study analysed the biodiesel produced from BSFL lipids achieved a yield of 97%, while physical properties such as the density, viscosity, and flash point of the biodiesel were comparable with palm waste, palm oil mill effluent (POME)-derived biodiesel, and fulfilled the requirements indicated in EN 14214 standards (Table 6.2) (Ishak and Kamari 2019). Currently, there are three possible processes that convert BSFL lipids into biodiesel, among them, (1) two-step

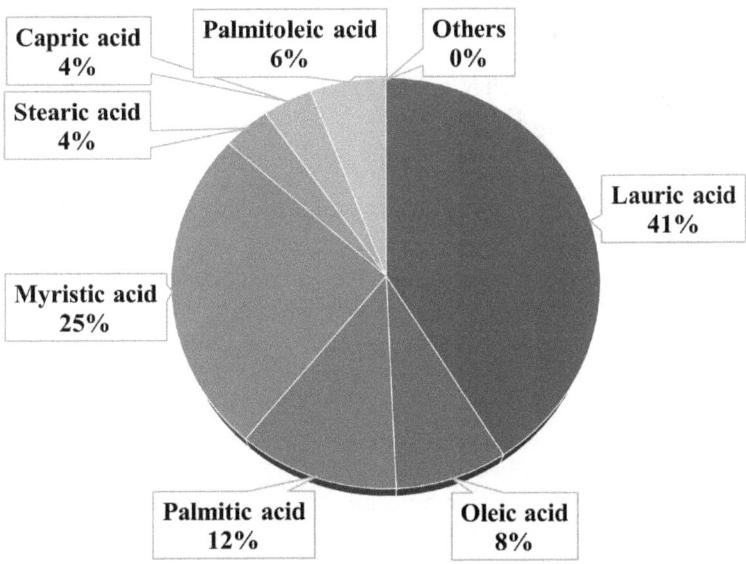

FIGURE 6.5 Proportions of fatty acid methyl esters analysed from BSFL lipids, data obtained from (Ishak and Kamari 2019).

TABLE 6.2
Comparisons in properties between POME-derived biodiesel and BSFL oil-transformed biodiesel reared on restaurant waste, according to EN 14214 standards (Eman and Cadence 2013; Ishak and Kamari 2019)

		Biodiesel	
Properties	**EN 14214 standards**	**POME-derived biodiesel**	**BSFL biodiesel**
Density at 15°C (kg/m³)	860–900	876	875
Viscosity at 40°C (mm²/s)	3.5–5.0	4.76	4.6
Water content (mg/kg)	<500	-	387
Flash point (°C)	>120	170	143
Cetane number	>51	62.8	49

transesterification, (2) one-step transesterification with co-solvent, and (3) biologically catalysed transesterification.

Two-step transesterification of BSFL lipids. The first step of this process is to esterify the BSFL fatty acids with an acidic catalyst. The second step would involve the transesterification of the converted esters into fatty acid methyl esters in the presence of an alkaline catalyst. In a study conducted, the lipids extracted from 1,000 BSFL reared with restaurant waste were used to produce biodiesel (Zheng, Li, et al. 2012). The two-step process was carried out at 75°C for an hour, using 1% H_2SO_4 as the catalyst and molar ratio of methanol-to-lipid at 8:1 for the first step. The second step involved 0.8% NaOH as the alkaline catalyst while the methanol-to-lipid ratio was maintained at 6:1. It was found that the conversion rate of free fatty acids obtained was recorded at 91.9%. About 23.6 g of biodiesel was produced from the BSFL reared with 1 kg of restaurant waste – equivalent to 2.4% of the total yield of biodiesel. While studies related to BSFL reared with FW are limited, the conversion of BSFL lipids reared with dairy manure into biodiesel via the same method was investigated (Q. Li, Zheng, Qiu, et al. 2011). After the two-step process, 15.8 g of biodiesel could be obtained from 1.2 kg of dairy manure fed. Analysis showed that biodiesel contained 58.2% saturated fatty acids with most properties satisfying the requirement from EN 14214 demonstrating the commercial feasibility of this biodiesel production method.

One-step transesterification with co-solvent was an alternative study to overcome the energy intensiveness of the conventional transesterification process by directly carrying out transesterification in one step. The use of excessive methanol would deteriorate the function of the transesterification catalyst, leading to the reduced yield of biodiesel. As such, it was proposed to combine co-solvents to reduce the proportions of methanol involved in the direct transesterification process (Nguyen, Liang, Li, et al. 2018). Co-solvents such as n-hexane, n-pentane, chloroform, acetone and petroleum ether were mixed with methanol at the volume ratio of 1.17:1. These solvents were incorporated as a homogenous catalyst to dissolve BSFL lipids and methanol. The use of n-hexane as the co-solvent yielded 63.37% of biodiesel, comparably better than acetone (54.83%), chloroform (48.50%), and petroleum ether (35.67%). However, the overall profile of the biodiesel produced was not analysed, and thus, the applicability of the biodiesel production method is yet to be validated.

Biologically catalysed transesterification. The conventional transesterification process of BSFL lipids used corrosive catalysts such as sulphuric acid to produce biodiesel which have issues related to separation and damage of equipment. A bio-catalyst, lipase has been proposed and studied as the alternative for these hazardous catalysts. The commercial lipases are Novozym 435, Lipozyme TL-IM, Lipozyme RM-IM, and other lipases produced *ex situ* by microorganisms (Nguyen et al. 2017). The transesterification process was carried out at 20°C for 4 hours, with a methanol-to-lipid ratio of 3:1 in the presence of 10% of the aforementioned lipase. The Novozyme 435 yielded the highest amount of biodiesel (56.78%), followed by Lipozyme RM-IM (42.18%) while the remaining tested lipases recorded yields lower than 30%. Though considerable yield of biodiesel was attainable with the incorporation of commercial lipase, this method still faces several challenges. The stability of lipase in the system is one of the major concerns as high amounts of alcohol would immobilise these enzymes and may result in deactivation. This is particularly concerning as the transesterification process involves the use of excessive alcohol such as methanol and ethanol. Additionally, the stability of enzymes is highly influenced by the operating temperature: overshooting of temperature above 40°C may cause the denature of lipase, and thus, could prematurely terminate the process (Nguyen et al. 2017). Lastly, in our opinion, the adoption of enzymes for biodiesel production is still not economically viable mainly due to the high cost of production of enzymes. Instead, incorporating lipase-producing microorganisms might be more feasible due to the reproducibility of the microbial strains. However, this suggestion necessitates further investigations and optimisation as in a study, the use of the strain-produced lipase led to a lower yield of biodiesel (Nguyen et al. 2017).

6.4.4 Black Soldier Fly Larvae Frass as Nutritious Fertiliser

The residual of the BSFL compost or known as the frass consists mainly of BSFL faeces and exoskeletons and residual substrates. The BSFL frass has good overall nutritional profiles making them good candidates to be upcycled as fertilisers and soil conditioners. Applying frass for agriculture use recaptures the essential elements such as nitrogen, phosphorus, and potassium to reduce the reliance on chemical fertilisers. On top of its nutritional profile, the shed-off exoskeletons of BSFL could serve as a source of chitin when applied to plants. Notably, BSFL frass is known to be the only plant-digestible chitin (Choi & Hassanzadeh 2019). These chitins produce antimicrobial peptides, and when applied, the peptides can help plants combat pathogenic microorganisms and act as a barrier for plant defences (Surendra et al. 2020). Furthermore, conditioning soils with BSFL frass would diversify the effective microbial consortium in the soil. These effective microorganisms could promote nitrogen-fixing and nitrifying, phosphorus, and potassium mobilisation for the plants. With all the aforementioned benefits, BSFL frass has been perceived as an emerging organic fertiliser.

Generally, a qualified fertiliser should contain at least 0.6%, 0.22%, and 0.25% of total nitrogen, phosphorus, and potassium, respectively. Many studies have shown that the frass resulted from various FW (Table 6.3) consist good N:P:K ratio – suggesting their feasibility in ebing used as organic fertiliser. On top of their N:P:K ratio, the C/N ratio of the frass was found to be favourable, of which all of them were below 30. This is because a C/N ratio higher than 30 would result in microbial immobilisation which impedes complex compound degradation. C/N ratio between 20 and 30, preferably at 24, favours the uptake of nutrients by the soil microbes (The agricultural European Innovation Partnership (EIP-AGRI) 2016). Lower C/N ratio, i.e. between 1 and 15, favours the release of N into the soil for plant uptake (Brust 2019). In other words, BSFL compost with a lower C/N ratio could more effectively disseminate their N for plant use. Enhanced growth was observed in crops such as lettuce, basil, and tomato following the application of BSFL frass (Setti et al. 2019). Additionally, it was found that the production of BSFL frass emits at least 12.5 kg less CO_2 compared to commercial fertiliser on mass basis (Gärttling and Schulz 2022). Consequently, substituting commercial fertiliser with BSFL frass not only fosters plant growth but also represents a greener alternative that benefits the environment with potentially lower carbon footprint.

Despite the benefits aforementioned, several challenges still pose ahead of the use of BSFL frass as fertiliser. The short composting period of BSFL (normally 20 days) often leads to immature compost of the frass. When immature frass was applied to crops, inhibited plant growth was observed primarily due to the undesirably high concentration of phytotoxins and pathogens present within the frass (S. Song et al. 2021). This is particularly concerning when household FW was used to produce BSFL frass, high relative abundance of *Xanthomonadaceae* sp. was identified (Kawasaki et al. 2020). Moreover, the elevated moisture content in FW poses another obstacle to the mature composting of BSFL. The excess moisture not only would enhance the release of ammonium but also diminishes the beneficial N levels in the frass. Applied BSFL frass derived from kitchen waste led to stunted growth of maize mainly due to the presence of phytotoxins and ammonium (Alattar, Alattar, and Popa 2016). Therefore, post-treatment of BSFL frass before fertilisation is highly recommended. In section 9.3.5, we will explore an innovative post-composting approach to address issues related to premature composting of BSFL as practised by a manufacturer in Malaysia. Figure 6.6 shows the actual raw BSFL frass to be managed by the manufacturer.

In our opinion, the immature BSFL frass could be transformed into biofertiliser, particularly by being treated with plant-growth-promoting strains (PGPM) that utilise phenols (the major phytotoxins) for growth and to produce catabolic enzymes. Hence, when being applied in the soils, the frass-transformed biofertiliser would have reduced concentration of phytotoxins while carrying a viable amount of PGPM which could synergise plant growths.

6.4.5 Case Study: Edafos Agro PLT's BSFL Frass Organic Fertiliser

Edafos Agro PLT, located in Johor, Malaysia, is a prominent manufacturer specialising in organic fertilisers, particularly all-natural biostimulants, pure black soldier fly (BSF) frass bio-organic fertilisers, and effective microorganisms (EM). These products are developed to address widespread agricultural issues such as slow crop

FIGURE 6.6 Pure BSF frass actual product managed by Edafos Agro PLT.

growth, low yields, poor crop quality, weak plant immunity, soil degradation, and reduced fertility. Edafos aims to provide sustainable solutions that enhance crop productivity and soil health. Figure 6.7 shows Edafos' composted BSFL frass piles, ready for packaging.

ASP Composting System for BSFL Frass at Edafos Agro PLT. Edafos Agro PLT employs the aerated static pile (ASP) composting system to process BSFL frass into a potent bio-organic fertiliser. This method is selected for its efficiency in controlling odours, maintaining consistent aeration, and eliminating pathogens, making it ideal for large-scale and industrial composting operations.

The ASP is an advanced composting method that involves placing organic material in static piles equipped with a network of perforated pipes. These pipes supply oxygen through compressed air, thereby eliminating the need for manual turning. The introduction of air supports the aerobic conditions necessary for efficient microbial activity and decomposition.

Several factors influence the efficiency and quality of the compost produced by the ASP system:

TABLE 6.3
The compositions of BSFL frass reared with different FW and its comparison with other compost and fertilisers

Fertilisers	Source of FW	C/N ratio	Moisture (%)	Total N (%)	Total P_2O_5 (%)	Total K_2O (%)	References
BSFL frass	Household FW	17:1	56	2.2	0.1	0.1	(Kawasaki et al. 2020)
	Fruit and vegetables	27:1	10	1.8	-	-	(Klammsteiner et al. 2020)
	Maize straw	-	38	0.6	2.5	2.1	(Gao et al. 2019)
	Okara and wheat bran	8:1	-	4.8	1.0	0.9	(S. Song et al. 2021)
		10:1	-	3.2	0.8	0.5	
	Brewery spent grain	17:1	30	2.1	1.2	0.2	(Beesigamukama et al. 2020)
Windrow compost	FW and yard waste	17:1	46	-	-	-	(Kamaruzzaman et al. 2018)
	Empty fruit bunch, POME and decanter cake	19:1	-	-	-	-	[245]
	POME biogas sludge and empty fruit bunch	16:1	-	-	-	-	
	Oil palm frond, chicken manure, rice	21:1	-	-	-	-	
Bin compost	Cafeteria FW and yard waste	-	-	0.9	0.8	0.4	(Hamid et al. 2019)
Aerated compost	Household FW	36:1	32	0.9	0.6	0.9	(Bhave and Kulkarni 2019)
Anaerobic digestate	Municipal and FW	11:1	85	9.6	2.4	2.3	(Manu et al. 2021)
Commercial organic fertiliser	-	-	-	2.3	2.3	2.3	(Hamid et al. 2019)
Commercial chemical fertiliser	-	-	-	16	16	16	(Hamid et al. 2019)

- **Aeration Rate**: Proper aeration is critical for maintaining aerobic conditions and ensuring effective microbial activity. Studies indicate that different aeration rates significantly affect the composting process. For example, Rasapoor et al. (2009) found that an aeration rate of 0.6 L/min/kg maintains optimal temperatures for microbial activity, resulting in a stable and nutrient-rich compost converted from MSW.
- **Temperature Control**: Regulating temperature through the various phases of composting ensures the elimination of pathogens and the stability of the

FIGURE 6.7 Edafos' Aerated Static Pile (ASP) composting technology.

final product as illustrated in Figure 6.8. The thermophilic phase, in particular, is vital for pathogen reduction (Keng et al. 2023).

- **Moisture Content**: Generally, maintaining a moisture level of 40–60%, or as high as 70% depending on the C:N ratio, is essential for microbial activity and decomposition (Keng et al. 2023). Excess moisture can lead to anaerobic conditions, while insufficient moisture can slow down microbial activity.
- **Carbon-Nitrogen Ratio**: An optimal C/N ratio is necessary for efficient composting. A balanced C/N ratio ensures that microorganisms have enough carbon for energy and nitrogen for protein synthesis.

Edafos' ASP system is designed in modules, each capable of handling a 50 tonne (wet weight) pile of fresh frass. Key components of the system include:

- **Compressed Air Distribution**: Air is delivered via a blower through four perforated 100 mm, SCH40 CPVC pipes. These pipes spaced 1.4 meters apart and each 18 meters long feature perforations with diameters of 4.8 mm, 6.35 mm, and 7.9 mm arranged in 6-meter segments from the blower. Removable caps at the ends of the pipes facilitate maintenance.
- **Aeration and Temperature Control**: The system maintains humidity levels between 50% and 60% and regulates temperature through the different composting phases: the initial heating phase at 40°C, the thermophilic phase at 55–65°C, the cooling phase at 35–40°C, and the maturation phase at ambient temperature.

FIGURE 6.8: Composting process with temperature monitoring.

Use of Effective Microorganisms (EM). Edafos integrates EM into the composting process to enhance the quality of the final product. EM consists of beneficial microbial strains that accelerate the decomposition of organic matter, increase nutrient availability, and improve compost quality. The use of EM ensures the production of a high-quality organic fertiliser that promotes healthy plant growth and robust soil health.

The combination of the ASP system and EM produces a stable, nutrient-rich fertiliser with a balanced profile of nitrogen, phosphorus, and potassium. The compost is free of pathogens and enriched with antimicrobial peptides from BSFL exoskeletons, which help strengthen plant immunity. This innovative composting approach helps farmers address challenges such as soil degradation and low crop yields, promoting sustainable and profitable farming practices. The advanced ASP system and the integration of EM result in a superior organic fertiliser that supports long-term agricultural sustainability and productivity.

From the investor's perspective, the ASP and EM systems at Edafos Agro PLT require an initial investment of approximately USD 20,000 to 50,000, depending on the capacity, to cover equipment like blowers, piping, and sensors. This system reduces operational costs by 15–30% by cutting labour and energy use. The cost of EM is about USD 5–10 per cubic meter of compost. Additionally, the ASP system accelerates the composting cycle from 40 days to 30 days, increasing production capacity. Together, these factors allow for premium pricing and operational savings, generating an estimated annual return on investment (ROI) of 30–40%, with a payback period of 2–3 years. Compared to traditional windrow composting, ASP systems are more efficient, particularly for processing BSFL frass, and are less capital and labour-intensive.

TABLE 6.4
Summary of FW management via BSFL composting

	Advantages	Disadvantages
Comparison with other composting methods	• Lower capital cost required. • Lower technological and operational barriers. • Lower space, water, and energy required.	• May result in the release of fault smell. • May require more rigorous segregation of FW from MSW as most wastes are not palatable for BSFL and will result in delayed pupation.
Applications in animal feeds	• Can be exploited to manage manures produced from the farm and turn them into nutritious insect meals for the animals. • Balanced nutritional profiles, particularly high levels of proteins in the insect meals.	• Animals especially the livestock fed on BSFL reared with manures may raise hygienic concerns among the consumers. • Poor fatty acid profiles for animal consumption due to low levels of polyunsaturated fatty acids that are crucial for animal health.
Applications in biodiesel	• Lipids extracted from BSFL consisthigh levels of saturated fatty acids promote oxidative stability when processed into biodiesel. • Lipids extracted from BSFL mainly consist of middle-chain fatty acids favourable for transesterification into biodiesel.	• The current extraction method for BSFL lipids requires further advancement to be commercially viable.
Applications in composted fertiliser	• Balanced nutritional profiles with comparable N, P and K levels compared to other organic fertilisers. • Contain shed-off exoskeletons of BSFL capable of producing antimicrobial peptides that help strengthen a plant's immune system.	• Short composting period results in immature compost. • May contain phytotoxins such as phenols and other pathogenic species (depending on the feeds reared) that may stunt the growth of plants when applied.

6.5 CONCLUSIONS

The bioconversion of FW by BSFL presents many benefits including reduced carbon emissions, lower capital and operating cost incurred, and smaller land required as compared to other FW management methods. The larval stage is the only feeding stage throughout the entire BSF cycle where BSFL can ingest a wide range of organic waste substrates. However, the palatability of the substrate for BSFL is highly influenced by their compositions. Particularly, substrates rich in lignocelluloses and greases were found to have lower palatability for BSFL. To overcome this, blending of substrates and fermentation were introduced as methods to improve highly recalcitrant substrates for BSFL in this chapter. In the latter part of this chapter, the applications of BSFL were briefly introduced. BSFL meals rich in proteins and lipids proved to be excellent sources of animal feed for fishes and poultries and potentially for human consumption. However, BSFL lipids contain high levels of unsaturated fats which may deteriorate the health of consumers. Hence, defatted BSFL meals were suggested as alternative feeds, while the extracted fats were deemed more appropriate for conversion into biodiesel, owing to their high oxidative stability. Furthermore, the residual of the compost or frass was deemed to be a suitable fertiliser with satisfying overall nutritional profiles. The immature compost may lead to undesirably high concentrations of phytotoxins and pathogens within the frass – essentially stunting the growth of plants. Further studies are required to study the characteristics of immature frass upon various post-treatment methods. In a nutshell, composting of FW via BSFL demonstrates an economically viable process with low technological barriers that could be adopted by developing countries with just minimal capital investment. A summary of this FW management via BSFL was provided in Table 6.4.

6.6 EXERCISE

1. What are the four main stages in the black soldier fly lifecycle? In which stage would the insect require food?
2. What are the components in substrates found to be highly recalcitrant to BSFL?
3. Ali has bought a pack of BSFL eggs from a farm near his school. He plans to cultivate it inside the lab, and so, he places the freshly acquired BSFL eggs on top of a sieve separating from okara in a plastic container. The small BSFL hatches from their eggs in approximately 4 days, and Ali allows them to grow using okara as the substrates. On the 22nd day, about 80% of the skin of larvae turned darker in which Ali justified that these larvae had already turned into prepupae. Therefore, he decided to harvest the population. Upon harvesting, he found that now only 800 larvae were left from the originally 1000 larvae inoculated. The weight of the larvae was improved to 100 mg/larval from 10 mg/ larval. He also measured the weight of the residual and found that only 35 g were left from the originally 1000 g of okara fed.

   Now, find the development period (80% pupation), the survival rate of larvae, substrate conversion rate, substrate overall degradation, and waste reduction index of this rearing process.

4. Despite having high crude proteins and lipids in BSFL meals, why is it recommended that defatted BSFL meals be used as animal feeds instead of raw BSFL meals?
5. What are the issues that may be caused by applying immature BSFL frass as fertiliser?

ACKNOWLEDGEMENT

This research is supported by the Institution of Chemical Engineers (IChemE) via Palm Oil Processing Special Interest Group (POPSIG) Research Bursary via two grants EENG0052 and EENG0039 with the project title "Biofertiliser production through insect frass degradation: Applications and production process modelling" and "Valorisation of blended palm oil mill effluent (POME) sediment and oil palm frond (OPF) as the substrate for black soldier fly larvae (BSFL)", respectively. This research is supported by the Ministry of Higher Education Malaysia through the Fundamental Research Grant Scheme (FRGS), project number "FRGS/1/2024/WAS04/XMU/02/1". This research is also supported by Xiamen University Malaysia Research Fund (XMUMRF) Project "XMUMRF/2024-C14/IENG/0075". Additionally, we extend our appreciation to *Kelip-kelip*! Center of Excellence for Light Enabling Technologies for providing the necessary laboratory spaces for the cultivation of BSFL and China-ASEAN College of Marine Science for the equipment and support. Lastly, we also would like to extend our appreciation to Edafos Agro PLT for providing us the critical information in ASP composting system that addresses the current issues facing immature BSFL frass composting and making this chapter more comprehensive.

BIBLIOGRAPHY

Alattar Manar, Fetheya Alattar, and Radu Popa. 2016. "Effects of Microaerobic Fermentation and Black Soldier Fly Larvae Food Scrap Processing Residues on the Growth of Corn Plants (Zea Mays)." *Plant Science Today* 3 (1): 57–62. doi:10.14719/pst.2016.3.1.179.

Barragan-Fonseca, K. B., M. Dicke, and J. J. A. van Loon. 2017. "Nutritional Value of the Black Soldier Fly (Hermetia Illucens L.) and Its Suitability as Animal Feed - a Review." *Journal of Insects as Food and Feed* 3 (2): 105–20. doi:10.3920/JIFF2016.0055.

Barragan-Fonseca, Karol B., Marcel Dicke, and Joop J. A. van Loon. 2018. "Influence of Larval Density and Dietary Nutrient Concentration on Performance, Body Protein, and Fat Contents of Black Soldier Fly Larvae (Hermetia Illucens)." *Entomologia Experimentalis et Applicata* 166 (9): 761–70. doi:10.1111/eea.12716.

Beesigamukama Dennis, Benson Mochoge, Nicholas K. Korir, Komi K.M. Fiaboe, Dorothy Nakimbugwe, Fathiya M. Khamis, Sevgan Subramanian, et al. 2020. "Exploring Black Soldier Fly Frass as Novel Fertilizer for Improved Growth, Yield, and Nitrogen Use Efficiency of Maize Under Field Conditions." *Frontiers in Plant Science* 11 (September): 574592. doi:10.3389/fpls.2020.574592.

Beniers, J. J. A., and R. I. Graham. 2019. "Effect of Protein and Carbohydrate Feed Concentrations on the Growth and Composition of Black Soldier Fly (Hermetia Illucens) Larvae." *Journal of Insects as Food and Feed* 5 (3): 193–99. doi:10.3920/JIFF2018.0001.

Bhave, Prashant P., and Bhargavi N. Kulkarni. 2019. "Effect of Active and Passive Aeration on Composting of Household Biodegradable Wastes: A Decentralized Approach." *International Journal of Recycling of Organic Waste in Agriculture* 8 (c). Springer Berlin Heidelberg: 335–44. doi:10.1007/s40093-019-00306-7.

Brust, Gerald E. 2019. *Management Strategies for Organic Vegetable Fertility. Safety and Practice for Organic Food*. Elsevier Inc. doi:10.1016/B978-0-12-812060-6.00009-X.

Cho Sam, Chul Hwan Kim, Min Ji Kim, and Haegeun Chung. 2020. "Effects of Microplastics and Salinity on Food Waste Processing by Black Soldier Fly (Hermetia Illucens) Larvae." *Journal of Ecology and Environment* 44 (1). doi:10.1186/s41610-020-0148-x.

Choi, S., and N. Hassanzadeh. 2019. "BSFL Frass: A Novel Biofertilizer for Improving Plant Health While Minimizing Environmental Impact." *Canadian Science Fair Journal* 2 (2): 41–46.

Chung, Arthur Y. C. 2010. "Edible Insects and Entomophagy in Borneo." In P.B. Durst, D.V. Johnson, R.L. Leslie, and K. Shono (Eds.), *Forest Insects as Food: Humans Bite Back, Proceedings of a Workshop on Asia-Pacific Resources and their Potential for Development*, pp. 131–40.

Cullere, M., G. Tasoniero, V. Giaccone, G. Acuti, A. Marangon, and A. Dalle Zotte. 2018. "Black Soldier Fly as Dietary Protein Source for Broiler Quails: Meat Proximate Composition, Fatty Acid and Amino Acid Profile, Oxidative Status and Sensory Traits." *Animal* 12 (3): 640–47. doi:10.1017/S1751731117001860.

Cullere, M., G. Tasoniero, V. Giaccone, R. Miotti-Scapin, E. Claeys, S. De Smet, and A. Dalle Zotte. 2016. "Black Soldier Fly as Dietary Protein Source for Broiler Quails: Apparent Digestibility, Excreta Microbial Load, Feed Choice, Performance, Carcass and Meat Traits." *Animal* 10 (12): 1923–30. doi:10.1017/S1751731116001270.

Cummins, Vaun C., Steven D. Rawles, Kenneth R. Thompson, Alejandro Velasquez, Yuka Kobayashi, Janelle Hager, and Carl D. Webster. 2017. "Evaluation of Black Soldier Fly (Hermetia Illucens) Larvae Meal as Partial or Total Replacement of Marine Fish Meal in Practical Diets for Pacific White Shrimp (Litopenaeus Vannamei)." *Aquaculture* 473. Elsevier B.V: 337–44. doi:10.1016/j.aquaculture.2017.02.022.

Dolezal Martin, Marc Helbling, and Swen Hutter. 2008. "Zwischen Gipfelkreuz Und Halbmond. Die Auseinandersetzung Um Den Islam in Österreich Und Der Schweiz, 1998–2007." *Osterreichische Zeitschrift Fur Politikwissenschaft* 37 (4): 401–17.

Erickson, Marilyn C., Mahbub Islam, Craig Sheppard, Jean Liao, and Michael P. Doyle. 2004. "Reduction of Escherichia Coli O157:H7 and Salmonella Enterica Serovar Enteritidis in Chicken Manure by Larvae of the Black Soldier Fly." *Journal of Food Protection* 67 (4): 685–90. doi:10.4315/0362-028X-67.4.685.

Finke, Mark D. 2015. "Complete Nutrient Content of Four Species of Commercially Available Feeder Insects Fed Enhanced Diets during Growth." *Zoo Biology* 34 (6): 554–64. doi:10.1002/zoo.21246.

Furman, Deane P., Robert D. Young, and Paul E. Catts. 1959. "Hermetia Illucens (Linnaeus) as a Factor in the Natural Control of Musca Domestica Linnaeus." *Journal of Economic Entomology* 52 (5): 917–21. doi:10.1093/jee/52.5.917.

Gao Zhenghui, Wanqiang Wang, Xiaoheng Lu, Fen Zhu, Wen Liu, Xiaoping Wang, and Chaoliang Lei. 2019. "Bioconversion Performance and Life Table of Black Soldier Fly (Hermetia Illucens) on Fermented Maize Straw." *Journal of Cleaner Production* 230 (May). Elsevier Ltd: 974–80. doi:10.1016/j.jclepro.2019.05.074.

Gärttling Daniel, and Hannes Schulz. 2022. "Compilation of Black Soldier Fly Frass Analyses." *Journal of Soil Science and Plant Nutrition* 22 (1). Springer International Publishing: 937–43. doi:10.1007/s42729-021-00703-w.

Hamid, Hazren A., Lim Pei Qi, Hasnida Harun, Norshuhaila Mohamed Sunar, Faridah Hanim Ahmad, and Mimi Suliza Muhamad. 2019. "Development of Organic Fertilizer from Food Waste by Composting in UTHM Campus Pagoh." *Journal of Design for Sustainable and Environment* 1 (1): 1–6. https://fazpublishing.com/jdse/index.php/jdse/article/view/3.

Hasnol Sabrina, Kunlanan Kiatkittipong, Worapon Kiatkittipong, Chung Yiin Wong, Cheng Seong Khe, Man Kee Lam, Pau Loke Show, Wen Da Oh, Thiam Leng Chew, and Jun Wei Lim. 2020. "A Review on Insights for Green Production of Unconventional Protein and Energy Sources Derived from the Larval Biomass of Black Soldier Fly." *Processes* 8 (5). doi:10.3390/PR8050523.

Ishak, S., and A. Kamari. 2019. "Biodiesel from Black Soldier Fly Larvae Grown on Restaurant Kitchen Waste." *Environmental Chemistry Letters* 17 (2). Springer International Publishing: 1143–50. doi:10.1007/s10311-018-00844-y.

Kamaruzzaman Farishya, Shahrom Md Zain, Nur Fatin, Mat Saad, Hassan Basri, Noor Ezlin, and Ahmad Basri. 2018. "Effective Use of Indigenous Microorganism (IMO) in Composting a Mixture of Food and Yard Wastes on an Industrial Scale (Penggunaan

Mikroorganisma Effektif (IMO) Dalam Pengkomposan Campuran Sisa Makanan Dan Sisa Taman Berskala Industri).” *Jurnal Kejuruteraan SI* 1 (5): 53–58. https://dx.doi.org/10.17576/jkukm-2018-si1(5)-08.

Kawasaki Kiyonori, Toshiya Kawasaki, Hirofumi Hirayasu, Yoshiki Matsumoto, and Yasuhiro Fujitani. 2020. “Evaluation of Fertilizer Value of Residues Obtained after Processing Household Organic Waste with Black Soldier Fly Larvae (Hermetia Illucens).” *Sustainability (Switzerland)* 12 (12). doi:10.3390/SU12124920.

Keng Zi Xiang, Jamie Jean Minn Tan, Bao Lee Phoon, Chee Chang Khoo, Ianatul Khoiroh, Siewhui Chong, Christinavimala Supramaniam, Ajit Singh, and Guan Ting Pan. 2023. “Aerated Static Pile Composting for Industrial Biowastes: From Engineering to Microbiology.” *Bioengineering* 10 (8): 1–16. doi:10.3390/bioengineering10080938.

Kim Chul Hwan, Junhee Ryu, Jongkeun Lee, Kwanyoung Ko, Ji Yeon Lee, Ki Young Park, and Haegeun Chung. 2021. “Use of Black Soldier Fly Larvae for Food Waste Treatment and Energy Production in Asian Countries: A Review.” *Processes* 9 (1): 1–17. doi:10.3390/pr9010161.

Kim Wontae, Sungwoo Bae, Kwanho Park, Sangbeom Lee, Youngcheol Choi, Sangmi Han, and Youngho Koh. 2011. “Biochemical Characterization of Digestive Enzymes in the Black Soldier Fly, Hermetia Illucens (Diptera: Stratiomyidae).” *Journal of Asia-Pacific Entomology* 14 (1). Korean Society of Applied Entomology, Taiwan Entomological Society and Malaysian Plant Protection Society: 11–14. doi:10.1016/j.aspen.2010.11.003.

Klammsteiner Thomas, Veysel Turan, Marina Fernández Delgado Juárez, Simon Oberegger, and Heribert Insam. 2020. “Suitability of Black Soldier Fly Frass as Soil Amendment and Implication for Organic Waste Hygienization.” *Agronomy* 10 (10 October). doi:10.3390/agronomy10101578.

Kroeckel, S., A. G. E. Harjes, I. Roth, H. Katz, S. Wuertz, A. Susenbeth, and C. Schulz. 2012. “When a Turbot Catches a Fly: Evaluation of a Pre-Pupae Meal of the Black Soldier Fly (Hermetia Illucens) as Fish Meal Substitute - Growth Performance and Chitin Degradation in Juvenile Turbot (Psetta Maxima).” *Aquaculture* 364–365. Elsevier B.V.: 345–52. doi:10.1016/j.aquaculture.2012.08.041.

Lailatul Ferdousi, and Sultana Nahid. 2021. “Impact of Daylight Duration on Pupation and Adult Emergence of Black Soldier Fly (Hermetia Illucens).” *International Journal of Entomology Research* 6 (1): 142–45. https://www.researchgate.net/publication/349588578_Impact_of_daylight_duration_on_pupation_and_adult_emergence_of_black_soldier_fly_Hermetia_Illucens.

Larouche Jennifer, Marie Hélène Deschamps, Linda Saucier, Yolaine Lebeuf, Alain Doyen, and Grant W. Vandenberg. 2019. “Effects of Killing Methods on Lipid Oxidation, Colour and Microbial Load of Black Soldier Fly (Hermetia Illucens) Larvae.” *Animals* 9 (4): 1–19. doi:10.3390/ani9040182.

Lensvelt Eveline J. S., and L. P. A. Steenbekkers. 2014. “Exploring Consumer Acceptance of Entomophagy: A Survey and Experiment in Australia and the Netherlands.” *Ecology of Food and Nutrition* 53 (5): 543–61. doi:10.1080/03670244.2013.879865.

Leong Siew Yoong, Shamsul Rahman Mohamed Kutty, Amirhossein Malakahmad, and Chew Khun Tan. 2016. “Feasibility Study of Biodiesel Production Using Lipids of Hermetia Illucens Larva Fed with Organic Waste.” *Waste Management* 47. Elsevier Ltd: 84–90. doi:10.1016/j.wasman.2015.03.030.

Li Qing, Longyu Zheng, Hao Cai, E. Garza, Ziniu Yu, and Shengde Zhou. 2011. “From Organic Waste to Biodiesel: Black Soldier Fly, Hermetia Illucens, Makes It Feasible.” *Fuel* 90 (4). Elsevier Ltd: 1545–48. doi:10.1016/j.fuel.2010.11.016.

Li Qing, Longyu Zheng, Ning Qiu, Hao Cai, Jeffery K. Tomberlin, and Ziniu Yu. 2011. “Bioconversion of Dairy Manure by Black Soldier Fly (Diptera: Stratiomyidae) for Biodiesel and Sugar Production.” *Waste Management* 31 (6). Elsevier Ltd: 1316–20. doi:10.1016/j.wasman.2011.01.005.

Li Senlin, Hong Ji, Binxin Zhang, Jingjing Tian, Jishu Zhou, and Haibo Yu. 2016. "Influence of Black Soldier Fly (Hermetia Illucens) Larvae Oil on Growth Performance, Body Composition, Tissue Fatty Acid Composition and Lipid Deposition in Juvenile Jian Carp (Cyprinus Carpio Var. Jian)." *Aquaculture* 465. Elsevier B.V.: 43–52. doi:10.1016/j.aquaculture.2016.08.020.

Li Shuhong, Ye Chen, Kejuan Li, Zhongfang Lei, and Zhenya Zhang. 2016. "Characterization of Physicochemical Properties of Fermented Soybean Curd Residue by Morchella Esculenta." *International Biodeterioration and Biodegradation* 109. Elsevier Ltd: 113–18. doi:10.1016/j.ibiod.2016.01.020.

Lim Jun Wei, Siti Nuraini Mohd-Noor, Chung Yiin Wong, Man Kee Lam, Pei Sean Goh, J. J.A. Beniers, Wen Da Oh, Khairulazhar Jumbri, and Noraini Abd Ghani. 2019. "Palatability of Black Soldier Fly Larvae in Valorizing Mixed Waste Coconut Endosperm and Soybean Curd Residue into Larval Lipid and Protein Sources." *Journal of Environmental Management* 231 (July 2018). Elsevier: 129–36. doi:10.1016/j.jenvman.2018.10.022.

Makkar Harinder P.S., Gilles Tran, Valérie Heuzé, and Philippe Ankers. 2014. "State-of-the-Art on Use of Insects as Animal Feed." *Animal Feed Science and Technology* 197 (November 2017). Elsevier B.V.: 1–33. doi:10.1016/j.anifeedsci.2014.07.008.

Manu, M. K., Dongyi Li, Luo Liwen, Zhao Jun, Sunita Varjani, and Jonathan W.C. Wong. 2021. "A Review on Nitrogen Dynamics and Mitigation Strategies of Food Waste Digestate Composting." *Bioresource Technology* 334 (January). Elsevier Ltd: 125032. doi:10.1016/j.biortech.2021.125032.

Maurer, V., M. Holinger, Z. Amsler, B. Früh, J. Wohlfahrt, A. Stamer, and F. Leiber. 2016. "Replacement of Soybean Cake by Hermetia Illucens Meal in Diets for Layers." *Journal of Insects as Food and Feed* 2 (2): 83–90. doi:10.3920/JIFF2015.0071.

Mekonnen, Mesfin M., and Arjen Y. Hoekstra. 2012. "A Global Assessment of the Water Footprint of Farm Animal Products." *Ecosystems* 15 (3): 401–15. doi:10.1007/s10021-011-9517-8.

Menino, R., and Murta, D. 2021. "BSF-time to change the flies." *Horticultural International Journal* 5(3), 114–117.

Mohana Devi, S., and In Ho Kim. 2014. "Effect of Medium Chain Fatty Acids (MCFA) and Probiotic (Enterococcus Faecium) Supplementation on the Growth Performance, Digestibility and Blood Profiles in Weanling Pigs." *Veterinarni Medicina* 59 (11). Veterinární Medicína: 527–35. doi:10.17221/7817-VETMED.

Mohd-Noor Siti Nuraini, Chung Yiin Wong, Jun Wei Lim, Mah Iazam Azuri Mah-Hussin, Yoshimitsu Uemura, Man Kee Lam, Anita Ramli, Mohammed J.K. Bashir, and Leony Tham. 2017. "Optimization of Self-Fermented Period of Waste Coconut Endosperm Destined to Feed Black Soldier Fly Larvae in Enhancing the Lipid and Protein Yields." *Renewable Energy* 111. Elsevier B.V.: 646–54. doi:10.1016/j.renene.2017.04.067.

Mutafela and Richard. 2015. "High Value Organic Waste Treatment via Black Soldier Fly Bioconversion." *Trita-Im-Ex 2015:*16, 1–64. https://www.diva-portal.org/smash/get/diva2:868277/FULLTEXT02.pdf.

Nguyen Hoang Chinh, Shih Hsiang Liang, Shang Sian Chen, Chia Hung Su, Jhih Huei Lin, and Chien Chung Chien. 2018. "Enzymatic Production of Biodiesel from Insect Fat Using Methyl Acetate as an Acyl Acceptor: Optimization by Using Response Surface Methodology." *Energy Conversion and Management* 158 (October 2017). Elsevier: 168–75. doi:10.1016/j.enconman.2017.12.068.

Nguyen Hoang Chinh, Shih Hsiang Liang, Thien Thanh Doan, Chia Hung Su, and Pei Cheng Yang. 2017. "Lipase-Catalyzed Synthesis of Biodiesel from Black Soldier Fly (Hermetica Illucens): Optimization by Using Response Surface Methodology." *Energy Conversion and Management* 145. Elsevier Ltd: 335–42. doi:10.1016/j.enconman.2017.05.010.

Nguyen Hoang Chinh, Shih Hsiang Liang, Sing Ying Li, Chia Hung Su, Chien Chung Chien, Yi Ju Chen, and Dinh Thi My Huong. 2018. "Direct Transesterification of Black Soldier Fly Larvae (Hermetia Illucens) for Biodiesel Production." *Journal of the Taiwan Institute of Chemical Engineers* 85. Elsevier B.V.: 165–69. doi:10.1016/j.jtice.2018.01.035.

Okey Aniebo Alphonsus, Ebere Erondu, and Onyema Joseph Owen. 2009. Replacement of Fish Meal with Maggot Meal in African Catfish (Clarias gariepinus) Diets. *Revista Cientifica UDO Agricola*, 9: 666–71.

Paritosh Kunwar, Sandeep K. Kushwaha, Monika Yadav, Nidhi Pareek, Aakash Chawade, and Vivekanand. 2017. "Food Waste to Energy: An Overview of Sustainable Approaches for Food Waste Management and Nutrient Recycling." *BioMed Research International* 2017. doi:10.1155/2017/2370927.

Pastor, B., Y. Velasquez, P. Gobbi, and S. Rojo. 2015. "Conversion of Organic Wastes into Fly Larval Biomass: Bottlenecks and Challenges." *Journal of Insects as Food and Feed* 1 (3): 179–93. doi:10.3920/JIFF2014.0024.

Purschke Benedict, Rafaela Scheibelberger, Sonja Axmann, Andreas Adler, and Henry Jäger. 2017. "Impact of Substrate Contamination with mycotoxins, Heavy Metals and Pesticides on the Growth Performance and Composition of Black Soldier Fly Larvae (*Hermetia illucens*) for use in the Feed and Food Value Chain." *Food Additives & Contaminants* 34(8): 1410–20. Doi: 10.1080/19440049.2017.1299946.

Ramos María Jesús, Carmen María Fernández, Abraham Casas, Lourdes Rodríguez, and Ángel Pérez. 2009. "Influence of Fatty Acid Composition of Raw Materials on Biodiesel Properties." *Bioresource Technology* 100 (1): 261–68. doi:10.1016/j.biortech.2008.06.039.

Rasapoor, M., T. Nasrabadi, M. Kamali, and H. Hoveidi. 2009. "The Effects of Aeration Rate on Generated Compost Quality, Using Aerated Static Pile Method." *Waste Management* 29 (2). Elsevier Ltd: 570–73. doi:10.1016/j.wasman.2008.04.012.

Rehman Kashif ur, Abdul Rehman, Minmin Cai, Longyu Zheng, Xiaopeng Xiao, Abdul Aziz Somroo, Hui Wang, Wu Li, Ziniu Yu, and Jibin Zhang. 2017. "Conversion of Mixtures of Dairy Manure and Soybean Curd Residue by Black Soldier Fly Larvae (Hermetia Illucens L.)." *Journal of Cleaner Production* 154: 366–73. doi:10.1016/j.jclepro.2017.04.019.

Salomone, R., G. Saija, G. Mondello, A. Giannetto, S. Fasulo, and D. Savastano. 2017. "Environmental Impact of Food Waste Bioconversion by Insects: Application of Life Cycle Assessment to Process Using Hermetia Illucens." *Journal of Cleaner Production* 140. Elsevier Ltd: 890–905. doi:10.1016/j.jclepro.2016.06.154.

Schiavone Achille, Marco Cullere, Michele De Marco, Marco Meneguz, Ilaria Biasato, Stefania Bergagna, Daniela Dezzutto, et al. 2017. "Partial or Total Replacement of Soybean Oil by Black Soldier Fly Larvae (Hermetia Illucens L.) Fat in Broiler Diets: Effect on Growth Performances, Feed-Choice, Blood Traits, Carcass Characteristics and Meat Quality." *Italian Journal of Animal Science* 16 (1). Informa Healthcare USA, Inc: 93–100. doi:10.1080/1828051X.2016.1249968.

Schiavone Achille, Michele De Marco, Silvia Martínez, Sihem Dabbou, Manuela Renna, Josefa Madrid, Fuensanta Hernandez, et al. 2017. "Nutritional Value of a Partially Defatted and a Highly Defatted Black Soldier Fly Larvae (Hermetia Illucens L.) Meal for Broiler Chickens: Apparent Nutrient Digestibility, Apparent Metabolizable Energy and Apparent Ileal Amino Acid Digestibility." *Journal of Animal Science and Biotechnology* 8 (1): 1–9. doi:10.1186/s40104-017-0181-5.

Sealey, Wendy M., T. Gibson Gaylord, Frederic T. Barrows, Jeffery K. Tomberlin, Mark A. McGuire, Carolyn Ross, and Sophie St-Hilaire. 2011. "Sensory Analysis of Rainbow Trout, Oncorhynchus Mykiss, Fed Enriched Black Soldier Fly Prepupae, Hermetia Illucens." *Journal of the World Aquaculture Society* 42 (1): 34–45. doi:10.1111/j.1749-7345.2010.00441.x.

Setti Leonardo, Enrico Francia, Andrea Pulvirenti, Silvia Gigliano, Massimo Zaccardelli, Catello Pane, Federica Caradonia, Sara Bortolini, Lara Maistrello, and Domenico Ronga. 2019. "Use of Black Soldier Fly (Hermetia Illucens (L.), Diptera: Stratiomyidae) Larvae Processing Residue in Peat-Based Growing Media." *Waste Management* 95. Elsevier Ltd: 278–88. doi:10.1016/j.wasman.2019.06.017.

Skrivanova, E., Milan Marounek, V. Benda, and P. Brezina. 2007. "Susceptibility of Escherichia Coli, Salmonella Sp. and Clostridium Perfringens to Organic Acids and Monolaurin." *Veterinarni Medicina* 51 (3). Veterinární Medicína: 81–88. doi:10.17221/5524-VETMED.

Somroo Abdul Aziz, Kashif ur Rehman, Longyu Zheng, Minmin Cai, Xiaopeng Xiao, Shencai Hu, Alexander Mathys, Moritz Gold, Ziniu Yu, and Jibin Zhang. 2019. "Influence of Lactobacillus buchneri on soybean curd residue co-conversion by black soldier fly larvae (Hermetia illucens) for food and feedstock production." *Waste Management*, 86: 114–22. https://doi.org/10.1016/j.wasman.2019.01.022

Song Shuang, Alvin Wei Liang Ee, Jonathan Koon Ngee Tan, Jia Chin Cheong, Zhongyu Chiam, Srishti Arora, Weng Ngai Lam, and Hugh Tiang Wah Tan. 2021. "Upcycling Food Waste Using Black Soldier Fly Larvae: Effects of Further Composting on Frass Quality, Fertilising Effect and Its Global Warming Potential." *Journal of Cleaner Production* 288. Elsevier Ltd: 125664. doi:10.1016/j.jclepro.2020.125664.

Spranghers Thomas, Matteo Ottoboni, Cindy Klootwijk, Ann Ovyn, Stefaan Deboosere, Bruno De Meulenaer, Joris Michiels, Mia Eeckhout, Patrick De Clercq, and Stefaan De Smet. 2017. "Nutritional Composition of Black Soldier Fly (Hermetia illucens) Prepupae Reared on Different Organic Waste Substrates" *Journal of the Science of Food and Agriculture*, 97(8): 2594–600. doi: 10.1002/jsfa.8081

Starčević Kristina, Aleksandar Gavrilović, Željko Gottstein, and Tomislav Mašek. 2017. "Influence of Substitution of Sunflower Oil by Different Oils on the Growth, Survival Rate and Fatty Acid Composition of Jamaican Field Cricket (Gryllus Assimilis)." *Animal Feed Science and Technology* 228: 66–71. doi:10.1016/j.anifeedsci.2017.04.007.

Surendra, K. C., Jeffery K. Tomberlin, Arnold van Huis, Jonathan A. Cammack, Lars Henrik L. Heckmann, and Samir Kumar Khanal. 2020. "Rethinking Organic Wastes Bioconversion: Evaluating the Potential of the Black Soldier Fly (Hermetia Illucens (L.)) (Diptera: Stratiomyidae) (BSF)." *Waste Management* 117. Elsevier Ltd: 58–80. doi:10.1016/j.wasman.2020.07.050.

The agricultural European Innovation Partnership (EIP-AGRI). 2016. "Soil Organic Matter Matters Investing in Soil Quality for Long-Term Benefits." *Soil Organic Matter Matters Investing in Soil Quality for Long-Term Benefits*, 8. https://ec.europa.eu/eip/agriculture/sites/agri-eip/files/eip-agri_brochure_soil_organic_matter_matters_2016_en_web.pdf.

Thrastardottir Runa, Hildur Thora Olafsdottir, and Ragnheidur Inga Thorarinsdottir. 2021. "Yellow Mealworm and Black Soldier Fly Larvae for Feed and Food Production in Europe, with Emphasis on Iceland." *Foods* 10 (11): 1–35. doi:10.3390/foods10112744.

Tomberlin, Jeffery K., and D. Craig Sheppard. 2001. "Lekking Behavior of the Black Soldier Fly (Diptera: Stratiomyidae)." *Florida Entomologist* 84 (4): 729–30. doi: 10.2307/3496413.

Tschirner, M., and A. Simon. 2015. "Influence of Different Growing Substrates and Processing on the Nutrient Composition of Black Soldier Fly Larvae Destined for Animal Feed." *Journal of Insects as Food and Feed* 1 (4): 249–59. doi:10.3920/JIFF2014.0008.

Wang Dou, Jing Ai, Fei Shen, Gang Yang, Yanzong Zhang, Shihuai Deng, Jing Zhang, Yongmei Zeng, and Chun Song. 2017. "Improving Anaerobic Digestion of Easy-Acidification Substrates by Promoting Buffering Capacity Using Biochar Derived from Vermicompost." *Bioresource Technology* 227. Elsevier Ltd: 286–96. doi:10.1016/j.biortech.2016.12.060.

Wang Yu Shiang, and Matan Shelomi. 2017. "Review of Black Soldier Fly (Hermetia Illucens) as Animal Feed and Human Food." *Foods* 6 (10). doi:10.3390/foods6100091.

Wong Chung Yiin, Muhammad Naeim Mohd Aris, Hanita Daud, Man Kee Lam, Ching Seong Yong, Hadura Abu Hasan, Siewhui Chong, et al. 2020. "In-Situ Yeast Fermentation to Enhance Bioconversion of Coconut Endosperm Waste into Larval Biomass of Hermetia Illucens: Statistical Augmentation of Larval Lipid Content." *Sustainability (Switzerland)* 12 (4): 1–10. doi:10.3390/su12041558.

Zheng, Longyu, Yanfei Hou, Wu Li, Sen Yang, Qing Li, and Ziniu Yu. 2012. "Biodiesel Production from Rice Straw and Restaurant Waste Employing Black Soldier Fly Assisted by Microbes." *Energy* 47 (1): 225–29. doi:10.1016/j.energy.2012.09.006.

Zheng, Longyu, Qing Li, Jibin Zhang, and Ziniu Yu. 2012. "Double the Biodiesel Yield: Rearing Black Soldier Fly Larvae, Hermetia Illucens, on Solid Residual Fraction of Restaurant Waste after Grease Extraction for Biodiesel Production." *Renewable Energy* 41. Elsevier Ltd: 75–79. doi:10.1016/j.renene.2011.10.004.

7 Valorisation of food waste into liquid biofuel

Gelyn L. Bongabong, Isaac Jerome C. Dela Cruz, and Don Nelson C. Potato

7.1 INTRODUCTION

Food waste (FW) valorisation is an emerging strategy and a promising sustainable approach for addressing the global FW problem by conversion into value-added products, including bio-plastics, peptides, essential amino acids, bioenergy, and biofuels. This chapter explores various sustainable engineering interventions in utilising FW and converting it into sustainable alternative products. Sustainable processes include gasification, torrefaction, fermentation, and transesterification. These processes prioritised efficiency, resource conservation, and environmental management. These methods extract energy from the organic waste while minimising ecological impact. Integrating advanced technologies like enzymatic hydrolysis, microbial fermentation, and algae cultivation enhances biofuel yields and reduces energy inputs and environmental footprint.

7.1.1 Conversion of food waste to bioenergy and biofuels

Currently, several practices and techniques are used in waste management alongside the production of energy from FW. Incineration is a traditional waste management method and is the lowest-demand technique for waste treatment (Mayer et al., 2019). The process involves the combustion of waste that produces heat. The energy recovered from the heat can be used mechanically to operate turbines or heat exchangers in process industries (Pham et al., 2015). Combustion typically operates at temperatures below 1,150°C to avoid corrosion due to high temperatures (Di Maria et al., 2018). Apart from grate furnaces, rotary kilns and fluidized bed boilers have emerged as the leading technologies for waste incineration (Mayer et al., 2019). Incineration can reduce the volume of solid waste for disposal by 80–85%. However, this method is said to negatively impact the environment due to excessive emissions and heavy metals that are air pollutants during the burning phase. Moreover, byproducts during incineration such as ashes need to be cautiously disposed of, as they contain inorganic materials (Sridhar et al., 2021).

Landfill is another traditional method of treating municipal solid waste (MSW) which includes FW. Landfills are mostly consist of food waste that possible to be used a raw materials to generates approximately 125 m^3 of greenhouse gases—primarily

 DOI: 10.1201/9781032718507-7

60–65% of CH_4 and about 40% of CO_2 (Adhikari et al., 2006). According to the US Environmental Protection Agency (USEPA), landfill gases (LFGs) are considered one of the largest sources of methane and are responsible for about 8% of global emissions and could potentially increase by 2025. Although landfills are economical, less complex, and convenient means of waste disposal, land use and availability are the main challenges of this method (Sridhar et al., 2021). Both incineration and landfill pose environmental impacts such as high global warming potential and human toxicity, thus not environmentally sustainable in the long term.

7.2 TYPES OF BIOFUELS PRODUCED FROM FOOD WASTE

Biofuel is defined as energy derived from biomass that can be used as fuel for heat, work, and electrical applications (Dhiman & Mukherjee, 2022). Biofuels have been considered renewable, clean, safe, and alternative fuels derived from various biomass conversion methods (Rodionova et al., 2017). Generally, biofuels can be classified into three based on their phase namely solid, liquid, and gas biofuels, as shown in Figure 7.1. Among the gas-type biofuels are biomethane, pyrogas, syngas, and biohydrogen. Biomethane or biogas is the product of a biological method known as the anaerobic digestion process. Pyrogas and syngas are products of thermochemical conversion methods called pyrolysis and gasification, respectively. Pyrolysis can also produce a liquid product known as bio-oil that can be further processed into advanced biofuels. Biohydrogen production can be done through both biological and thermochemical treatments. Liquid biofuels include bioethanol and biodiesel produced from fermentation and transesterification processes, respectively. Solid biofuel is known as biochar and can be derived from biomass through torrefaction or mild pyrolysis. Thorough discussions of the conversion processes are presented in the succeeding sections.

7.2.1 Biohydrogen

Biohydrogen can be considered as a clean biofuel. This is because it does not contain carbon, as the only by-product produced when it is used as fuel in fuel cells or heat engines is water. Biohydrogen is a non-toxic, colourless, tasteless, and odourless gas that can produce more energy than any other fuel at about 122 KJ/g and has a calorific value of 143 GJ/ton (Sekoai & Gueguim Kana, 2013; Show et al., 2011). The heating of lignocellulosic biomass alongside its alkaline or acid pretreatment has been determined to be beneficial for biohydrogen manufacturing. The most widely used acid and alkali are H_2SO_4 and NaOH, respectively. Heating is an essential step for pretreatment, and pH 5.5–6.5 is suitable for biohydrogen production (Zhang et al., 2016). Ultrasonic pretreatment gave better biohydrogen yields. Co-culture of microorganisms could increase biohydrogen yield if specific microorganisms are in symbiosis (Gadhe et al., 2014).

7.2.2 Bioethanol

Bioethanol is the most common biofuel produced commercially, primarily from corn and sugarcane. According to the Renewable Fuels Association (RFA), the global

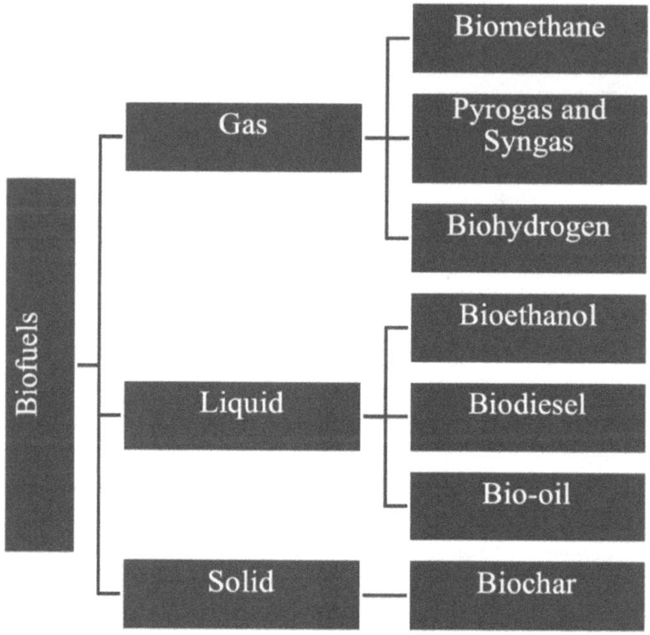

FIGURE 7.1 Classifications of biofuels based on product phase.

annual ethanol production reached 16,219 million gallons in 2024 (Annual Ethanol Production: U.S. and World Ethanol Production, 2025). Ethanol can be blended with gasoline because it can reduce greenhouse gas emissions, unburned hydrocarbons, and carcinogens. It can also reduce sulfur dioxide emission, which is the main cause of acid rain. There is ample potential for bioethanol production from various wastes and agricultural residues such as raw wood materials, corn stalks, corn sticks, rapeseed waste, sugar cane bagasse, citrus peels, leaves, and straw (Nigam & Singh, 2011). Additionally, carbohydrate-rich FW could be a great raw material for bioethanol production (Singh et al., 2022). FW rich in organic components, particularly polysaccharides, can be used directly to produce bioethanol since their hydrolytic products, such as glucose and xylose, are easily converted to ethanol (Prasoulas et al., 2020). Figure 7.2 presents the conversion of glucose and xylose to bioethanol via fermentation.

7.2.3 Biodiesel

Fractions of FW are oils and fats which can be utilised in the production of biodiesel. Biodiesel is a liquid biofuel produced from the reaction of oil (vegetable or animal oils) with a short-chain alcohol in the presence of a catalyst called the transesterification process (Chen et al., 2023). Biolipids extracted from FW and produced by oleaginous microbes are a promising feedstock for biodiesel production (Carmona-Cabello et al., 2021). Most of the FW components are difficult for oleaginous microbes to directly ingest; hence, the pretreatment process is needed to boost the availability of

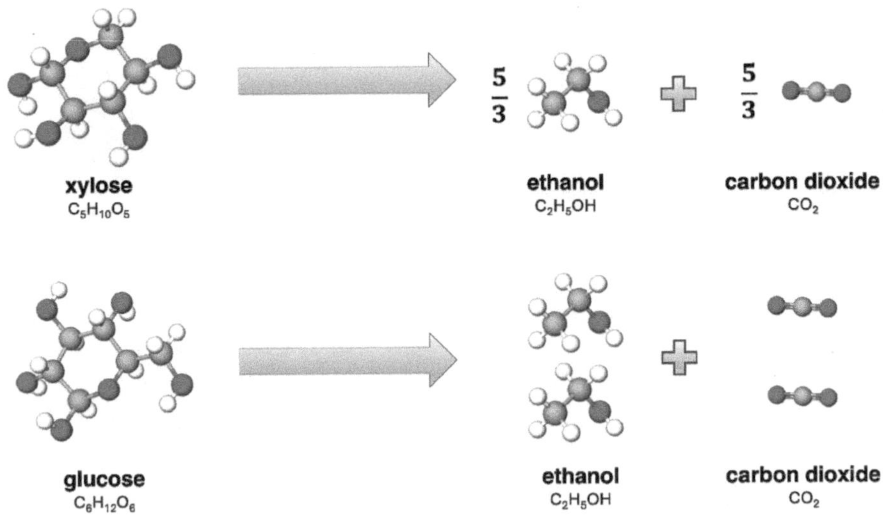

FIGURE 7.2 The conversion of glucose and xylose to bioethanol via fermentation.

the constituent nutrients. Apart from anaerobic pretreatment of FW for biolipid production, recent studies have shown that aerobic treatment is effective for the in situ treatment of FW with a short process time (Liu et al., 2022). In addition to aerobic treatment, hydrolysis can reduce FW weight in a short period of 1–35 h while also effectively recovering nutrients from FW into hydrolysate (Chen et al., 2023).

Short-chain alcohol can be classified as methanol, ethanol, or other suitable alcohols. Catalyst (homogeneous, heterogeneous, and biocatalyst) plays a significant role in the transesterification reaction by increasing the speed of the slow reaction, thereby increasing the biodiesel product yield (Ulukardesler, 2023). Homogenous catalysts such as sulfuric acid and sodium hydroxide are used in industrial applications because the reaction times are shorter, and the conditions are preferable. A large amount of heterogenous catalyst is used in a reaction condition of high temperature, high pressure, and high molar ratio of alcohol-to-oil are needed. Biocatalysts are biological substances that can be homogenous or heterogeneous such as algae, fungi, microbial bacteria, and enzymes (Mohiddin et al., 2021).

7.3 CLASSIFICATIONS OF BIOFUEL CONVERSION PROCESSES

FW contains various nutrients that make it an ideal source for biofuel production. FW is rich in organic compounds such as carbohydrates, lipids, and proteins. With the help of well-structured conversion technologies, the valorisation of FW into biofuels can be considered a sustainable solution to address the global energy crisis (Dhiman & Mukherjee, 2022). Various technologies have been used to convert FW into high-value biofuel, and these are categorized as thermochemical, biological, biochemical, and chemical conversion. As presented in Table 7.1, the primary biomass conversion processes include incineration, anaerobic digestion, fermentation, transesterification, pyrolysis, torrefaction, and gasification (Simonyan & Fasina, 2013;

TABLE 7.1
Comparative description of the different conversion techniques of biomass to produce biofuels

Classifications	Conversion process	Products
Thermal	Incineration (traditional)	Heat
Biological	Anaerobic digestion	Biogas/biomethane
Biochemical	Fermentation	Bioethanol
Chemical	Transesterification	Biodiesel
Thermochemical	Gasification	Syngas
	Pyrolysis	Bio-oil, pyrogas, biochar (by-product)
	Torrefaction	Biochar, torgas (by-product)

Adapted and modified from Simonyan and Fasina (2013) & Tumuluru et al. (2021).

Tumuluru et al., 2021). These conversion methods have advantages and disadvantages, and their suitability is dependent on FW characteristics.

7.3.1 Gasification

Given that FW includes fixed carbon, moisture, ash, and around 80% volatile components, it is considered a good source of energy. Gasification is among the thermochemical conversion methods that remove volatile components and recover energy in a short period. As a result, it has been determined that gasifying FW that contains moisture is an efficient energy recovery method that produces fewer pollutants (Murugesan et al., 2022). In addition to volatiles and moisture content, the ash content must be considered because it has a significant impact on the kinetics of the reaction between the char and the gasifying medium (Ahmed & Gupta, 2010).

Gasification is a low-polluting thermochemical process that treats solid carbonaceous materials at high temperatures (700–1,200°C) in a gasifier using various gasifying media (oxygen, steam, air, CO_2, etc.). The gasification process consists of four stages: feedstock drying, pyrolysis (decomposition), oxidation (combustion), and reduction (gasification). The gasification process produces syngas, a combustible combination of gases containing H_2, CO, CO_2, and CH_4, which can then be utilised as fuel or substrate in boiler combustion, fuel cells, and chemical product synthesis. Chars, oils, ash, and tars are byproducts of the gasification process (Murugesan et al., 2022). Various factors such as temperature, resistance time, precursor composition, and catalyst type have a significant impact on the composition of gasification products. Notably, using O_2 as a gasifying medium at a lower temperature produces a greater amount of CO_2 (Burra et al., 2016). Gasification can be classified based on the type of gasifying medium, namely steam gasification, subcritical water gasification (SbWG), and supercritical water gasification (SpWG). Different gasifying media can influence the H_2 concentration in syngas, gas yield, and the calorific value of the extracted gas fraction (Singh & Yadav, 2021).

7.3.2 Steam Gasification

Compared to conventional gasification which uses air as a gasifying medium, steam gasification can produce more H_2 from FW. H_2 and CO_2 are produced during the oxidation of the carbonaceous fuel in the steam gasification process, as seen in Eq. (7.1). In this process, the temperature and steam flow rate (SFR) affect the syngas properties. The methane reformation and water-gas shift (WGS) reaction are accelerated by an increase in SFR, which improves syngas quality (Murugesan et al., 2022).

$$C_nH_m + 2nH_2O \leftrightarrow \left(2n + \left(\frac{m}{2}\right)\right) H_2 + nCO_2 \tag{7.1}$$

The temperature and gasifying medium flow rate affect syngas concentration. Singh and Yadav (2021) found that the maximum H_2 production occurred at a steam flow rate (SFR) of 0.5 mL/min (Singh & Yadav, 2021). The steam gasification process can convert kitchen FW into H_2. The maximum H_2 yield can be as high as 1.2 m^3/kg at 0.5 mL/min SFR process at a temperature of 800°C. The syngas characteristics were unchanged by the increase in SFR up to 0.625 mL/min (Singh et al., 2020).

7.3.3 Hydrothermal Gasification

Hydrothermal gasification is a hydrothermal treatment to produce syngas from FW at a temperature and pressure range of more than 374 °C and 22.1 MPa, respectively. This condition is employed in the supercritical water gasification (SCWG) process (Wang et al., 2022). It is considered to have less impact on the environment as confirmed by a life cycle assessment study by Gasafi et al. (2003). Supercritical water gasification of FW could produce hydrogen synthesis gas. The hydrothermal temperature of FW has a positive effect on gas yield. Increasing temperature promoted the degradation of insoluble fractions resulting in high gas yield (Nanda et al., 2016). SCWG has been used for the hydrothermal conversion of FW to produce hydrogen-rich gas products. The alkali salts of FW influence the catalysis of the WGS reaction (Eq. (7.2)) causing an increase in hydrogen yield and a decrease in the CO_2 yield (Kruse, 2009).

$$CO + H_2O \rightarrow H_2 + CO_2 \tag{7.2}$$

7.3.4 Fermentation

Types of substrates for ethanol production can be divided into three categories. The first category includes sugar (molasses) and fine sugar that can be directly converted into ethanol by microbes without complex pretreatment; however, this type demands a high cost for industrial quantity. The second category is high starch crops such as corn, wheat, and rice, which have issues with food and feed competition. The third category is agricultural residue which can be used to produce 2nd generation bioethanol (Singh et al., 2022). Alternatively, FW that falls under agricultural residue classification can be used as substrate for bioethanol production. FW is affordable, abundant, renewable, and, most importantly, following pretreatment and enzymatic hydrolysis, glucose productivity is higher than agricultural residues (Kazemi Shariat Panahi et al., 2022).

Bioethanol production from FW can be produced using a fermentation process. Prior to fermentation process, FW undergoes several stages such as enzyme production and enzymatic hydrolysis. These processes can be done separately, simultaneously, or in combination. The development of ethanol fermentation to produce hydrolysate from the enzymatic hydrolysis of FW is another effective technique for developing economically viable bioconversion technology (Hafid et al., 2017).

7.3.5 Separate Hydrolysis and Fermentation

This method is commonly employed when the hydrolysis or saccharification process is considered. It offers optimized conditions for both catalyst and fermenting organisms which are used for saccharification and fermentation processes, respectively. In separate hydrolysis and fermentation (SHF), the saccharification/hydrolysis process comes after enzyme production, and when completed, optionally recovers sugar via centrifugation or filtration before introducing microbes for fermentation (Offei et al., 2018). Separate vessels are required for each process, as depicted in Figure 7.3. However, in SHF, the cellobiose and glucose produced during enzymatic hydrolysis inhibit cellulase activity, thus contributing a significant negative effect on the reaction rate (Sasaki et al., 2014).

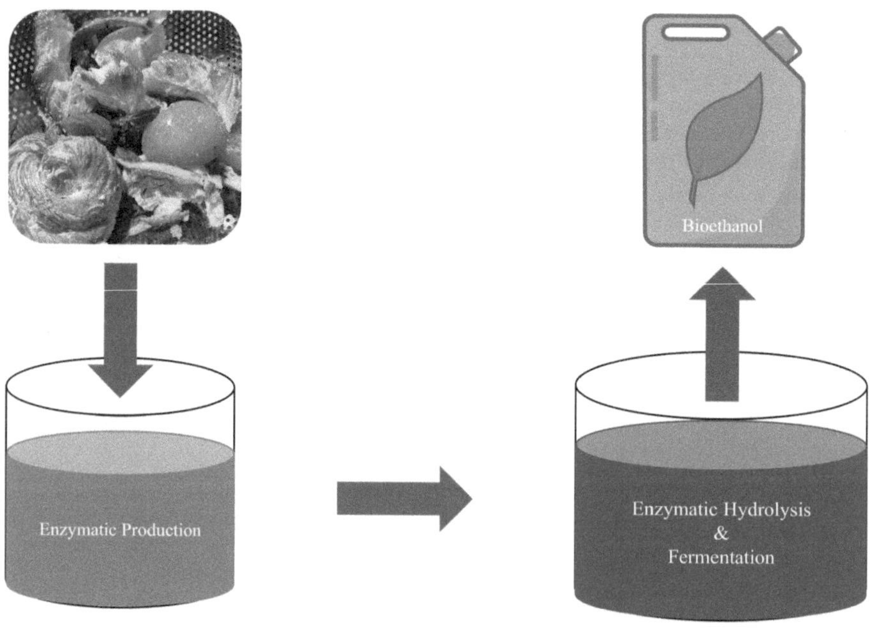

FIGURE 7.3 Simplified schematic diagram of food waste separate hydrolysis and fermentation (SHF) process.

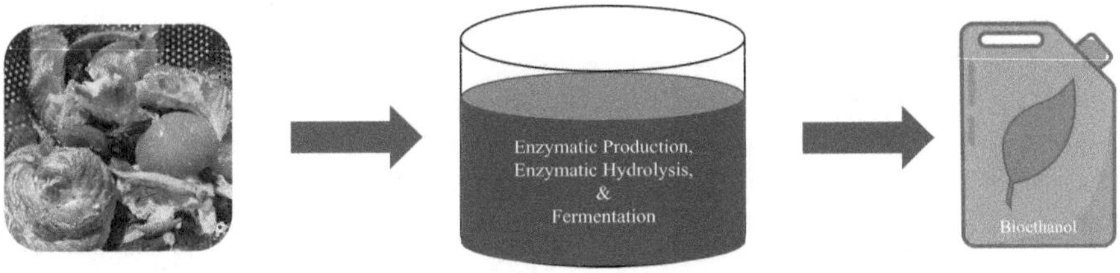

FIGURE 7.4 Simplified schematic diagram of food waste simultaneous saccharification and fermentation (SSF) process.

7.3.6 Simultaneous saccharification and fermentation

In SSF, saccharification enzymes and yeast or other fermenting microorganisms are added concurrently into the reactor with the same optimized operational conditions (Singh et al., 2022). SSF process is cost-effective in terms of reactor requirement since both hydrolysis and fermentation happen in one reactor at the same time, as shown in Figure 7.4. However, it is difficult to obtain optimal conditions for each phase during the simultaneous process (Offei et al., 2018). Simultaneous saccharification and fermentation (SSF) eliminate end-product inhibition, resulting in higher ethanol yield, since microorganisms can use the sugar generated throughout the process for both growth and ethanol production (Saha et al., 2011).

7.3.7 Consolidated bioprocessing

The consolidated bioprocessing (CBP) technique is a single-step bioprocessing of FW to produce bioethanol. In this strategy, a consortium of microorganisms is employed and three steps (i.e., enzyme production, saccharification, and fermentation) are carried out in a single reactor. Thus, the cost of infrastructure is reduced, and feedback inhibition is minimised. However, process efficiency is still a challenge (Singhania et al., 2022). A study on non-sterile CBP of FW was carried out by Peinemann et al. (2020) using *Saccharomyces cerevisiae.* At unregulated pH, 80% of the substrate was converted into ethanol (900–1000 mM), whereas at regulated pH (6), sugars produced were evenly converted to ethanol and lactic acid (400–500 mM) (Peinemann et al., 2020). Process efficiency also depends on solid loading of biomass for enzyme hydrolysis, which is vital and affects the economy of bioethanol production overall because solid loading of biomass is necessary for high sugar concentration in the broth, which may eventually result in higher bioethanol concentration in the broth (Singhania et al., 2022).

7.3.8 Transesterification

Transesterification, or alcoholysis, is a chemical process widely used to convert vegetable and animal oil into biodiesel (Win & Trabold, 2018). The process involves the combination of oil with alcohol with the aid of a catalyst, such as acids, bases, and enzymes, to produce an alkyl ester, commonly referred to as biodiesel (Jain, 2023). This step is crucial in biodiesel production, encompassing reaction mechanisms, catalysts, influencing factors, and optimization challenges (Baskar et al., 2018). During this process, triglycerides and alcohol react to form fatty acid alkyl esters (FAAE) and glycerol. The typical transesterification reaction is shown in Figure 7.5 (Ejikeme et al., 2010). Initially, triglycerides and alcohol combine to form diglycerides, which further convert into monoglycerides and glycerol, yielding alkyl esters (Quader & Ahmed, 2017). Factors influencing biodiesel yield include reaction time, temperature, catalyst and concentration, feedstock oil, and alcohol-to-oil ratio (Farouk et al., 2024). Excess alcohol is required to promote product formation, and alcohols with varying chain lengths are preferable. In addition, the transformation from oil to ester reduces molecular weight, viscosity, density, and increased volatility (Ahmed Elgharbawy et al., 2021; Farouk et al., 2024). Methanol and ethanol are frequently chosen alcohols for transesterification due to their widespread availability and cost-effectiveness. Meanwhile, methanol is particularly favoured because of its high reactivity, lower boiling point, and affordability. The effectiveness of catalysts, such as NaOH or KOH, in accelerating the equilibrium adjustment and enhancing ester yields is influenced by the free fatty acid content in the feedstock (Ejikeme et al., 2010; Tchuifon et al., 2020). Moreover, transesterification has two main methods: catalytic transesterification and non-catalytic transesterification. Catalytic transesterification encounters several challenges, notably the time-consuming nature of the process and the necessity for separating the oil, alcohol, catalyst, and saponified impurities mixture from the resulting biodiesel. Further to this, biodiesel purification is simplified in the non-catalytic supercritical transesterification process, as it eliminates the need for a catalyst, thereby preventing soap formation or saponification (Ahmed Elgharbawy et al., 2021; Interlenghi, de Medeiros, & Fernandes Araújo, 2023; Mumtaz et al., 2017; Quader & Ahmed, 2017).

7.4 FOOD WASTE-TO-BIOFUELS AND BIOECONOMY: A CASE STUDY

With the world's rapid population growth, continual rising energy demand, and depleting fossil fuel resources, the quest for alternative energy sources has been

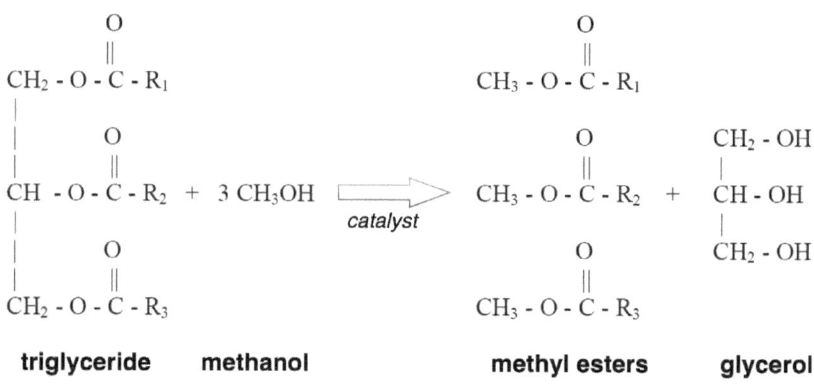

FIGURE 7.5 Transesterification reaction for biodiesel production.

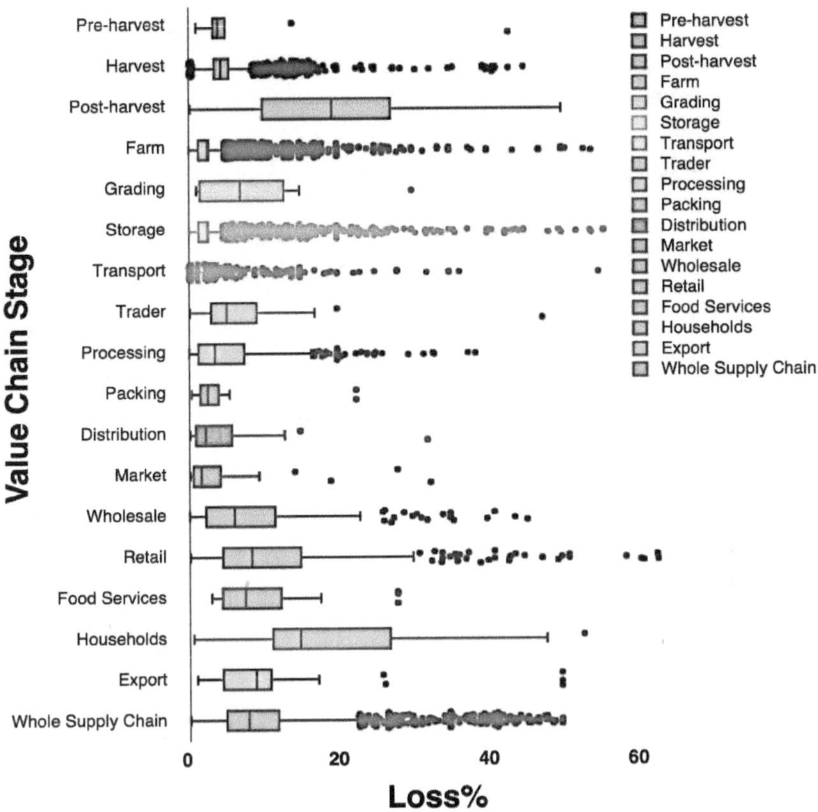

FIGURE 7.6. Global food wastage per food value chain stage from 2020 to 2022. Adapted from (Food and Agriculture Organization of the United Nations, 2022).

increasingly receiving attention. This has led to the abrupt transition of research focus from a fossil-based unsustainable economy to a sustainable biobased economy (Dahiya et al., 2018). The bioeconomy concept deals with the sustainable utilisation of biomass to generate a range of biobased products such as biofuels employing the multidisciplinary disciplines of science, engineering, and management. Feedstock selection is a vital step in biofuel production. Features of a sustainable feedstock include abundance, economics, and no food and feed competitions. Thus, nonedible feedstocks such as FW are potential sustainable biofuel sources (Dhiman & Mukherjee, 2021).

The amount of wasted food is increasing drastically at an alarming rate parallel to the population and economic growth. FW is the wastage results from various food value chain stages. Depicted in Figure 7.6 is the global relative food wastage at different food supply chain stages as reported by the Food and Agricultural Organization from 2020 to 2022, (FAO, 2022). The FW from households and food services have significant shares in the total food wastage. With this substantial amount, FW can be a sustainable feedstock for biofuel production with the use of efficient conversion technology. Among biofuel conversion technologies that are commonly used are bioethanol and biodiesel technologies. Bioethanol and biodiesel have policies on blending with commercial gasoline and diesel, respectively. Hence, in this chapter, data from previous studies on the conversion of household FW to produce these two liquid biofuels are presented.

7.4.1 Estimation of Bioethanol and Biodiesel Yield from Household FW

This section deals with the estimations of bioethanol and biodiesel yields and the theoretical energy generation from global household food waste (HFW). Figure 7.7 shows the comparative valorisation of FW to produce bioethanol and biodiesel. The carbohydrate fraction of HFW is needed to produce bioethanol, while the lipid constituent is required by the production of biodiesel (Karmee & Lin, 2014). Table 7.2 presents the data from previous studies on the production of bioethanol and biodiesel from household/kitchen FW.

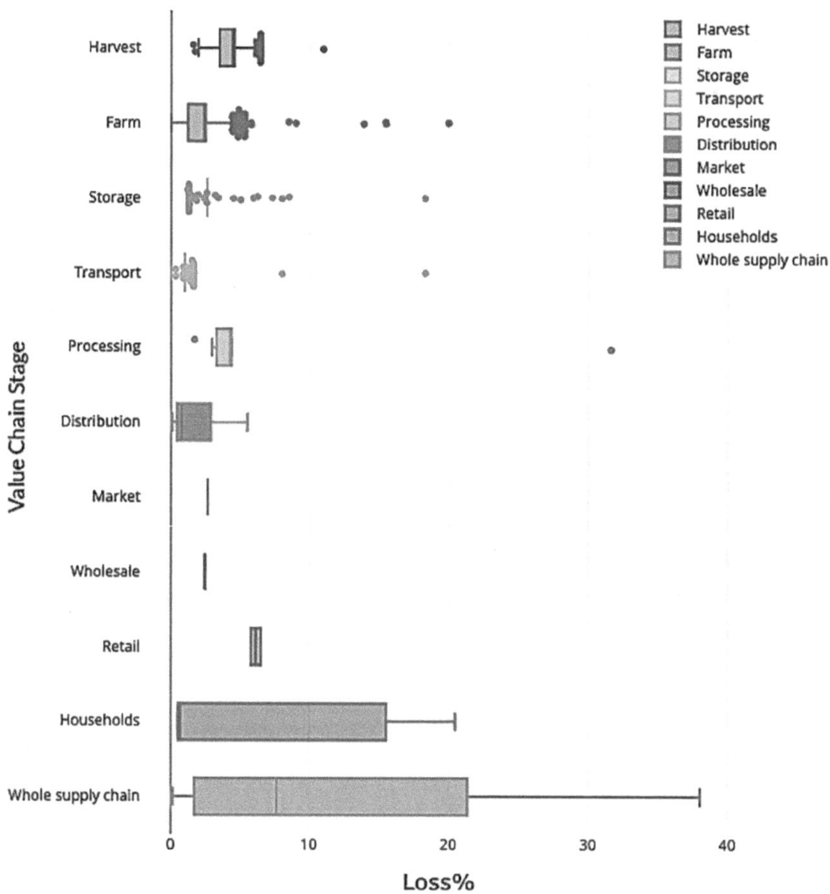

FIGURE 7.7 Valorisation of HFW to produce bioethanol and biodiesel. Adapted from (Food and Agriculture Organization of the United Nations, 2022).

Matsakas et al. (2014) worked on bioethanol production from HFW through liquefaction pretreatment and fermentation process with *Saccharomyces cerevisiae* as the catalyst. A hydrothermal pretreatment was employed for solid residues retrieved after the first fermentation run. The pretreated residues underwent 2nd fermentation process. The overall ethanol yield from soluble and cellulose sugars of 45% dry matter concentration is 63.64% (Matsakas et al., 2014). However, their study did not include ethanol characterisation; thus, there is no data on the calorific value of obtained bioethanol. Kannah et al. (2020) reported that the calorific value of FW bioethanol is 26.7 MJ/kg (Kannah et al., 2020).

Barik et al. (2018) studied the transesterification of kitchen FW to produce biodiesel using methanol as a solvent and sulfuric acid as a catalyst. Lipid extraction using methanol solvent was performed prior to transesterification. The transesterification reaction was performed in the ratio 11:1 (methanol:lipid molar ratio) at 60°C for 2–3 h and catalyst of 2.5 wt. % was added. The obtained biodiesel yield is 32.3% with a calorific value of 31.38 MJ/kg.

With the above yield and calorific values of bioethanol and biodiesel and based on annual HFW generated per country by the UNEP Food Waste Index Report (UNEP, 2021), the estimated theoretical bioethanol and biodiesel produced with their respective bioenergy generation are presented in Table 7.3. China being the top producer of HFW has the highest bioenergy potential for both bioethanol and biodiesel produced. Since the reported bioethanol yield is higher than the biodiesel yield from the same feedstock, it is expected that the bioenergy yield from bioethanol is higher than that of biodiesel. Overall, a significant amount of energy can be generated from the global FW. A categorized FW characteristics and composition and optimized operating pretreatment methods and operating parameters can increase the bioenergy yield. In addition, online measurement of operating data and product output of the system can utilise different FW.

7.5 CONCLUSIONS

FW is a great potential alternative biomass source for bioenergy production. Conversion of FW to produce value-added products such as biofuels (solid, liquid, and gaseous) is a promising valorisation strategy that would benefit both the environment and the economy. FW-derived biofuels such as bioethanol and biodiesel can address the issue of fossil fuel reliance. The existing policy on bioethanol-gasoline and biodiesel-diesel blends supports the FW valorisation focus of the researchers and scientists. The policies on solid and gaseous biofuels can further be strengthened to maximise sustainability strategies on the conversion of FW into these valuable products.

EXERCISES

1. How do FW-derived biofuels benefit the economy and environment?
2. Which is the most advantageous among liquid biofuel conversion technologies?
3. Given 90 tonnes of FW with a household fraction of 60%, estimate the amount of bioethanol that can be derived from household FW if the liquid yields are 65%. The energy demand of a specific region is about 1,000 GJ, is the total energy production from bioethanol able to cover this demand if its energy value is 27 MJ/kg?

TABLE 7.2
Conversion of household/kitchen food waste into bioethanol and biodiesel

Biofuel type	Pretreatment	Conversion process	Product yield (% w/w)	Calorific value (MJ/kg)	References
Bioethanol	Liquefaction and hydrothermal for solid residue	Fermentation using *S. cerevisiae*	63.64	26.7*	Matsakas et al. (2014)
Biodiesel	Lipid extraction	Transesterification using methanol as solvent and sulfuric acid as catalyst	32.3	31.38	Barik et al. (2018)

*Kannah et al. (2020).

TABLE 7.3
Estimated HFW-derived bioethanol and biodiesel produced and their generated energy per country

Country	Annual HFW produced[a] (Million tonnes)	Estimated bioethanol produced (M. tonnes)	Estimated bioenergy from bioethanol (GJ)	Estimated biodiesel produced (M. tonnes)	Estimated bioenergy from biodiesel (GJ)
China	91.64	58.32	1.56×10^9	29.60	9.29×10^8
India	68.76	43.76	1.17×10^9	22.21	6.97×10^8
US	19.36	12.32	3.29×10^8	6.25	1.96×10^8
Japan	8.16	5.19	1.39×10^8	2.64	8.27×10^7
Germany	6.26	3.98	1.06×10^8	2.02	6.34×10^7
France	5.52	3.51	9.38×10^7	1.78	5.59×10^7
UK	5.20	3.31	8.84×10^7	1.68	5.27×10^7
Russia	4.87	3.10	8.28×10^7	1.57	4.94×10^7
Spain	3.61	2.30	6.13×10^7	1.17	3.66×10^7
Australia	2.56	1.63	4.35×10^7	0.83	2.59×10^7

[a]Based on UNEP Food Waste Index Report 2021 (Forbes et al., 2021).

REFERENCES

Adhikari, B. K., Barrington, S., & Martinez, J. (2006). Predicted growth of world urban food waste and methane production. *Waste Management and Research*, *24*(5), 421–433 https://doi.org/10.1177/0734242X06067767

Ahmed, I. I., & Gupta, A. K. (2010). Pyrolysis and gasification of food waste: syngas characteristics and char gasification kinetics. *Applied Energy*, *87*(1), 101–108. https://doi.org/10.1016/j.apenergy.2009.08.032

Ahmed Elgharbawy, A., Sadik, W., Sadek, O., & Kasaby, M. (2021). Transesterification reaction conditions and low-quality feedstock treatment processes for biodiesel production-a review. *Journal of Petroleum and Mining Engineering*, *0*(0), 98–103.

Annual Ethanol Production: U.S. and World Ethanol Production (2025). https://ethanolrfa.org/markets-and-statistics/annual-ethanol-production. Accessed on 6 March 2025.

Barik, S., Paul, K.K., Priyadarshi, D. Utilization of kitchen food waste for biodiesel production, July 2018, *IOP Conference Series Earth and Environmental Science 167*(1), 012036, DOI:10.1088/1755-1315/167/1/012036

Baskar, G., Kalavathy, G., Aiswarya, R., & I. Abarnaebenezer Selvakumari. (2018). Advances in eco-fuels for a sustainable environment. *Advances in bio-oil extraction from nonedible oil seeds and algal biomass*. Elsevier Ltd. https://dx.doi.org/10.1016/B978-0-08-102728-8.00007-3

Burra, K. G., Hussein, M. S., Amano, R. S., & Gupta, A. K. (2016). Syngas evolutionary behavior during chicken manure pyrolysis and air gasification. *Applied Energy*, *181*, 408–415. https://doi.org/10.1016/j.apenergy.2016.08.095

Carmona-Cabello, M., García, I. L., Papadaki, A., Tsouko, E., Koutinas, A., & Dorado, M. P. (2021). Biodiesel production using microbial lipids derived from food waste discarded by catering services. *Bioresource Technology, 323*(December 2020), 124597. https://doi.org/10.1016/j.biortech.2020.124597

Chen, N., Xie, Y., Liang, Z., & Shim, H. (2023). Biodiesel production and properties estimation from food waste and domestic wastewater by Rhodosporidium toruloides. *Journal of Environmental Management, 348*(June), 119368. https://doi.org/10.1016/j.jenvman.2023.119368

Dahiya, S., Kumar, A. N., Shanthi Sravan, J., Chatterjee, S., Sarkar, O., & Mohan, S. V. (2018). Food waste biorefinery: Sustainable strategy for circular bioeconomy. *Bioresource Technology*, *248*(August 2017), 2–12. https://doi.org/10.1016/j.biortech.2017.07.176

Dhiman, S., & Mukherjee, G. (2021). Present scenario and future scope of food waste to biofuel production. *Journal of Food Process Engineering*, *44*(2). https://doi.org/10.1111/jfpe.13594

Dhiman, S., & Mukherjee, G. (2022). Utilization of food waste for biofuel production: a biorefinin perspective. *Materials Today: Proceedings*, *xxxx*. https://doi.org/10.1016/j.matpr.2022.12.009

Di Maria, F., Sisani, F., & Contini, S. (2018). Are EU waste-to-energy technologies effective for exploiting the energy in bio-waste? *Applied Energy*, *230*(September), 1557–1572. https://doi.org/10.1016/j.apenergy.2018.09.007

Ejikeme et al. (2010). Catalysis in biodiesel production by transesterification processes-an insight | Ikechukwu Anyaogu - Academia.Edu. *E-Journal of Chemistry*, *7*(4), 1120–1132. https://www.academia.edu/796041/Catalysis_in_Biodiesel_Production_by_Transesterification_Processes-An_Insight

FAO. (2022). *Food loss and waste database*. Food and Agriculture Organization of the United Nations. https://www.fao.org/platform-food-loss-waste/flw-data/en/

Farouk, S. M., Tayeb, A. M., Abdel-Hamid, S. M. S., & Osman, R. M. (2024). Recent advances in transesterification for sustainable biodiesel production, challenges, and prospects: a comprehensive review. *Environmental Science and Pollution Research*, *31*(9), 12722–12747. https://doi.org/10.1007/s11356-024-32027-4

Forbes, H., Quested, T., & O'Connor, C. (2021). United Nations Environment Programme (2021). Food Waste Index Report 2021. Nairobi.

Gadhe, A., Sonawane, S. S., & Varma, M. N. (2014). Ultrasonic pretreatment for an enhancement o biohydrogen production from complex food waste. *International Journal of Hydrogen Energy*, *39*(15), 7721–7729. https://doi.org/10.1016/j.ijhydene.2014.03.105

Gasafi, E., Meyer, L., & Schebek, L. (2003). Using life-cycle assessment in process design: supercritical water gasification of organic feedstocks. *Journal of Industrial Ecology, 7*(3–4), 75–91. https://doi.org/10.1162/108819803323059415

Hafid, H. S., Rahman, N. A. A., Shah, U. K. M., Baharuddin, A. S., & Ariff, A. B. (2017). Feasibility of using kitchen waste as future substrate for bioethanol production: a review. *Renewable and Sustainable Energy Reviews, 74*(January), 671–686. https://doi.org/10.1016/j.rser.2017.02.071

Interlenghi, S. F., de Medeiros, J. L., & de Queiroz Fernandes Araújo, O. (2023). Chapter 4: Sustainability of methylic and ethylic biodiesel production routes. In *Fatty Acid Methyl Ester and Fatty Acid Ethyl Ester Inventories and Sustainability Indicators, 1st Edition.* Woodhead Publishing

Jain, S. (2023). Biodiesel production from food waste using insitu transesterification method. *Sustainable Energy Technologies and Assessments, 58*(November 2022), 103380. https://doi.org/10.1016/j.seta.2023.103380

Kannah, R. Y., Sivashanmugham, P., Kavitha, S., & Banu, J. R. (2020). Valorization of food waste for bioethanol and biobutanol production. In *Food waste to valuable resources: applications and management* (Vol. 1). INC. https://doi.org/10.1016/B978-0-12-818353-3.00003-1

Karmee, S. K., & Lin, C. S. K. (2014). Valorisation of food waste to biofuel: current trends and technological challenges. *Theoretical Chemistry Accounts, 2*(1), 4–7. https://doi.org/10.1186/s40508-014-0022-1

Kazemi Shariat Panahi, H., Dehhaghi, M., Guillemin, G. J., Gupta, V. K., Lam, S. S., Aghbashlo, M., & Tabatabaei, M. (2022). Bioethanol production from food wastes rich in carbohydrates. *Current Opinion in Food Science, 43*, 71–81. https://doi.org/10.1016/j.cofs.2021.11.001

Kruse, A. (2009). Hydrothermal biomass gasification. *Journal of Supercritical Fluids, 47*(3), 391–399. https://doi.org/10.1016/j.supflu.2008.10.009

Liu, D., Ma, X., Huang, J., Shu, Z., Chu, X., Li, Y., & Jin, Y. (2022). Investigation of the aerobic biochemical treatment of food waste: a case study in Zhejiang and Jiangsu provinces in China. *Science of the Total Environment, 806*, 150414. https://doi.org/10.1016/j.scitotenv.2021.150414

Matsakas, L., Kekos, D., Loizidou, M., & Christakopoulos, P. (2014). Utilization of household food waste for the production of ethanol at high dry material content. *Biotechnology for Biofuels, 7*(1), 1–9. https://doi.org/10.1186/1754-6834-7-4

Mayer, F., Bhandari, R., & Gäth, S. (2019). Critical review on life cycle assessment of conventional and innovative waste-to-energy technologies. *Science of the Total Environment, 672*, 708–721. https://doi.org/10.1016/j.scitotenv.2019.03.449

Mohiddin, M. N. Bin, Tan, Y. H., Seow, Y. X., Kansedo, J., Mubarak, N. M., Abdullah, M. O., Chan, Y. S., & Khalid, M. (2021). Evaluation on feedstock, technologies, catalyst and reactor for sustainable biodiesel production: a review. *Journal of Industrial and Engineering Chemistry, 98*, 60–81. https://doi.org/10.1016/j.jiec.2021.03.036

Mumtaz, M. W. et al. (2017). Clean energy for sustainable development: comparisons and contrasts of new approaches. In: *Biodiesel production through chemical and biochemical transesterification: trends, technicalities, and future perspectives.* Elsevier Inc. https://dx.doi.org/10.1016/B978-0-12-805423-9.00015-6

Murugesan, P., Raja, V., Dutta, S., Moses, J. A., & Anandharamakrishnan, C. (2022). Food waste valorisation via gasification – a review on emerging concepts, prospects and challenges. *Science of the Total Environment, 851*(July), 157955. https://doi.org/10.1016/j.scitotenv.2022.157955

Nanda, S., Isen, J., Dalai, A. K., & Kozinski, J. A. (2016). Gasification of fruit wastes and agro-food residues in supercritical water. *Energy Conversion and Management, 110*, 296–306. https://doi.org/10.1016/j.enconman.2015.11.060

Nigam, P. S., & Singh, A. (2011). Production of liquid biofuels from renewable resources. *Progress in Energy and Combustion Science, 37*(1), 52–68. https://doi.org/10.1016/j.pecs.2010.01.003

Offei, F., Mensah, M., Thygesen, A., & Kemausuor, F. (2018). Seaweed bioethanol production: a process selection review on hydrolysis and fermentation. *Fermentation, 4*(4), 1–18 https://doi.org/10.3390/fermentation4040099

Peinemann, J. C., Rhee, C., Shin, S. G., & Pleissner, D. (2020). Non-sterile fermentation of food waste with indigenous consortium and yeast – effects on microbial community and product spectrum. *Bioresource Technology, 306*(March), 123175. https://doi.org/10.1016/j.biortech.2020.123175

Pham, T. P. T., Kaushik, R., Parshetti, G. K., Mahmood, R., & Balasubramanian, R. (2015). Food waste-to-energy conversion technologies: current status and future directions. *Waste Management, 38*(1), 399–408. https://doi.org/10.1016/j.wasman.2014.12.004

Prasoulas, G., Gentikis, A., Konti, A., Kalantzi, S., Kekos, D., & Mamma, D. (2020). Bioethano production from food waste applying the multienzyme plasma at laboratory and bench-scale level and its application as a starter culture in a meat product. *Fermentation, 6*(2). https://doi.org/10.3390/FERMENTATION6020039

Quader, M. A., & Ahmed, S. 2017. Clean energy for sustainable development: comparisons and contrasts of new approaches. *Bioenergy with carbon capture and storage (BECCS): future prospects of carbon-negative technologies*. Elsevier Inc. https://dx.doi.org/10.1016/B978-0-12-805423-9.00004-1

Rodionova, M. V., Poudyal, R. S., Tiwari, I., Voloshin, R. A., Zharmukhamedov, S. K., Nam, H. G., Zayadan, B. K., Bruce, B. D., Hou, H. J. M., & Allakhverdiev, S. I. (2017). Biofuel production: challenges and opportunities. *International Journal of Hydrogen Energy, 42*(12), 8450–8461. https://doi.org/10.1016/j.ijhydene.2016.11.125

Saha, B. C., Nichols, N. N., Qureshi, N., & Cotta, M. A. (2011). Comparison of separate hydrolysis and fermentation and simultaneous saccharification and fermentation processes for ethanol production from wheat straw by recombinant *Escherichia coli* strain FBR5. *Applied Microbiology and Biotechnology, 92*(4), 865–874. https://doi.org/10.1007/s00253-011-3600-0

Sasaki, C., Kushiki, Y., Asada, C., & Nakamura, Y. (2014). Acetone-butanol-ethanol production by separate hydrolysis and fermentation (SHF) and simultaneous saccharification and fermentation (SSF) methods using acorns and wood chips of Quercus acutissima as a carbon source. *Industrial Crops and Products, 62*, 286–292. https://doi.org/10.1016/j.indcrop.2014.08.049

Sekoai, P. T., & Gueguim Kana, E. B. (2013). A two-stage modelling and optimization of biohydrogen production from a mixture of agro-municipal waste. *International Journal of Hydrogen Energy, 38*(21), 8657–8663. https://doi.org/10.1016/j.ijhydene.2013.04.130

Show, K. Y., Lee, D. J., & Zhang, Z. P. (2011). Production of biohydrogen: current perspectives and future prospects. Biofuels*:* Alternative Feedstocks and Conversion Processes, 467–479. https://doi.org/10.1016/B978-0-12-385099-7.00021-8

Simonyan, K. J., & Fasina, O. (2013). Biomass resources and bioenergy potentials in Nigeria. *African Journal of Agricultural Research, 8*(40), 4975–4989. DOI: 10.5897/AJAR2013.6726

Singh, A., Singhania, R. R., Soam, S., Chen, C. W., Haldar, D., Varjani, S., Chang, J. S., Dong, C. Di, & Patel, A. K. (2022). Production of bioethanol from food waste: status and perspectives. *Bioresource Technology, 360*(July), 127651. https://doi.org/10.1016/j.biortech.2022.127651

Singh, D., & Yadav, S. (2021). Steam gasification with torrefaction as pretreatment to enhance syngas production from mixed food waste. *Journal of Environmental Chemical Engineering, 9*(1), 104722. https://doi.org/10.1016/j.jece.2020.104722

Singh, D., Yadav, S., Bharadwaj, N., & Verma, R. (2020). Low temperature steam gasification to produce hydrogen rich gas from kitchen food waste: influence of steam flow rate and temperature. *International Journal of Hydrogen Energy*, *45*(41), 20843–20850 https://doi.org/10.1016/j.ijhydene.2020.05.168

Singhania, R. R., Patel, A. K., Singh, A., Haldar, D., Soam, S., Chen, C. W., Tsai, M. L., & Dong, C. Di. (2022). Consolidated bioprocessing of lignocellulosic biomass: technological advances and challenges. *Bioresource Technology, 354*(April), 127153. https://doi.org/10.1016/j.biortech.2022.127153

Sridhar, A., Kapoor, A., Senthil Kumar, P., Ponnuchamy, M., Balasubramanian, S., & Prabhakar, S (2021). Conversion of food waste to energy: a focus on sustainability and life cycle assessment. *Fuel*, *302*(May), 121069. https://doi.org/10.1016/j.fuel.2021.121069

Tchuifon, D. R. T. et al. (2020). Production of biodiesel by transesterification reaction of waste cooking oil. *Chemical Science International Journal 29*(July): 1–12.

Tumuluru, J. S., Ghiasi, B., Soelberg, N. R., & Sokhansanj, S. (2021). Biomass torrefaction process, product properties, reactor types, and moving bed reactor design concepts. *Frontiers in Energy Research*, *9*(September), 1–20. https://doi.org/10.3389/fenrg.2021.728140

Ulukardesler, A. H. (2023). Biodiesel production from waste cooking oil using different types of catalysts. *Processes, 11*(7). https://doi.org/10.3390/pr11072035

United Nations Environment Programme (UNEP) (2021).*Food Waste Index Report 2021*, Nairobi.

Wang, C., Wang, Z., Wang, X., Li, N., Tao, J., Zheng, W., Yan, B., Cui, X., Cheng, Z., & Chen, G. (2022). A review on the hydrothermal treatment of food waste: processing and applications. *Processes, 10*(11). https://doi.org/10.3390/pr10112439

Win, S. S., & Trabold, T. A. (2018). Sustainable food waste-to-energy systems. In *Sustainable waste-to-energy technologies: transesterification*. Elsevier Inc. https://dx.doi.org/10.1016/B978-0-12-811157-4.00006-1

Zhang, S. C., Lai, Q. H., Lu, Y., Liu, Z. D., Wang, T. M., Zhang, C., & Xing, X. H. (2016). Enhanced biohydrogen production from corn stover by the combination of Clostridium cellulolyticum and hydrogen fermentation bacteria. *Journal of Bioscience and Bioengineering*, *122*(4), 482–487. https://doi.org/10.1016/j.jbiosc.2016.03.014

8 Maximising biogas production from different food waste categories through anaerobic digestion

Collen Yang, Ryan Zhe Hse Soh, and Yi Jing Chan

8.1 INTRODUCTION

Food waste (FW) constitutes the largest portion (44%) of the total waste composition, whose sources include food production, processing, and consumption from catering services and households (Negri et al. 2020). According to The World Bank Group (2024), FW is roughly 30% of all food globally, amounting to 1.3 billion tonnes per year. Proper management of FW is crucial due to its potentially severe impact on the environment, including the release of harmful gases into the atmosphere, the leaching of pathogenic bacteria, and the spread of foul odours (Liang et al. 2022). Landfills are commonly used to treat municipal solid waste, including FW (Chakraborty et al. 2018). However, landfill treatment leads to secondary pollution and requires significant land area, making it an unsustainable waste management method. The increasing legal and regulatory restrictions, rising land costs, and challenges associated with urbanisation make FW management even more difficult (Vergara and Tchobanoglous 2012).

For a significant period, fossil fuels such as coal, oil, and natural gas have been the primary global energy sources. As of 2023, approximately 84% of the world's energy consumption relies on fossil fuels (European Investment Bank 2023). However, fossil fuels are currently depleting rapidly, and their combustion leads to the release of pollutants, which have detrimental effects on both the environment and human health (Soares et al. 2020). To address these issues, researchers are extensively exploring renewable energy sources like solar, wind, geothermal, hydro-energy, and bioenergy as viable alternatives to fossil fuels (Nazir et al. 2019). The International Energy Agency emphasises that global investments in clean energy must triple by 2030 and reach over $4 trillion annually to achieve carbon neutrality by the mid-century (European Investment Bank 2023). One such renewable energy resource is bioenergy, which is environmentally friendly and sustainable. It primarily involves harnessing hydrogen (H_2) and methane (CH_4) from perishable organic waste materials. These gases can provide heat and power, therefore such facilities that process these gases for that purpose are known as combined heat and power (CHP) facilities.

DOI: 10.1201/9781032718507-8

Anaerobic digestion (AD) is a promising technology for extracting bioenergy from household waste. This biological process occurs in the absence of oxygen and involves the decomposition and stabilisation of organic matter. AD offers several benefits, including minimising environmental footprint, reduction of greenhouse gas emissions, and a net negative impact on global warming potential. Furthermore, it generates valuable products such as electricity, heat, and biomethane by converting biodegradable materials into biogas. As a result, AD can be considered a sustainable waste management technology. FW holds great potential for biogas production due to its high biochemical composition. The composition of FW, including its macromolecular structure, total solids, and organic matter contents, as well as the presence of inorganic and organic compounds, makes it highly suitable for conversion into biogas through AD. The identification of optimal parameters and pretreatment methods are the main methods in enhancing biogas production through AD. Research efforts in the past decade normally constitute different FW mixtures and different inoculum types, i.e. different bacteria species/mixtures to digest the FW anaerobically (see Section 8.4).

As indicated by the most recent global and company data in 2024 presented in Tables 8.1 to 8.3, AD technology has made significant progress and is now widely adopted commercially. This technology effectively converts large quantities of FW into valuable new or green products, such as biofuels, fertilisers, animal feed, and biogas. Table 8.1 provides an overview of the processes and product descriptions used by the top 20 global AD solution providers. The impacts of AD are measured using key parameters such as waste processing capacity, greenhouse gas (GHG) reductions, landfill diversion rates, and the production of thermal energy, electrical power, and biomethane. These metrics, sourced from company websites and other reliable sources are summarised in Table 8.2. Some companies operate multiple plants locally or globally. Table 8.3 highlights successful projects for each company, specifying the types of FW processed, facility capacities, and power outputs measured in MW, MMBtu/year, or Nm^3/h CH_4.

8.2 ANAEROBIC DIGESTION PROCESS

Addressing the treatment of FW is a critical global challenge that demands an effective solution. Proper monitoring and management of FW are essential due to their harmful effects on soil pollution and environmental disruption. One highly efficient method for treating FW is AD, which involves the breakdown of organic matter by bacteria in the absence of oxygen. This process utilises microbial communities to digest the waste, resulting in the generation of renewable energy in the form of biogas and a residual product known as digestate. These byproducts can be utilised for various purposes, as depicted in Figure 8.1, showcasing the range of digestible materials and the potential applications of the resulting products.

Biogas consists mainly of CH_4 at a composition of 50–70 vol%, along with carbon dioxide (CO_2), hydrogen sulphide (H_2S), water vapour (H_2O), and small amounts of other gases. Similar to natural gas, biogas can be utilised for heating, electricity generation, and cooling systems. By removing low-value components like CO_2, H_2S, and H_2O, renewable natural gas can be obtained. This purified gas can be compressed and used as a fuel for vehicles or as an energy source for various applications. It can also serve as a feedstock to produce advanced biochemicals including plastics.

TABLE 8.1
Process/product descriptions of the top 20 anaerobic digestion solution companies in the world

No	Company	Location	Brief process/product description	References
1	Anaergia	Burlington, Ontario, Canada	1. Organics extraction and digestion OREX™ technology - Negates the need for waste separation 2. Omnivorous AD - 60% less CAPEX than conventional AD technology 3. Biogas upgrading system to create pipeline-quality RNG 4. Water and residue treatment via PyroSys™ technology	(Anaergia 2024c)
2	Biogest	Vienna, Austria	PowerRing™ modular tank-in-tank/ring-in-ring digestor - Consists of primary and secondary digestors, complete with agitation and heating systems	(Biogest 2024a)
3	BTS Biogas/ Bioenergy Devco	Affi, Trentino-Alto Adige, Italy	1. BIOaccelerators™ and BIOseparators™ to optimise decomposition, size reduction and material separation of organics 2. Fermentation of pretreated organics. Process is monitored by a Plug & Produce™ modular technical centre. 3. Post-treatment of slurry via drying and pelleting into fertiliser	(BTS Biogas 2024b)
4	Severn Trent Green Power Limited	Coventry, West Midlands, United Kingdom	1. Collected food waste is separated from packaging and macerated to <12 mm in particle size 2. Macerated food waste is digested at 37 °C for 80–90 days 3. Biogas is upgraded/combusted in a CHP engine to provide energy 4. Slurry is pasteurised at 70 °C and output as digestate/soil amendment	(Severn Trent 2024a)
5	Agrivert	Gagingwell, Oxfordshire, United Kingdom	Conventional process (depackaging, size reduction, digestion, biogas combustion in CHP engine and digestate output as fertiliser)	(Agrivert 2024a, 2024b)
6	ReFood UK	Doncaster, South Yorkshire, United Kingdom	1. Food waste is separated from packaging in hoppers and added with liquid to facilitate pumping 2. Separated food waste is pasteurised at 70 °C, cooled to 40 °C and held in a mixing tank for 3 days. 3. Processed food waste is anaerobically digested in digester tanks for 30–35 days. 4. Collected biogas is scrubbed to remove hydrogen sulphide (H_2S). 5. Clean biogas is used to fuel a CHP engine.	(ReFood 2024b)
7	GPS Renewables/ proweps envirotec GmBH	Bangalore, Karnataka, India	1. Food waste is pretreated and hydrolysed before undergoing AD. 2. Biogas is upgraded. Biomethane is compressed and stored in cascades. 3. Digestate is composted and wastewater is treated.	(proweps envirotec 2024)

(Continued)

TABLE 8.1 (Continued)

No	Company	Location	Brief process/product description	References
8	Green Create	Amsterdam, Noord-Holland, Netherlands	1. Industry organic waste is screened and blended before being fed into AD tanks. 2. Biogas is formed by microbial communities breaking down food waste without oxygen. 3. Biogas is upgraded to biomethane in a biogas upgrading plant to remove CO_2, H_2S and moisture. 4. Biomethane is sent to the natural gas grid to supply electricity for common use.	(Green Create 2024a)
9	PlanET Biogastechnik GmbH	Gescher, North Rhine-Westphalia, Germany	1. Food industry waste is pretreated conventionally (unpacking and shredding). 2. Food waste is subject to two-stage fermentation: the primary digesters are three 825 m^3 tanks mixed with "paddle giants" and the secondary fermenter is a 3,070 m^3 tank mixed using a submersible agitator. 3. Fermentation residue is stored in two 3,435 m^3 tanks mixed using submersible motor agitators. 4. A CHP Otto engine generates power by combusting product biogas. 5. Biogas is also upgraded to biomethane by membrane processes or PSA to be used in power grids.	(Plan ET 2024a, 2024d)
10	Group Op de Beeck	Olen, Antwerp, Belgium	Conventional process (pretreatment, digestion, biogas combustion in CHP engine and digestate output as fertiliser)	(Group Op der Beeck 2024a)
11	Olleco	Northampton, Northamptonshire, United Kingdom	Conventional process (pretreatment, digestion, biogas combustion in CHP engine and digestate output as fertiliser)	(Olleco 2024a)
12	Prodeval	Chateauneuf-Sur-Isère, France	Prodeval specialises not in the AD process itself, but in biogas upgrading via: 1. Chilling raw biogas to 5 °C in a VALOGAZ™ unit to separate liquid fraction. 2. Chilled biogas is filtered in a VALOPACK™ unit to remove H_2S, siloxanes and volatile organic compounds (VOCs). 3. Three-stage membrane separation to upgrade biogas to biomethane in a VALOPUR™ unit with 99.5% efficiency. 4. Odourisation of biomethane with tetrahydrothiophene in a VALODOR™ unit.	(Prodeval 2024a)

TABLE 8.1 (Continued)

No	Company	Location	Brief process/product description	References
13	Asja	Rivoli, Piedmont, Italy	Conventional process (collection, digestion, biogas upgraded to biomethane and sent to grid, and digestate sent for composting)	(Asja 2020)
14	Keenan Recycling	Bridge of Don, Aberdeen, Scotland	1. Collected waste is depackaged and sorted by turbo separator technology and water injection to produce optimal AD feedstock with 20% dry matter content. 2. Depending on the plant, pretreated feedstock is digested in AD tanks for about 40 days. 3. Biogas is collected for CHP and digestate for fertiliser.	(Keenan Recycling 2024a)
15	ESI	Sisters, Oregon, US	ESI's speciality is in AD biogas *utilisation.* AD biogas is upgraded using a Paques Thiopaq Gas Pretreatment system, a Wärtsilä Puregas CA80 Gas Upgrading system, and a ANGI Compression system.	(ESI 2020)
16	Vanguard Renewables	Weston, Massachusetts, United States	Conventional process (depackaging, pretreatment, digestion, biogas directly used/upgraded to biomethane for use in CHP engine and digestate output as fertiliser)	(Vanguard Renewables 2024c)
17	Weltec Biopower	Vechta, Lower Saxony, Germany	Conventional process (depackaging, pretreatment, digestion, biogas directly used/upgraded to biomethane for use in CHP engine and digestate output as fertiliser) - Weltec's MULTIMix™ technology ensures optimal shredding and mixing of material.	(Weltec Biopower 2024b)
18	Brocklesby	North Cave, England	Conventional process (depackaging, pretreatment, digestion, biogas directly used/upgraded to biomethane for use in CHP engine and digestate output as fertiliser) - The facility uses Permastore glass-fused-to-steel tanks with bolted containment technology and proprietary glass coating for AD tanks requiring little to no maintenance.	(Permastore 2024)
19	GENeco	Bristol, Bournemouth, United Kingdom	Conventional process (depackaging, pretreatment, digestion, biogas directly used/upgraded to biomethane for use in CHP engine and digestate output as fertiliser) - The produced biomethane is also used to power Bio-Buses, buses that are powered solely by waste-derived gases.	(GENeco 2024a)

TABLE 8.2
Sustainability figures for the top 20 anaerobic digestion solution companies in the world

No	Company	Waste processing capacity	GHG reductions	Landfill diversion	Power generation		Biomethane production		References
		(kton/year)	(ton CO_{2e}/year)	(ton/year)	(MW)	(equivalent homes)	(MMBtu/year)	(Nm³/h CH_4)	
1	Anaergia		>46,0000	~8,000,000	134	100,000	5,000,000		(Anaergia 2024a)
2	Biogest				122[1]		12,031,065[1]	10,925[1]	(Biogest 2024b)
3	BTS Biogas/ Bioenergy Devco								
4	Severn Trent Green Power Limited	500			31.8	77,000			(Severn Trent 2024b)
5	Agrivert								
6	ReFood UK		93,822	203,963	11.9	24,000			(ReFood 2024c)
7	GPS Renewables		493,000	120,000				9,100,000 m³	(GPS Renewables 2024)
8	Green Create								
9	PlanET Biogastechnik GmbH		~1,450,000						(Plan ET 2024c)
10	Group Op de Beeck	350			8.5	23,000			(Group Op der Beeck 2024b)
11	Olleco		96%[2]			12,000			(Olleco 2024a)
12	Prodeval		880,000						(Prodeval 2024b)
13	Asja		1,500,000		68.5				(Asja 2024b)

TABLE 8.2 (Continued)

No	Company	Waste processing capacity	GHG reductions	Landfill diversion	Power generation	Biomethane production	References
14	Keenan Recycling	39.55	27,328		7,220		(Keenan Recycling 2024b)
15	ESI						
16	Vanguard Renewables	76.03 + 44,620 (food waste + manure)	3,167,542.5			81,675	(Vanguard Renewables 2024b)
17	Weltec Biopower		600,000			18,835 (biogas)	(Weltec Biopower 2024c)
18	Brocklesby	60		3.5	10,500		
19	GENeco	35			8,300	1,941[3]	(GENeco 2024b)

Notes

1. Data cleaned up and calculated from https://www.biogest.at/products/references.
2. AD-based power emits 96% less CO_2 relative to conventional electricity sources in the UK.
3. Also 17 million m^3 of biomethane per year.

TABLE 8.3
Success cases of the top 20 anaerobic digestion solution companies in the world

No	Company	Samples of successful project								References
		Facility			Input		Output			
		Plant name	Location	Startup year	(kton/y)	Content	(MW)	(MMBtu/ year)	(Nm³ /h CH_4)	
1	Anaergia	Enerbio	Italy	2012	-	- Whey - Cattle manure - Maize Silage	~1			(Anaergia 2024b)
2	Biogest	Frogmary Biomethane Plant	Frogmary, UK	2016	70	- cattle slurry - farmyard manure - poultry manure - beet - maize silage - grass silage - whey	7.6	19,9838	550	(Biogest 2024c)
3	BTS Biogas/ Bioenergy Devco	Maryland Bioenergy Center	Jessup, Maryland, US	2021	110	- fruit /vegetable processing waste - meat - baked goods, - oil and fats - food industry wastewater	10.4 16.6 kton of soil amendment	312,000		(Bioenergy Devco 2024)

TABLE 8.3 (Continued)

No	Company	Samples of successful project								References
		Facility			Input		Output			
		Plant name	Location	Startup year	(kton/y)	Content	(MW)	(MMBtu/ year)	(Nm³ /h CH_4)	
4	Severn Trent Green Power Limited	Severn Trent Green Power Cassington AD Facility	Worton, Wiltshire, UK		50	- food and beverage production waste - restaurant/bakery / domestic food waste - fats/oils /liquid waste		2.1		(Severn Trent 2024c)
5	Agrivert	Agrivert North London AD Facility	Colney, UK	2016	64+33 (solid + liquid)	- source segregated food waste from commercial production and local authority		3		(Agrivert 2024a, 2024b)
6	ReFood UK	ReFood Doncaster Plant	Doncaster, South Yorkshire, UK	2011	160	- packaged/unpackaged food/ drink waste - food production organic residue - bones, fat, meat - fruits and vegetables			1,598	(ReFood 2024a)
7	GPS Renewables	BioCNG plant	Indore, India	2021	220	Source separated municipal organic waste		17 tonnes/day (6.205 ktonne/year) of compressed biogas		(GPS Renewables 2024; proweps envirotec 2024)

(Continued)

TABLE 8.3 (Continued)

No	Company	Samples of successful project								References
		Facility			Input		Output			
		Plant name	**Location**	**Startup year**	**(kton/y)**	**Content**	**(MW)**	**(MMBtu/ year)**	**(Nm^3 /h CH_4)**	
8	Green Create	Wijster plant	Wijster, Netherlands	2022	136.5	Poultry manure from broiler and egg-laying farms	33.1[1]			(Green Create 2024b)
9	PlanET Biogastechnik GmbH	Groot Zevert Vergisting BV	Groot Zevert, Netherlands	N/A (extended in 2016)	-	Coferments, manure, glycerin, food residues	1.2			(Plan ET 2024b)
10	Group Op de Beeck	Kallo Site	Kallo, Belgium	–	350	Packed & unpacked organic waste	7.5 MWh			(Group Op der Beeck 2024c)
11	Olleco	Westcott Park	Buckinghamshire, UK	2018	96	General food waste	8[2]			(Olleco 2024b)
12	Prodeval	Project: Barczewo	Barczewo, Poland	2023	–	Biowaste, agricultural waste			220 (biogas)	(Prodeval 2023)
13	Asja	Tuscania Plant	Tuscania, Viterbo, Italy	2019	60	Organic fraction of municipal solid waste (OFMSW)	0.74			(Asja 2024a)
14	Keenan Recycling	NIL (The company specialises in preparing feedstock rather than carrying out AD)								(Keenan Recycling 2024a)
15	ESI	Threemile Canyon Farms	Boardman, Oregon, US	2012	3,100 scfm biogas input		4.8	985,500		(ESI 2020)

TABLE 8.3 (Continued)

No	Company	Samples of successful project								References
		Facility			Input		Output			
		Plant name	Location	Startup year	(kton/y)	Content	(MW)	(MMBtu/ year)	(Nm³ /h CH_4)	
16	Vanguard Renewables	Goodrich Family Farm	Salisbury, Vermont, US	–	65.7 + 36.5	Food waste + manure			453[3]	(Vanguard Renewables 2024a)
17	Weltec Biopower	Arneburg Biomethane Refinery	Arneburg, Germany	2013	70	- Maize - Cereal silage - Cereals			700	(Weltec Biopower 2024a)
18	GENeco	Avonmouth facility / Bristol sewage treatment works	Avonmouth, Bristol, UK		35	Inedible food waste			1,941	(GENeco 2024b)

Notes:

1. Biogas produces 3.0 MW of electricity and 3.1 MW of thermal energy, while 27.0 MW of biomethane is sent to local gas grids.
2. 3.2 MW of electricity and 4.8 MW of heat.
3. or 140,000 Mcf/year.

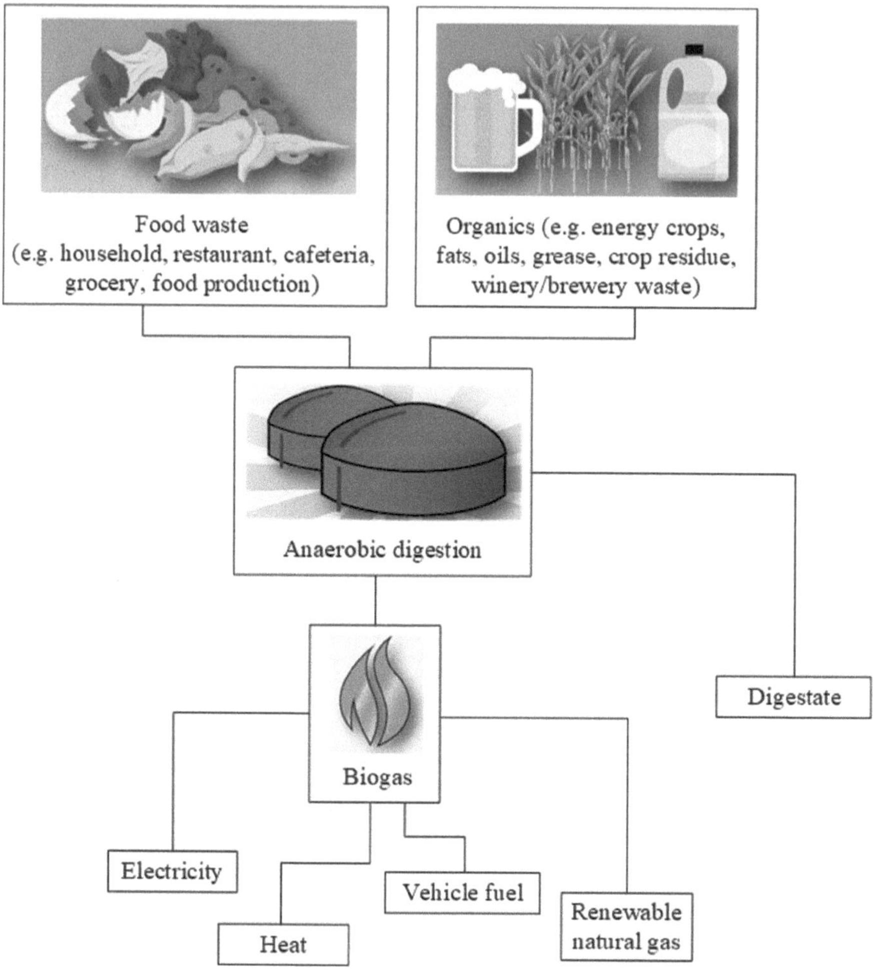

FIGURE 8.1 Flow of feedstocks through the anaerobic digestion to produce biogas and digestate (EPA 2023).

Digestate refers to the remaining substance following the AD process, comprising both liquid and solid components. These components are typically separated and managed separately due to their distinct post-processing potential. Each portion holds a value that can be maximised through appropriate treatment. The liquid and solid portions of digestate offer various advantageous applications, including but not limited to serving as animal bedding, nutrient-dense fertiliser, material for bioplastics, organic-rich compost, and soil amendment. Utilising digestate products can yield financial gains or cost reductions, contributing to the overall economic and environmental benefits of a biogas project. While digestate may be a type of useful byproduct, it will not be the focus of this chapter.

8.2.1 Basic Principles of Anaerobic Digestion

The process of AD unfolds through four consecutive stages, namely hydrolysis, acidogenesis, acetogenesis, and methanogenesis. In the case of single-stage batch reactors, all waste materials are loaded simultaneously into the reactor, and all four processes occur sequentially within the same reactor (Meegoda et al. 2018). These stages necessitate certain conditions to ensure the proliferation of AD bacteria and the continuation of the AD process.

Hydrolysis is carried out by specialised hydrolytic bacteria release enzymes to break down carbohydrates, lipids, and proteins into sugars, long-chain fatty acids (LCFAs), and amino acids, respectively (Li, Park, and Zhu 2011). Enzymes are sometimes added to enhance the hydrolysis of carbohydrates such as lignin, cellulose, and hemicellulose that are challenging to degrade and remain inaccessible to microbes due to their complex structures (Lin et al. 2010). Typically, hydrolysis works optimally at temperatures between 30–50 °C and pH levels of 5–7 (Azman 2016).

Acidogenesis is the process of hydrolysis product metabolism by acidogenic bacteria, which convert simpler compounds into volatile fatty acids (VFAs) (acetates, propionates, and butyrates) and other substances such as ethanol and lactate (Van Lier, Mahmoud, and Zeeman 2008). The specific concentrations of these intermediates can vary depending on the digester conditions, especially in digesters operating at different pH levels (Wu, Yao, and Zhu 2010). This process results in a mixture of VFAs, along with some CO_2 and H_2. Excessive acidification of VFAs can lead to digester failure, making it a critical consideration in AD processes (Akuzawa et al. 2011).

Acetogenesis is the process of converting higher VFAs into acetic acid (CH_3COOH), concurrently generating H_2. This conversion is controlled to a large extent by the partial pressure of H_2 in the mixture since excessive H_2 partial pressure could be harmful to the acetogenic microorganisms (Dinopoulou, Rudd, and Lester 1988). Fortunately, H_2 can be swiftly consumed by the hydrogenotrophic methanogens, ensuring that the partial pressures remain favourable for acetogenesis through an exergonic reaction (Stams and Plugge 2009).

Methanogenesis is the production of methane by specialised microorganisms called methanogens, which are accessible intermediates such as acetic acid and other VFAs produced in earlier stages, producing CH_4 (Ferry 2010). Methanogens are highly sensitive to the presence of oxygen (O_2), as evidenced by studies showing that exposure to O_2 leads to the death of 99% of *Methanococcus voltae* and *Methanococcus vannielli* cells within ten hours (Kiener and Leisinger 1983). Approximately two-thirds of CH_4, arises from acetoclastic methanogenesis (acetate to CH_4), while the remaining one-third comes from hydrogenotrophic methanogenesis (Belay, Sparling, and Daniels 1986). Additionally, methanogens require specific environmental conditions, including a higher pH range (6.5–8) compared to earlier stages of AD. To assess the extent of digestion in sludge, factors like its volatile solids (VS) content and dewatering capability are commonly evaluated (Clarke 1999).

8.3 KEY FACTORS INFLUENCING ANAEROBIC DIGESTION PERFORMANCE

Various crucial parameters affect the rates of the different steps of the digestion process such as temperature, pH, retention time, substrate characteristics and reactor types.

8.3.1 Temperature

Temperature variations significantly impact the growth rate of microorganisms and the degradation of organic matter during AD, leading to varying CH_4 production levels (Shang et al. 2023). Digesters are designed to operate within two main temperature ranges: mesophilic (around 35 °C) and thermophilic (around 55 °C) (Meegoda et al. 2018). Each temperature range supports different populations of anaerobic microbes.

Mesophilic digestion occurs at lower temperatures, resulting in slower digestion and lower biogas yields, but it has the advantage of reduced heater energy costs compared to thermophilic digesters (Moset et al. 2015). On the other hand, thermophilic digestion operates at higher temperatures, which increases reaction rates, allowing for higher loading rates and greater biogas production (Hartmann and Ahring 2006). Additionally, thermophilic AD is suitable for achieving "Class A Biosolids", a designation for treated sewage sludge that meets U.S. EPA guidelines, allowing its safe use as fertiliser on farms and gardens.

Both temperature conditions are suitable for the AD of FW (Rajagopal, Massé, and Singh 2013). However, thermophilic methanogenic bacteria are sensitive to temperature changes and require optimal temperatures to handle higher organic loading rates (OLR). Excessive operating temperatures can lead to increased ammonia concentration, which inhibits the essential activity of acetic acid-type methanogenic bacteria, negatively affecting biogas production (Li, Chen, and Li 2010). Therefore, anaerobic co-digestion of FW typically shows higher microbial diversity at medium temperatures, enhancing stability and shock resistance throughout the AD process (Liu et al. 2017).

8.3.2 pH

The stability of the AD process is significantly influenced by pH levels. Extremely low or high pH values can hinder the process as microorganisms involved in the digestion of FW function optimally within different pH ranges (Panigrahi and Dubey 2019). Initially, hydrolytic acidifying bacteria produce organic acids, leading to a pH decrease in the digestive system (Shang et al. 2023). Methanogenesis, on the other hand, involves methanogenic bacteria consuming VFAs and CO_2 to produce CH_4 and increase pH (Paranjpe, Saxena, and Jain 2023). Fermentative bacteria are active across a broad pH range of 4.0–8.0, whereas methanogenic bacteria function between pH 6.5–7.5. Maintaining the pH within 6.8–7.2 is crucial for efficiently converting VFAs to CH_4 and CO_2 (Blasius et al. 2020). In the pre-digestion period, a slightly acidic environment enhances ammonia protonation and increases accumulation. However, as the pH increases, the accumulated NH_4^+ is reversibly converted to harmful NH_3 by methanogenic bacteria (Dai et al. 2017). The performance of AD is significantly dependent on the system's buffering capacity, which can be enhanced by the addition of buffering agents such as sodium bicarbonate ($NaHCO_3$) and potassium bicarbonate ($KHCO_3$).

8.3.3 Retention time

The solid retention time (SRT) and hydraulic retention time (HRT) play crucial roles in biological treatment processes. SRT refers to the average duration that biomass remains in the digester, while HRT represents the average time liquids stay in the digester (Chatterjee and Mazumder 2019; Meegoda et al. 2018). Normally, SRT exceeds HRT except in continuous stirred tank reactors. HRT (often symbolised as θ) can be determined by dividing the digester volume (V) by the digester's flow rate (Q) (Cooper 2014). A shorter HRT indicates a higher loading rate, which can result in VFA acidification and inhibitory effects (Kim et al. 2013). However, it also leads to improved process efficiency and reduced capital costs. Conversely, the digestion of lignocellulosic wastes necessitates longer HRTs (Shi et al. 2017). Process conditions, especially temperature and substrate type, also influence HRT. Thermophilic conditions have shorter retention times but offer superior AD performance compared to mesophilic conditions. In cost-benefit analyses of municipal wastes, digesters with low loading rates and longer HRT are found to yield the highest benefits (Hartmann and Ahring 2006).

8.3.4 Total solids (TS)

Total solids represent the dry matter content within sludge, regardless of its organic or inorganic composition, and can be measured as a percentage or concentration (Eaton et al. 2005). The process of measuring TS involves gradually drying the sludge sample at temperatures ranging from 103 to 105 °C until a consistent weight is achieved (U.S. Environmental Protection Agency 2001). In the realm of digester operations, TS content plays a pivotal role as it impacts the system's efficiency. Higher TS content in AD requires smaller digester sizes and reduces the energy needed for heating, as concentrated dry matter results in a more effective conversion of organic materials into biogas (Yi et al. 2014). Notably, recent research examining the effects of increasing TS content on AD of FW revealed that continuous high TS digesters yield more biogas than those with lower TS content, even when under the same retention times (Duan et al. 2012). This underscores the advantages of managing higher TS levels in AD processes for enhanced biogas production, waste treatment, and sustainable energy generation. By optimising TS content, we can promote greener and more efficient waste management strategies and energy production through AD.

8.3.5 Organic loading rate

The OLR is a crucial parameter in continuous digesters, representing the amount of organic matter introduced daily. It significantly impacts the stability of VFA production and overall AD. Maintaining an appropriate OLR is essential to provide an adequate carbon supply for the process (Wainaina et al. 2019). Higher OLR values are generally associated with increased treatment capacity. However, overloading the digester with organic matter can lead to detrimental effects. The rapid hydrolysis and acidification of waste at high OLRs result in the rapid formation of metabolic intermediates, particularly VFAs. These VFAs may accumulate excessively and surpass the rate at which they can be consumed, ultimately causing inhibitory effects on methanogenesis and inducing process instability (Franke-Whittle et al. 2014; Nagao et al. 2012). One challenge in the literature is inconsistent presentations of OLR units, such as kg COD/(m^3·day) or kg VS/(m^3·day), which hinders standardisation due to the lack of conversion factors. For instance, increasing the loading rate in olive mill waste digesters led to adverse effects on pH, chemical oxygen demand (COD) reduction, and biogas production rates (Rincon et al. 2007). Furthermore, high solids batch digesters typically have half of the loading rates compared to other single-stage reactors, resulting in a larger land footprint (Verma 2002).

8.3.6 Inoculum

An inoculum refers to a population of microorganisms or cells introduced into a fermentation medium or suitable environment (Sood et al. 2011). In the context of AD of FW, selecting an appropriate inoculum is crucial for efficient performance (Cremonez et al. 2021). The two stages of Thermophilic Sequential AD (TSAD) involve different dominant microorganisms, with acidogenesis typically carried out by *Thermoanaerobacterium sp.* and methanogenesis by *Methanoculleus sp. and Methanosarcina sp.* This variation is due to different operating parameters (Mamimin et al. 2015). Inoculum can be obtained in either pure cultures or mixed forms. Pure cultures tend to produce biogas at a higher rate compared to mixed forms (Mamimin et al. 2019). However, for engineering applications of AD systems, mixed cultures are more suitable due to several advantages including not requiring sterile operating conditions, reduced dependence on specific microorganism strains, increased substrate consumption range by microorganisms, and their availability at low costs. The main components of mixed culture are typically anaerobic sludge, organic compost, and bovine manure. It is important to note that the inoculum may contain inhibitors, such as hydrogen-consuming bacteria, which can limit hydrogen production during the initial stage of AD. As a result, pretreatment becomes necessary to eliminate these inhibitors (Kan 2013).

8.3.7 Reactor types used in industrial anaerobic digestion systems

8.3.7.1 Continuous stirred tank reactors

Bioreactors play a crucial role in influencing the growth and activity of microorganisms. Various bioreactors have been developed to enhance the AD process. Among them, the continuous stirred tank reactor (CSTR) is widely used due to its efficient substrate blending systems, ensuring high mixing levels (Gianico et al. 2015). This makes CSTRs preferable for AD of FW, whether in batch or continuous conditions, as they can handle high solid loading rates. For example, Zorg Biogas GmbH manufactures industrial-grade CSTRs as shown in Figure 8.2 with the following distinctive features (Zorg Biogas 2024a):

1. heating from piping installed on reactor walls
2. sulphur removal using a biofilter colonised by aerobic sulphur-consuming bacteria and air supply positioned in the reactor centre
3. flexible gasholder to hold biogas above the reactor
4. inclined, submersible mixers for mixing operations

Such CSTRs are made from either bolted steel panels or reinforced concrete. It functions well at moisture contents of 93–97% (and in turn, solids content from 3–7%)

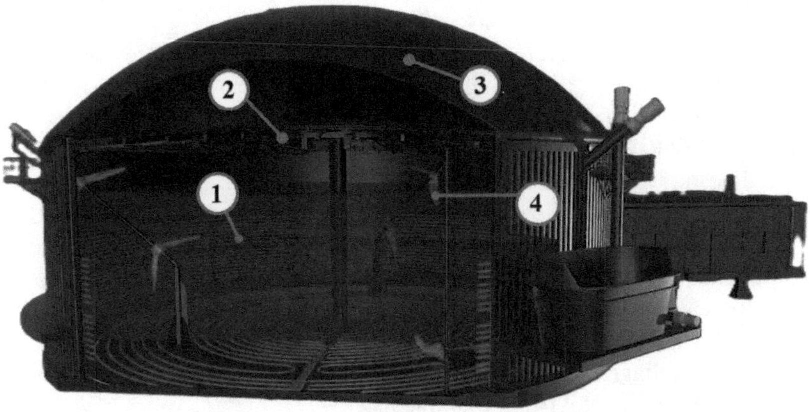

FIGURE 8.2 A Zorg Biogas CSTR with the distinctive features labelled (Zorg Biogas 2024a).

and organic dry matter loads of 2–5 kg/(m^3 CSTR · day). Suitable substrates for CSTRs include mostly animal manure. Its ease of operation and maintenance makes it ideal for rural applications, but more effective and more sophisticated technologies are available to increase efficiency (Zorg Biogas 2024a).

An instance of such a CSTR is used in an AD plant digesting 100 tonnes per day of sewage sludge, 50 tonnes per day of dung and 20 tonnes per day of slaughterhouse waste in Voroshnevo, Russia. Four such CSTRs at 3,600 m^3 can produce 2 MW (Zorg Biogas 2024b).

8.3.7.2 Upflow anaerobic sludge reactor (UASB)

Another suitable choice for AD is the upflow anaerobic sludge blanket reactor (UASB). This reactor includes a feeding system in the lower part, allowing the substrate to flow upwards through a sludge blanket containing suspended beds and granules. UASBs are effective for high OLRs and have low HRTs. However, they are not suitable for treating organic wastes with high solid loads (Cremonez et al. 2021). For example, Veolia's Biothane® Advanced UASB can handle COD loading rates of 8–14 kg COD/(m^3 day) and is suitable for areas with height limitations. Their patented advanced three-phase, double-baffle settler enhances separation efficiency between biomass and biogas (BiogasWorld 2024; Veolia 2024a). Such a reactor has been successfully installed in food-processing facilities (Veolia 2024b).

Another successful instance of a UASB reactor is employed by Gills Onions. Before employing AD as an onion waste treatment solution, Gills Onions had to haul 250,000 pounds of waste off-site every day, and when transport was infeasible (e.g. on rainy days), the onions were left to decompose. Using juice from ground onion waste as feedstock, biogas was recovered from a UASB reactor and reformed using steam to run two 300 kW electrochemical fuel cells with 47% electrical efficiency at 480 V. This utilises up to 930 Btu per cubic feet of biogas. Overall, this reduced Gills Onions' waste production by 99%, effectively eliminating a waste stream and decreasing their carbon footprint (Gills Onions 2024).

8.3.7.3 Plug flow digesters

Plug flow digesters are a preferable alternative to CSTRs in cases of large solid concentration variations (U.S. Environmental Protection Agency 2024). Plug flow digesters for AD of rich substrates with 15–50% dry matter content have seen industrial use under STRABAG, whose patented dry, continuous and mesophilic or thermophilic digestion process has been in use for over 15 years. AD takes place in a horizontal LARAN® plug flow digester under local mixing by agitators arranged transversely to the flow direction, whose schematic diagram is shown in Figure 8.3.

Components of the digester such as feed units and biogas collection systems are designed for easy external access and maintenance. The digestate is discharged using a wear-resistant STRABAG vacuum discharge system (STRABAG 2024a). A cross-sectional diagram of the plug flow digestor is shown in Figure 8.4 with the description of each aspect written in Table 8.4.

A successful dry digestion plant using STRABAG's LARAN® plug flow digesters has its place in Helsinki, Finland. This plant has a capacity of 44 kilotonnes per year of biowaste (i.e. biodegradable food and kitchen waste) and produces 6 MW of biogas as of October 2023 (STRABAG 2024b).

Several other reactor models are available, such as anaerobic packed-bed reactors, anaerobic baffled reactors, expanded granular sludge blanket (EGSB) reactors, leachate bed reactors, internal circulation (IC) reactors, or sequential batch reactors (SBRs) (Cremonez et al. 2021; Van et al. 2020). The choice of reactor should be based on the specific requirements of the application.

8.4 PERFORMANCE OF ANAEROBIC DIGESTION OF DIFFERENT CATEGORIES OF FW

Table 8.5 provides a comprehensive overview of diverse FW and expired food categories and their corresponding biogas yields. This tabulation serves as a valuable

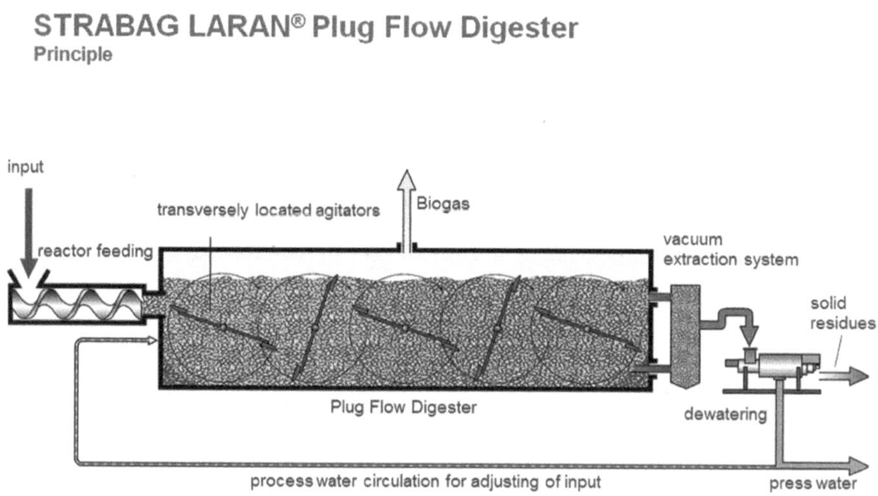

FIGURE 8.3 Schematic diagram of a STRABAG LARAN® Plug Flow Digester (STRABAG 2024a).

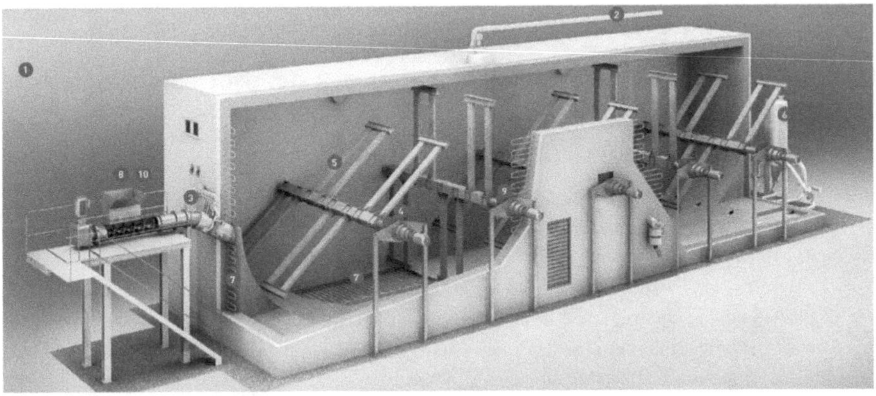

FIGURE 8.4 Cross-sectional diagram of a STRABAG LARAN® Plug Flow Digester (STRABAG 2016).

reference point for forthcoming enhancements in the realm of AD, as well as the identification of pivotal elements influencing the efficacy of this process.

It is worth noting that many of the studies listed in Table 8.5 focus on small-scale AD of FW, typically with digester volumes ranging from 0.25 to 37 L under mesophilic conditions (35–37 °C). An exception is found in the treatment of household and restaurant wastes, where a relatively larger volume of 4 m^3 is reported. Most studies indicate relatively low to average total VS (TVS) and COD removal, spanning from 42.7 to 78%. However, certain cases, such as the anaerobic co-digestion (ACD) of household and restaurant wastes, peach pulp waste, and apple pulp waste, demonstrate higher COD removal rates, ranging from 83 to 97.6%.

The AD for FW and expired food can be classified as either wet or dry AD depending on the TS content of the feedstock. While FW is primarily composed of solids, it is often mixed with water before undergoing anaerobic digestion to achieve the desired consistency for either wet or dry AD processes. However, it is important to note that the specific solid-to-water ratios are not always specified in most of the research studies presented in Table 8.5. Instead, many studies focus on reporting the OLR, a critical parameter influencing the efficiency and stability of the AD process (Section 8.3). Focusing on OLR has advantages as it directly relates to the digester's performance and stability, covering various operational scenarios, including different solid contents and moisture levels without the need for specifying solid-to-water ratios. Standardising OLR as a key parameter ensures comparability and relevance across different studies and operational conditions. This approach allows for a more flexible and adaptable operation, where operators can adjust the solid-to-water ratio based on specific circumstances while maintaining the target OLR to optimise digester performance. The OLR in these studies falls within the range of 0.9 to 4.34 kg COD/(m^3·day), strategically adjusted to prevent overloading of anaerobic bacteria, as discussed in Section 8.3.

Methane yields vary between 5.7 and 768 L/kg COD consumed or L/kg VS. Particularly, studies involving FW with polylactic acid bioplastic, as reported by Zhu et al. (2023), and sunflower meal and wheat straw, reported by Rajput and Sheikh (2019), achieved the highest methane yield. This is mainly attributed to the key

TABLE 8.4
Description of STRABAG LARAN® plug flow digester process characteristics (STRABAG 2016)

Item	Description
1	Industry-approved and patented 15-year-old process
2	High biogas yield from plug-flow characteristics
3	Quasi-steady feeding system, providing continuous biogas production
4	Multiple, robust, independently driven agitators with short, crosswise shafts
5	Localised mixing controls scum layer sedimentation and transport to digestate section
6	Unique, wear-resistant vacuum digestate system
7	Wall and floor-attached heating elements allow seamless heat transfer
8	Substrates with varying dry matter content flexibility
9	Individually controlled, intermittently operated agitators provide low electrical energy consumption
10	Safe operation even with substrates containing high dry matter content

factors as discussed in Section 8.3 such as an appropriate OLR (2 kg VS/(m^3·day)), HRT of 45–46 days, temperature of 35–37 °C and a pH of 7.0.

These findings highlight the importance of carefully considering operational parameters in optimising AD processes. This is particularly crucial as the reported COD and VS removal rates (Table 8.5) tend to be low. Further efforts are necessary to enhance these removal rates. Moreover, it is apparent that the current studies (Table 8.5) are confined to laboratory-scale investigations, highlighting the essential need for expanding research into large-scale AD of FW.

8.5 ANAEROBIC DIGESTION PERFORMANCE MONITORING AND OPTIMISATION SOFTWARE

AD is inherently volatile since it is a biological process, necessitating stringent process control protocols and accurate process parameter monitoring software. Issues in industrial-scale anaerobic digestion may arise from, for instance, excess feedstock, excess biogas /sulphur/AD-inhibiting substance accumulation, and suboptimal feedstock ratios which will impact biogas yield, or in more severe cases, cause equipment wear and structural failure. In compliance with Industry 4.0 to realise digital transformation in industrial processes and real-time decision-making, commercially available AD process monitoring and optimisation software is capable of mitigating such quantifiable issues through the Internet of Things (IoT) that links sensors attached to equipment with computers in control rooms, sometimes even located remotely and a large distance away from the plant.

An instance of retention time optimisation software is dinaMETAN, developed by BTS Biogas. Feed is characterised using near-infrared spectroscopy (NIRS) in real-time using 26 parameters, including starch levels, carbohydrate levels and protein levels. Through graphical analysis, dinaMETAN can also calculate the optimum mixture ratios and biological efficiency of different feedstock mixtures, along with cost estimations. As a result, dinaMETAN can calculate the exact substrate residence time for maximum biogas production (Douard 2014).

TABLE 8.5
Performance of anaerobic digestion of different categories of FW and expired food

Substrates	Inoculum source	Reactor volume (L)	Temperature (°C)	pH	HRT (days)	OLR Value	OLR Unit	Removal efficiency Value	Removal efficiency Unit	Biomethane yield H_2	Biomethane yield CH_4	Biomethane yield Unit	References
Simulated food waste	Anaerobic sludge	4	35	6	42	1.9	kg/(m³·day)	–		31.1	460	L/kg	(Feng et al. 2020)
Pineapple waste	Pig farm digestate	5	37	6	3	328.8	kg COD/(m³·day)	42.7	% COD	4.1	167.9	L/kg COD	(Chu, Vo, and Chen 2020)
Food waste	Anaerobic sludge	4.5	45	8	0.708	20	kg COD/(m³·day)	74.1	% COD	22.7	30	L/kg $COD_{consumed}$	(Hassan, Hemdan, and El-Gohary 2020)
Fast food restaurant leftovers	Anaerobic sludge	4.6	35	8	0.708	–	–	47.5	% COD	4.1	5.7	L/kg $COD_{consumed}$	(Hassan et al. 2021)
Artificial food waste	Anaerobic sludge	10	37	7.2	–	–	–	59.7	% VS	30.3	190.3	L/kg VS_{added}	(Luo et al. 2021)
Citrus peel waste	UASB reactor sludge	0.25	30	8.5	–	–	–	–	–	13.29	50.2	mmol/L	(Camargo et al. 2021)
Household and restaurant wastes	Cow farm sludge	4000	35	7.5	7	2.5–3.8	kg VS / (m³·day)	83–87	% tCOD	189-263	–		(Nguyen et al. 2021)
Food waste	Maize straw digestate and municipal sewage sludge	4.5	35	7	9	18	kg VS / (m³·day)	56.6	% tCOD	135	510	L/kg VS_{in}	(Algapani et al. 2019)
Food waste and activated sludge	Activated sludge	37	37	7.4	12	18	kg VS / (m³·day)	54.5	% TVS	8.6	428.3	NL/(kg TVS·day)	(Baldi, Pecorini, and Iannelli 2019)
Food and paper waste	Thermophilic sludge, anaerobic sludge	12	37.5	7.3	24	4.34	kgCOD / (m³·day)	78.4	% VS	79	329	NL/kg VS_{fed}	(Li et al. 2019)

(Continued)

TABLE 8.5 (Continued)

Substrates	Inoculum source	Reactor volume (L)	Temperature (°C)	pH	HRT (days)	OLR Value	OLR Unit	Removal efficiency Value	Removal efficiency Unit	Biomethane yield H_2	Biomethane yield CH_4	Biomethane yield Unit	References
Peach pulp waste	Anaerobic sludge	5	30	7.4	5	0.9	kg COD / (m^3·day)	97.6	% COD	–	320	L/kg COD	(Carvalheira et al. 2018)
Apple pulp waste	Anaerobic sludge		37	–	2.5	7.4	kg COD / (m^3·day)	91.9	% COD	–	300	L/kg COD	
Food and vegetable wastes	Food waste slurry	1.8	–	7	–	–	–	54.1	% COD	–	226.86	L/kg VS	(Chakraborty and Venkata Mohan 2018)
Organic market waste	Cow manure	13.4	35	7	15	1.27–1.95	kg VS/ (m^3·day)	–	–	0.05–0.07	120–180	NL/kg VS	(Gómez Camacho et al. 2019)
Sunflower meal and Wheat Straw	Acclimatised sludge	0.225	35	7	45	2	kg VS/ (m^3·day)	78	% VS	–	768	NL/kg VS	(Rajput and Sheikh 2019)
Food waste	Synthetic microbial consortia (E(C2)Tx)	0.2	37	6	4	2.5	kg VS/ (m^3·day)	77.5	% VS	–	368	L/ kg VS_{added}	(Thakur et al. 2023)
Food waste and algae	Anaerobic sludge	20	35	7.5	25	1–2	kg COD / (m^3·day)	–	–	–	334	L/kg COD_{Input}	(Khanthong et al. 2023)
Food waste with polylactic acid bioplastic	Anaerobic digestion sludge	0.4	35	7	46	2	kg VS / m^3 reactor	–	–	–	659	L/kg VS	(Zhu et al. 2023)
Expired food waste-used disposable adult nappies	Anaerobic sludge from wastewater treatment plant	4	37 ± 0.5	–	10.6	4.03	kg COD / (m^3·day)	67	% COD	–	10.03 ± 0.63	L/L feed	(Tsigkou et al. 2021)

METANload, a feed optimisation software developed also by BTS Biogas, is an example of commercially available OLR monitoring and optimisation software. Using integrated hardware components, METANload can display and control feed filling by providing real-time information about substrate flow rates for operators, and as a result, the flow rate of which substrate must be changed to optimise mixing ratios. The METANload system stands out with its simple controls and interface, early warning systems at low biomass levels, along with retrofitting capabilities for existing plants (BTS Biogas 2024a).

Hitachi Zosen Inova has also developed its artificial intelligence-based Digester Performance Monitoring (DPM AI) system for its in-house Kompogas® dry anaerobic digestion technology, used in their FW AD plant in Jönköping, Sweden. The system was developed to replace conventional sampling and laboratory analysis, a process that may take days to diagnose the issue since the process is biological-based. With DPM AI, laboratory analysis is rendered unnecessary, as plant data collected from existing measurements and available online is evaluated to detect potential upcoming disturbances within a timescale of seconds. The results are then presented neatly in a user-friendly visual report to allow immediate operator response. Implementation of this detection system enables processes to be operated closer to biological limits and increases productivity by up to 15% (Hitachi Zosen Inova 2023, 2024).

8.6 STRATEGIES FOR MAXIMISING BIOGAS PRODUCTION OF FW

Enhancing biogas production through anaerobically digesting FW encompasses various strategies focused on refining the digestion process and maximising gas output. Researchers are delving into more intricate approaches to address challenges like ammonia-induced inhibition of the digestion process and system breakdown caused by fatty acid accumulation (Liu et al. 2023). To bolster operational stability and effectiveness, several studies propose modifications aimed at elevating the performance of AD. These proposed adjustments encompass pretreatment methods, the integration of co-digestion practices, and the introduction of additives.

8.6.1 Pretreatment

Before undergoing AD, it is crucial to pretreat the substrate properly to optimise biogas yield (Chen et al. 2023). Appropriate pretreatment techniques can preserve organic content, accelerate hydrolysis, and prevent inhibitory chemical formation. By carefully selecting the optimal pretreatment conditions, the rate-limiting hydrolysis step can be enhanced and consequently, methane yield can be increased. Methods used to pretreat FW are mainly categorised into mechanical and chemical pretreatments.

8.6.2 Mechanical pretreatment

Mechanical pretreatment involves the fragmentation and grinding of solid substrate particles to release cellular components and significantly increase total surface area

(TSA). Increased TSA promotes interactions between substrate and anaerobic bacteria, thereby substantially improving the AD process (Elliott and Mahmood 2012; Ariunbaatar et al. 2015). It is worth noting that larger particle sizes are associated with decreased degradation of COD and a reduction in the rate of methane production. Consequently, methods like sonication, liquid shear, collision, high-pressure homogenisation, and liquefaction are employed as mechanical pretreatment techniques to reduce the size of substrate particles.

In the case of sonication, electrical energy is transformed into mechanical vibrations, leading to the phenomenon of cavitation. Ultrasonic treatment harnesses this cavitation to disrupt cell structures and induce flocculation by generating bubbles within aqueous or slurry media (Rowan et al. 2022). However, while ultrasonic treatment is effective, its notable energy consumption presents economic challenges. Thus, it is imperative for future research to prioritise the reduction of energy requirements associated with ultrasonic and microwave pretreatment methods.

Another mechanical strategy, high-pressure homogenisation (HPH), involves pressurising heterogeneous sludge components for 3 to 30 minutes at 30 to 150 MPa. This application of pressure induces homogenisation through shear forces generated upon the impact of pressurised sludge against a ring. The resultant cavitation from this impact disrupts cell membranes, leading to a heightened reduction of VS during AD (Barjenbruch and Kopplow 2003). For example, an observable 25% increase in VS reduction was recorded in the AD of HPH-pretreated sewage sludge. This improvement was attributed to the elevated concentration of soluble proteins, lipids, and carbohydrates (Engelhart et al. 2000).

8.6.3 Chemical Pretreatment

Chemical pretreatment methods are mainly divided into acid and alkaline pretreatment. Acid pretreatment involves using substances like HCl, H_2SO_4, H_3PO_4, and HNO_2 to break down the polymeric structures and cells in waste-activated sludge. This process occurs at a pH range of 1 to 5.5. As an example, Wang et al. discovered that the flocculation of microbial extracellular polymeric substances (EPS) to increase their surface area, and subsequently accessibility for AD, were the most efficient at pH 4.8 (Wang et al. 2012). However, excessive use of strong acidic pretreatment can generate inhibitory byproducts like furfural and hydroxyl-methyl furfural, which is why it's typically avoided (Mussoline et al. 2013). Instead, mild acidic pretreatment is combined with thermal methods.

Alkaline pretreatment has its advantages. It is useful for processes like lipid saponification to enhance lipid availability and prepare FW for AD (Carrere et al. 2012). Different alkali agents like NaOH, KOH, $Mg(OH)_2$, and $Ca(OH)_2$ can be used for sludge solubilisation, with NaOH being the most effective followed by KOH, $Mg(OH)_2$, and $Ca(OH)_2$. However, excessive concentrations of Na^+ or K^+ must be avoided as they might hinder the AD process. For example, Hamzawi et al. found that at an alkali content of 150 mg/L, a TS concentration of 23% and a sewage sludge particle size of 8 mm, alkaline pretreatment resulted in a 31% increase in biogas generation and a 32% increase in TS elimination within the digestion system (Hamzawi, Kennedy, and McLean 1998). Alkaline pretreatment also has benefits related to reducing the production of toxic hydrogen sulphide (H_2S) during AD. Dai et al. have shown that after alkali pretreatment, cumulative production of H_2S decreased by 54.2% compared to without pretreatment (Dai et al. 2017).

When comparing the efficacy of acidic and alkaline pretreatment methods in terms of soluble COD and short-chain fatty acids from excess sludge, alkaline pretreatment came out as more effective. However, it is worth noting that using these pretreatment methods requires additional pH adjustments during the AD process, which can increase operational costs and raise environmental concerns due to the production of acidic or alkaline waste which can prove either corrosive to materials or detrimental to water and soil systems if discharged without neutralisation.

8.6.4 Co-digestion

Co-digestion refers to the simultaneous AD of multiple organic wastes in one digester to increase methane production from low-yielding or difficult-to-digest materials (EPA 2012). Co-digestion techniques can balance the nutrients in FW and dilute toxic substances, thereby improving biohydrogen yield and energy recovery (Xu et al. 2018). In a co-digestion process, care must be taken to select compatible co-digestion feedstocks that enhance methane production while avoiding the production of methanogenic inhibitors. For example, digestion time can be shortened by implementing the co-digestion of animal waste, phytomass, crop residues, and sewage sludge (Solé-Bundó et al. 2019).

8.6. Additives

Additives can have a significant impact on anaerobic digestive systems by aiding in the initial hydrolytic acidification phase, maintaining pH stability, and accelerating the direct interspecies electron transfer pathway, thus reducing lag time. For example, enzymes can be introduced during the pre-hydrolysis of FW to rapidly break down complex molecules (Yan et al. 2011). Enzymes such as amylase and glycosylase cleave glycosidic bonds in starch, while proteases break down peptide chains, accelerating the release of ammonia due to increased acidity. Further to this, fungal sludge contains a diverse range of enzymes, notably accelerating the hydrolytic acidification phase compared to using individual enzymes (Meng et al. 2017).

pH levels within the mixture considerably influence microbial growth, subsequently impacting the overall performance of AD. To regulate the system's pH, natural solutions such as neutralisers, bicarbonates, and buffering agents are introduced. By adding micronutrient and macronutrient supplements, methane production is not only stimulated but process stability is also enhanced. This is achieved by maintaining pH within the optimal range, preventing the accumulation of VFAs, and ensuring minimal alkalinity levels. Even when low levels of neutralisers are added, high VS removal rates can be achieved (Wang et al. 2016). Bicarbonate addition is particularly favoured as a method for pH management due to its effectiveness with just a single application, akin to a neutraliser.

Conductive materials are incorporated to bolster the stability of co-digestion systems by increasing the capacity for direct interspecies electron transfer (DIET) and balancing microbial community structures (Li et al. 2021). DIET involves swift electron transfer facilitated by microbial cell structures and exogenous materials, reducing the dependence on biological enzymes (Wang et al. 2017). Capitalising on their robust electrical conductivity, conductive materials promote DIET, potentially surpassing the efficiency of bacterial flagella. These materials can also concentrate on specific microorganisms with extracellular electron transfer capabilities through adsorption (Zhao et al. 2021).

8.7 CASE STUDIES

8.7.1 Laboratory Case Study 1: Anaerobic Digestion of Sunflower Meal and Wheat Straw Using Acclimatised Sludge

Rajput and Sheikh (2019) conducted batch experiments by manipulating inoculum type and OLR to study their effects on the biogas yield of sunflower meal and wheat straw. This specific case study has been chosen due to its highest methane yield, as discussed in Section 8.4. This case study aims to investigate the factors contributing to the high achievement observed in this study.

Initially, in this study, the OLR was kept constant at 2 g VS/L, maintaining a feed-to-inoculum ratio of 1 on a g VS basis, and a temperature of 35 °C to maximise degradation. Sets of anaerobic digesters were conducted using digested manure, septic tank sludge and acclimatised sludge for sunflower meal and wheat straw respectively, totalling for $3 \times 2 = 6$ reaction sets. The best results were obtained for the reaction set containing digested manure and sunflower meal at 768 NmL biogas/g VS and 58% methane content, followed by that containing digested manure and wheat straw at 556 NmL biogas/g VS and 56% methane content. The performance of acclimatised sludge, while weaker than digested manure, was better than that of septic tank sludge (Rajput and Sheikh 2019). Table 8.6 summarises the biogas yield, methanol content and VS removal for different feed and inoculum.

Compared to another study, biogas yields for wheat straw were higher in this work. The authors cited inoculum preparation, pre-incubation, degassing, and a sufficiently low OLR (2 g VS/L) as possible reasons for stable operation and increased biogas volume. Inoculum preparation involves conditioning microbial communities to ensure robust activity. Pre-incubation acclimatised these microbes to the wheat straw substrate, enhancing their efficiency. It also helps to decrease biogas production from endogenous material in the inoculum, minimising gas production from the inoculum itself and maximising gas production from the wheat straw substrate. Degassing removed background methane and other dissolved gases, particularly oxygen, to maintain strict anaerobic conditions and ensure accurate biogas measurements. These pretreatment steps combined with the low OLR can prevent overloading and support stable microbial activity, thus resulting in higher biogas yields.

The accumulation of VFA as a result of septic tank sludge digested in the tank it was taken from may be the reason why reaction sets involving septic tank sludge had low performance. Other reasons cited for the high performance of AD by digested manure include the provisions of initial microbial population by inoculum, inoculum nutrient content in itself, inoculum enzyme activity, and stable buffer (VFA and total alkalinity, TA) production (Rajput and Sheikh 2019).

TABLE 8.6
Performance of reaction sets for various inoculum (Rajput and Sheikh 2019)

Feed	Inoculum	Biogas yield (NmL/g VS)	Methane content (%)	VS removal (%)
Sunflower meal	Digested manure	768	58	78
	Septic tank sludge	381	53	31
	Acclimatised sludge	691	55	72
Wheat straw	Digested manure	556	56	70
	Septic tank sludge	297	52	26
	Acclimatised sludge	397	53	52

Second, the inoculum with the best results (which was digested manure) was selected to be combined with OLRs of 2, 6, 10 and 14 g VS/L. The best results were obtained, still, for the reaction set containing digested manure and sunflower meal at OLR of 2 g VS/L at 768 NmL biogas/g VS, and the poorest for the reaction set containing digested manure and sunflower meal at OLR of 14 g VS/L at 768 NmL biogas/g VS (Rajput and Sheikh 2019). Table 8.7 summarises the biogas yield, feed inoculum, pH, VFA/TA at various OLR.

A strong positive trend between OLR and VFA/TA and a strong negative trend between OLR and biogas yield were observed. It is suggested that at high OLRs, VFAs are produced too quickly to be consumed, resulting in low biogas yield. A noticeable jump between biogas yield, pH and VFA/TA values for sets containing 10 and 14 g VS/L OLR can be observed, suggesting there is an upper limit for OLR before its buffering capacity decreases drastically (Rajput and Sheikh 2019).

In summary, the findings from this case study highlight the critical role of maintaining a sufficiently low OLR at 2 g VS/L. This is essential for ensuring stable operation and increasing biogas volume in AD.

8.7.2 Laboratory Case Study 2: Anaerobic Co-Digestion of FW and Algae

The second case study focuses on the ACD of FW and algae. Realising that AD of FW exhibits poor process stability at high OLRs, and AD of algae biomass has a low biogas yield rate and is incapable of inhibiting ammonia production, Khanthong et al. investigated the optimisation of FW to algae ratio for efficient ACD by anaerobic sludge through measuring biochemical methane potential (BMP). Reaction sets were homogenised using magnetic stir bars. pH was kept constant by adding bicarbonate and oxygen was removed by purging the reaction sets with nitrogen (Khanthong et al. 2023). This study reported that the carbon content of both algae and FW was similar, but algae had a higher nitrogen content, leading to a lower C/N ratio for algae. Importantly, the C/N ratio of the mixtures improved as the proportion of FW increased, reaching a maximum of 10.6 with a 2:8 algae to FW ratio.

Figure 8.5 shows the primary results of the batch experiment. The optimum algae to FW ratio was determined to be 2:8 (by volume), probably due to its highest C/N ratio compared to other algae to FW ratios. Besides, the high methane yield using

TABLE 8.7
Performance of reaction sets at various OLR (Rajput and Sheikh 2019)

Feed + inoculum	OLR (g VS/L)	Biogas yield (NmL/g VS)	pH		VFA/TA	
			Initial	Final	Initial	Final
Sunflower meal + digested manure	2	768	7.1	6.9	0.38	0.34
	6	636	7.3	6.8	0.36	0.32
	10	518	7.3	6.2	0.29	0.39
	14	226	7.2	3.9	0.34	1.11
Wheat straw + digested manure	2	509	7.4	7.1	0.28	0.31
	6	477	7.2	6.9	0.27	0.36
	10	434	7.4	6.3	0.25	0.39
	14	198	7.3	4.6	0.27	1.27

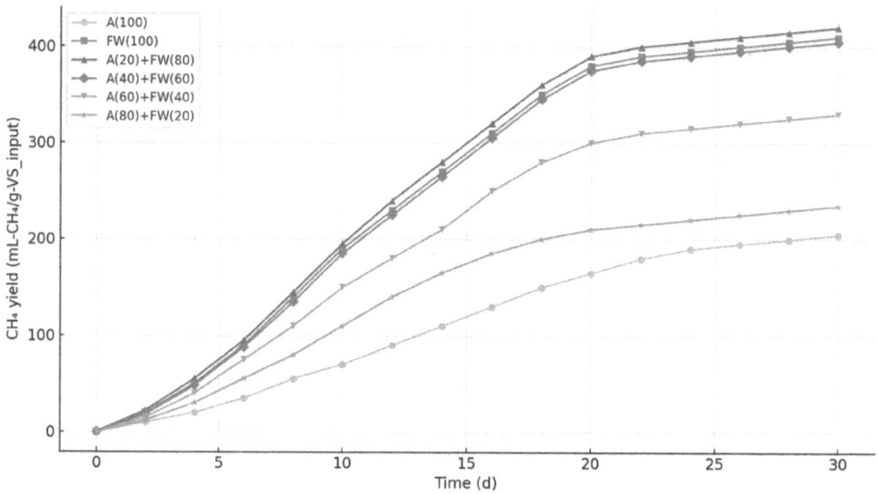

FIGURE 8.5 Algae (A) to FW ratio (by volume) effects on methane production in mL per g VS (Khanthong et al. 2023).

this ratio was explained by a high methane generation constant of 0.036 ± 0.008 d^{-1} and a short lag phase. Increasing FW will increase the methane generation constant and decrease the lag phase. The co-digestion performance index (as described by Ebner et al. 2016) of the 2:8 mixture was also the highest among other mixtures at 1.21 compared to 1.01, 1.05 and 1.18 at mixtures of 8:2, 6:4 and 4:6 respectively, indicating good synergy between algae and FW at a 2:8 ratio (Khanthong et al. 2023).

The experiment was succeeded by continuous operation by anaerobically digesting FW, algae and a 2:8 by volume mixture of algae to FW in three semi-continuous stirred tank reactors (SCSTRs), while varying OLR from 1 to 4 kg COD/(m^3·day) in 1 kg COD/(m^3·day) increments (Khanthong et al. 2023). In the primary results of the continuous experiment. The reactors were operated for 120 days, with initial OLR at 1 kg COD/(m^3·day) OLR was increased by 1 kg COD/(m^3·day) every 30 days until the AD process failed (indicated by lowered methane yield). The reactors containing pure algae and pure FW failed when OLR was increased to 3 kg COD/(m^3·day), but that did not occur for the mixed algae-FW substrate. CPI for the co-digestion reactor was valued at 1.24, 1.20 and 3.30 for OLRs of 1, 2 and 3 kg COD/(m^3·day) respectively, indicating high performance of co-digestion even at higher-than-usual OLRs. For the failure of the reactor at 4 kg COD/(m^3·day), the accumulation of VFAs was cited as the reason (Khanthong et al. 2023).

Algae is a poor substrate for AD due to low hydrolysis and acidogenesis rates. However, when combined with FW, algae are capable of converting this disadvantage into an advantage by lowering hydrolysis and acidogenesis rates of FW, thereby acting as a buffer to impede VFA accumulation and keep pH levels constant. This provides a balance in hydrolysis, acidogenesis, and methanogenesis rates and therefore maximises methane yield. The TVFA/alkalinity ratio (equivalent to VFA/TA in laboratory case study 1) of the different reactor sets reported that the ratio skyrocketed for FW substrates at 3 kg COD/(m^3·day), and while the ratios for algae and co-substrates remained fairly stable, the poor performance of algae by itself as a substrate was said to be due to high ammonia nitrogen levels as a result of high protein levels in algae which were broken down to produce ammonia that reduces biogas yield (Khanthong et al. 2023). Additionally, algae provide a rich source of macronutrients (potassium, magnesium, calcium) and micronutrients (iron, iodine) that are essential for microbial growth in AD (Koyande et al. 2019). These nutrients enhance microbial activity and stability in the digester, contributing to improved biogas yield and process efficiency.

In summary, the co-digestion of algae and FW at a 2:8 by volume ratio exhibited much better methane yield and process stability compared to the digestion of a single substrate of either material. This case study highlights the potential of ACD of FW with other biomass, emphasising the possible synergistic effects among the substrates.

8.7.3 Industrial Case Study 1: Hong Kong Organic Waste Recycling Facility

This section encourages the application of case study methodology to investigate the AD process facility in Hong Kong, developed under a contract with the Government of Hong Kong and managed by Agrivert. It focuses on the waste management challenges faced by Hong Kong, such as high municipal solid waste (MSW) production, and outlines the solutions implemented by the Environmental Protection Department (EPD).

Under a contract with the Government of Hong Kong Special Administrative Region, Agrivert is the principal joint venture partner responsible for designing, constructing, and operating Organic Resources Recovery Centre Phase 2 (O·PARK2), an AD facility as shown in Figure 8.6. As one of the largest organic waste treatments in the world, O·PARK2 processes 300 tonnes of FW daily. Utilising AD technology, the facility efficiently transforms organic wastes into biogas, subsequently producing over 45 GWh of electricity annually (Agrivert 2024a).

Hong Kong's waste management problems demonstrate the complexities of densely populated urban areas. With a yearly production of 5.67 million tonnes of

FIGURE 8.6 Hong Kong Anaerobic Digestion Food Waste Facility, O.PARK 2.

MSW, FW significantly contributes to the city's waste management infrastructure issues. The high volume, potential for odour, and methane emissions make FW a particularly problematic component. Despite efforts to recycle, only one-third of the city's waste is successfully recycled, with the rest destined for landfills. Analysts predict that Hong Kong's three major landfills will reach capacity by the late 2020s. The city's high population density of over 7 million people within just 1,106 km^2 exacerbates waste generation and complicates waste management. The limited and unsustainable use of landfills highlights the need for more advanced waste treatment technologies.

The Government of Hong Kong has outlined environmental protection policies through the EPD. These policies aim to reduce environmental pollution, encourage energy conservation and waste reduction, promote green buildings and environmental technology, and improve environmental education (Greater Bay Area n.d.). The EPD's long-term objective is to establish waste-to-energy facilities, reducing dependency on landfills. Waste-to-energy offers an alternative method for disposing of MSW while generating electricity (Abomohra, Wang, and Huang 2022). The EPD invested in designing and constructing an organic waste treatment facility that processes 300 tonnes of organic waste per day through AD.

Several essential systems are included in the facility. The waste reception and pretreatment system aim to remove waste solids in the water, using chemical pretreatments such as pH adjusters, coagulants, and flocculants (Chem READY 2020). AD decomposes organic matter in the absence of oxygen, producing biogas and digestate (Us EPA 2019). Biogas, mainly methane and carbon dioxide, is similar to natural gas and can produce heat and energy (Us EPA 2019). The digestate, a mixture of solid and liquid materials, undergoes dewatering to remove the liquid fraction, making the processing of the solid fraction more efficient. The solid fraction is converted into granulate fertiliser for agricultural and landscaping use (Us EPA 2019). This conversion process ensures the sustainability of the waste management system.

The facility also includes a wastewater treatment plant designed to remove impurities from wastewater through a series of treatment processes, including pretreatments, primary and secondary treatments. Membrane filtration, advanced oxidation processes, and tertiary treatment enhance effluent quality (Sheriff et al. 2024). A centralised air pollution control system manages and minimises pollutant emissions, ensuring compliance with air quality regulations. Biogas conditioning systems eliminate contaminants while maintaining constant pressure, flow, temperature, and quality (Petrogas 2017). A heat and power cogeneration plant (CHP) produces both heat and electricity, operating at about 75% more efficiency than traditional systems. The CHP plant significantly lowers greenhouse gas emissions, promoting environmental sustainability. The biogas generated by AD provides electricity for the facility and heat for process use.

O·PARK2 is anticipated to reduce the use of fossil fuels for electricity generation and divert approximately 110,000 tonnes of FW from landfills, minimising the release of about 67,000 tonnes of greenhouse gases annually. These benefits align with the United Nations Sustainable Development Goals (SDGs), particularly Goal 7: Affordable and Clean Energy, and Goal 13: Climate Action (SDSN Hong Kong). Through these benefits, O·PARK2 not only addresses Hong Kong's waste problems but also contributes to global efforts for sustainability and environmental conservation.

O·PARK2 has been meticulously designed to minimise its visual impact and integrate seamlessly with its surroundings. It features advanced technology such as hammermills for removing non-organic materials and processes for blending FW with liquids, ensuring effective digestion and efficient operation. By addressing the growing waste management demands and environmental concerns, O·PARK2 represents a pivotal advancement in Hong Kong's sustainability efforts.

8.7.4 Industrial case study 2: North London anaerobic digestion facility

The North London AD facility, managed by Agrivert in the United Kingdom, serves as a pioneering example of sustainable waste management (Figure 8.7). FW disposal is a severe environmental problem, with the UK producing approximately 9.5 million tonnes of FW in 2018, 15% less than in 2007 (Business Waste 2024). The North London AD facility addresses these issues by processing source segregated FW from local authorities and commercial producers. Using advanced AD technology, the facility converts organic waste into biogas and digestate, significantly reducing landfill dependency and promoting sustainable practices. This process helps mitigate methane emissions, a potent greenhouse gas that is 25 times more effective at trapping heat in the atmosphere than carbon dioxide over a 100-year period, significantly contributing to climate change (Us EPA 2023).

The facility tackles multiple challenges associated with FW management. One primary concern is the release of methane, which is approximately 100 times more potent in causing global warming than carbon dioxide over the first 20 years in the

FIGURE 8.7 Food Waste Anaerobic Digestion Facility in North London.

atmosphere (Reuters 2023). By diverting FW from landfills and processing it anaerobically, the facility prevents methane emissions and captures the gas for energy production. This biogas, primarily composed of methane and carbon dioxide, is purified and used to generate renewable electricity, which is fed into the National Grid, reducing reliance on fossil fuels. Additionally, the excess heat produced during digestion is utilised to warm the digesters and heat the pasteurisation process, enhancing overall energy efficiency. The facility also prevents valuable nutrients from being lost in landfills, as the digestate is used as a biofertiliser on local farmland, replacing synthetic fertilisers that are often derived from fossil fuels and contribute to nitrous oxide emissions, another potent greenhouse gas (Department of Energy and Environment n.d.).

The facility's engineering solutions include an advanced depackaging system that ensures high recycling rates and effective processing of peak waste deliveries. The digester tanks are designed to optimise biogas production, and the facility handles an average of 175 tonnes of solid waste and 90 tonnes of liquid waste daily. Robust control systems monitor and regulate the AD process, ensuring optimal conditions for biogas production and maintaining high-quality output.

The operating conditions at the North London AD facility are designed to maximise efficiency and sustainability. The robust depackaging system handles various types of FW, including packaged food, ensuring a high purity of organic feedstock for the digester tanks. The recessed reception bunker increases the facility's capacity to temporarily store incoming waste, enabling it to handle peak waste deliveries without compromising the processing flow. FW is stored for up to three days before undergoing AD, providing a continuous feed to the digester tanks. Each digester tank holds 5,800 m^3 of material, and the waste undergoes thorough pasteurisation before digestion, involving heating to kill pathogens and ensure the safety of the final digestate product. The facility uses a wet scrubber and biofilter on the building ventilation system to manage odour. Control systems monitor and regulate temperature, pH, and microbial activity to ensure optimal biogas production conditions. The biogas produced is captured and purified to remove impurities such as hydrogen sulphide, ensuring high-quality gas for electricity generation. The heat recovery system captures excess heat from the digestion process, used to maintain digester temperatures and for pasteurisation, enhancing overall energy efficiency.

By converting FW into renewable energy and valuable fertilisers, the North London AD facility demonstrates significant environmental and economic benefits. It produces enough renewable electricity to power over 6,000 homes and captures 5.6 million cubic metres of methane annually. The facility's operations align with the United Nations Sustainable Development Goals, particularly affordable and clean energy (SDG 7) and climate action (SDG 13) (SDG n.d.). Overall, the North London AD facility reduces greenhouse gas emissions and promotes sustainable agricultural practices, showcasing the potential of advanced waste management technologies in addressing global environmental challenges.

8.8 CHALLENGES AND RECOMMENDATIONS

The AD of FW and expired food faces several significant challenges that must be addressed to enhance its effectiveness and sustainability. One of the primary challenges in the AD of FW is maintaining process stability. Fluctuations in feedstock composition and loading rates can disrupt the balance of microbial communities, leading to process upsets and reduced biogas production. Besides, FW feedstock can be highly variable in terms of its composition, moisture content, and organic content. Such variability can impact the efficiency of biogas production as certain substrates might be more readily digestible by the microbial community than others. Another critical aspect is the quality of the digestate, the residual material left after AD. For digestate to be effectively utilised as a fertiliser or soil amendment, it must be free from pathogens, heavy metals, and other contaminants. Ensuring the digestate's safety and quality is essential for its safe application and for meeting regulatory standards.

The reported studies in Table 8.5 highlighted the crucial need to improve the COD and VS removal rate, which currently ranges from 42.7 to 78%. Numerous studies have also provided optimal parameters for a specific set of substrates and inoculum. However, the existing pool of information concerning optimal ratios and operational parameters derived from experimental research remains inadequate for direct application in practical, real-world projects. This insufficiency of comprehensive guidelines necessitates a more refined approach to implementing AD in actual settings.

Future research should focus on emerging technologies and innovative strategies to enhance AD efficiency. Key areas for development include microbial community engineering, which involves manipulating microbial communities to boost efficiency and stability through techniques such as microbial inoculation, co-digestion with specific substrates, and bioaugmentation (Khanthong et al. 2023). Furthermore, advanced monitoring systems should be implemented by integrating real-time monitoring systems that measure parameters like pH, temperature, methane production, and VFAs concentrations can provide valuable insights into process dynamics. This approach allows for timely interventions to maintain stability and optimise biogas yield. In addition to monitoring, the utilisation of sophisticated process modelling, and simulation tools can help to predict the effects of different operational parameters and feedstock compositions on performance. This predictive capability enables operators to fine-tune conditions for maximum biogas production, reducing the likelihood of process failures and inefficiencies (Chen et al. 2023). Moreover, it is critical to consider the broader environmental and economic impacts of AD. For instance, optimising the digestion process to produce a high-quality digestate can contribute to a circular economy by providing a sustainable source of fertiliser. Additionally, improving biogas yield can enhance the economic viability of AD facilities, making them more attractive investments.

The specific details about the use of AI, cloud computing, and IoT at O·PARK2 and the North London AD facility, as outlined in industry case studies (Section 8.7), are not available on the facilities' websites. However, incorporating these technologies could significantly enhance the efficiency and performance of the AD process, optimising biogas production and overall waste management operations. AI could be used to analyse data from the digestion process, adjusting operational parameters such as temperature and mixing rates to maximise methane output and improve system reliability. Cloud computing would enable the centralisation and real-time analysis of data from various sensors, facilitating remote monitoring and management of the facility, and allowing for data-driven decision-making. IoT sensors placed throughout the facility could continuously monitor critical variables like waste composition, temperature, and gas production rates, providing valuable insights that help to adjust processing conditions dynamically. Together, these technologies would improve the efficiency of biogas production, reduce operational costs, and enhance the facility's ability to handle varying waste inputs effectively.

Addressing these challenges requires a multidisciplinary approach, combining insights from microbiology, engineering, environmental science, and economics. Collaborative efforts between researchers, industry stakeholders, and policymakers will be essential to drive innovation and implement effective solutions. By focusing on these critical areas, the AD of FW can become a more stable, efficient, and sustainable process, contributing to waste reduction, renewable energy production, and environmental protection

8.9 CONCLUSIONS

AD emerges as a promising and sustainable avenue for harnessing renewable energy, particularly in its capacity to effectively address the issue of FW within communities. By integrating AD systems, the proper management of FW becomes achievable. The factors that exert influence on the AD process for FW were thoroughly examined, including temperature, pH, retention time, and substrate characteristics, all of which can be monitored in real-time by utilising Industry 4.0-ready technology including IoT and cloud computing for storing historical performance data online.

The performance of FW AD, both on the laboratory scale and industry scale, has been comprehensively investigated. Laboratory-scale FW AD was evaluated via its solid's removal efficiency and biomethane yield while manipulating substrate and inoculum sources and noting reactor volumes, temperatures, pH, HRT and OLR. Industry-scale FW AD plants were evaluated based on their waste processing capacities, greenhouse gas reductions, landfill diversions, power generation capacities and biogas/biomethane production, with success cases provided for the top 20 AD companies in the world, proving that AD is a sufficiently practical technology to produce renewable energy from municipal waste.

While numerous strategies are available for maximising biogas production, the field still experiences many challenges in making anaerobic digestion both industry-mature and profitable, such as inadequacy of parameter optimisation, lack of standardisation, instability of the FW AD process and the need to improve COD and VS removal rate. Nevertheless, the success of multiple companies in making AD a process that can deliver usable electricity to local communities, as presented in the case studies, cannot be overlooked. As such, AD of FW could be considered an emerging industry worth investing in to produce more sophisticated, more energy efficient and more cost-effective technologies and processes to achieve sustainability goals and valorise waste as an energy source to meet global energy demands.

EXERCISES

1. Briefly explain the basic principles of anaerobic digestion and each digestion stage for food waste and expired food.
2. What are the main factors affecting the performance of anaerobic digestion of food waste and expired food?
3. List three strategies that can maximise the performance of anaerobic digestion.
4. Which food waste pair as listed in Table 8.5 produced the maximum yield of biogas?
5. Aside from CSTRs, UASB reactors and plug flow digesters, several other reactors/digestion methods were listed. Select one and decide whether scale-up is possible for the selected reactor/digestion method. If not, explain one (1) inhibiting factor that makes scale-up difficult.

REFERENCES

Abomohra, A. E.- F., Q. Wang, and J. Huang (Eds.). 2022. *Waste-to-energy: recent developments and future perspectives towards circular economy* (1st ed.). Springer Nature.

Agrivert. 2024a. "Hong Kong anaerobic digestion food waste facility O.PARK 2". https://www.agrivert.co.uk/projects/organic-waste-anaerobic-digestion-plant

Agrivert. 2024b. "Food waste anaerobic digestion facility North London." https://www.agrivert.co.uk/projects/food-waste-anaerobic-digestion-facility-north-london

Akuzawa, Masateru, Tomoyuki Hori, Shin Haruta, Yoshiyuki Ueno, Masaharu Ishii, and Yasuo Igarashi. 2011. "Distinctive responses of metabolically active microbiota to acidification in a thermophilic anaerobic digester." *Microbial Ecology* 61:595–605.

Algapani, Dalal E., Wei Qiao, Marina Ricci, Davide Bianchi, Simon M. Wandera, Fabrizio Adani, and Renjie Dong. 2019. "Bio-hydrogen and bio-methane production from food waste in a two-stage anaerobic digestion process with digestate recirculation." *Renewable Energy* 130:1108–1115. doi: https://doi.org/10.1016/j.renene.2018.08.079

Anaergia. 2024a. "Anaergia." https://www.anaergia.com

Anaergia. 2024b. "Enerbio." https://www.anaergia.com/reference-facilities/enerbio/

Anaergia. 2024c. "Technology." https://www.anaergia.com/technology/

Ariunbaatar, Javkhlan, Antonio Panico, Daniel H. Yeh, Francesco Pirozzi, Piet N. L. Lens, and Giovanni Esposito. 2015. "Enhanced mesophilic anaerobic digestion of food waste by thermal pretreatment: Substrate versus digestate heating." *Waste Management* 46:176–181.

Asja. 2020. "BIOMETANO: da rifiuto a energia." https://www.asja.energy/wp-content/uploads/2020/03/Brochure_Biometano-web.pdf

Asja. 2024a. "entra in funzione l'impianto forsu di tuscania." https://www.asja.energy/news/entra-in-funzione-limpianto-forsu-di-tuscania/

Asja. 2024b. "Environmental responsibility." https://www.asja.energy/en/sustainability/environmental-responsibility/

Azman, Samet. 2016. *Anaerobic digestion of cellulose and hemicellulose in the presence of humic acids*. Wageningen University and Research.

Baldi, F., I. Pecorini, and R. Iannelli. 2019. "Comparison of single-stage and two-stage anaerobic co-digestion of food waste and activated sludge for hydrogen and methane production." *Renewable Energy* 143:1755–1765. doi: https://doi.org/10.1016/j.renene.2019.05.122

Barjenbruch, M., and O. Kopplow. 2003. "Enzymatic, mechanical and thermal pre-treatment of surplus sludge." *Advances in Environmental Research* 7(3):715–720.

Belay, Negashi, R. Sparling, and L. Daniels. 1986. "Relationship of formate to growth and methanogenesis by Methanococcus thermolithotrophicus." *Applied and Environmental Microbiology* 52(5):1080–1085.

Bioenergy Devco. 2024. "Maryland bioenergy center." https://bioenergydevco.com/maryland-bioenergy-center/

BiogasWorld. 2024. "Biothane UASB from Veolia." https://www.biogasworld.com/product/anaerobic-digestion/biothane-uasb-veolia/

Biogest. 2024a. "Products: PowerRing." https://www.biogest.at/products/powerring

Biogest. 2024b. "Products: references." https://www.biogest.at/products/references

Biogest. 2024c. "Success stories: Frogmary, UK." https://www.biogest.at/products/success-stories/frogmary-uk

Blasius, Jandir Pereira, Ronan Cleber Contrera, Sandra Imaculada Maintinguer, and Marcus Cesar Avezum Alves de Castro. 2020. "Effects of temperature, proportion and organic loading rate on the performance of anaerobic digestion of food waste." *Biotechnology Reports* 27:e00503.

BTS Biogas. 2024a. "METANload - calculation software." https://www.energy-xprt.com/software/metanload-calculation-software-513513

BTS Biogas. 2024b. "Plants development." https://bts-biogas.com/en/plants-development/

Business Waste. 2024. "Food waste 2024 - the facts." https://www.businesswaste.co.uk/food-waste-the-facts/

Camargo, Franciele Pereira, Isabel Kimiko Sakamoto, Tiago Palladino Delforno, Mahendra Mariadassou, Valentin Loux, Cédric Midoux, Iolanda Cristina Silveira Duarte, Edson Luiz Silva, Ariane Bize, and Maria Bernadete Amâncio Varesche. 2021. "Microbial and functional characterization of an allochthonous consortium applied to hydrogen production from Citrus Peel Waste in batch reactor in optimized conditions." *Journal of Environmental Management* 291:112631. doi: https://doi.org/10.1016/j.jenvman.2021.112631

Carrere, H., Y. Rafrafi, A. Battimelli, M. Torrijos, J. P. Delgenes, and C. Motte. 2012. "Improving methane production during the codigestion of waste-activated sludge and fatty wastewater: impact of thermo-alkaline pretreatment on batch and semi-continuous processes." *Chemical Engineering Journal* 210:404–409.

Carvalheira, Mónica, Joana Cassidy, João M. Ribeiro, Bruno A. Oliveira, Elisabete B. Freitas, Christophe Roca, Gilda Carvalho, Adrian Oehmen, and Maria A. M. Reis. 2018. "Performance of a two-stage anaerobic digestion system treating fruit pulp waste: The impact of substrate shift and operational conditions." *Waste Management* 78:434–445. doi: https://doi.org/10.1016/j.wasman.2018.06.013

Chakraborty, Debkumar, Obulisamy Parthiba Karthikeyan, Ammaiyappan Selvam, and Jonathan W. C. Wong. 2018. "Co-digestion of food waste and chemically enhanced primary treated sludge in a continuous stirred tank reactor." *Biomass and Bioenergy* 111:232–240.

Chakraborty, Debkumar, and S. Venkata Mohan. 2018. "Effect of food to vegetable waste ratio on acidogenesis and methanogenesis during two-stage integration." *Bioresource Technology* 254:256–263. doi: https://doi.org/10.1016/j.biortech.2018.01.051.

Chatterjee, Biswabandhu, and Debabrata Mazumder. 2019. "Role of stage-separation in the ubiquitous development of anaerobic digestion of organic fraction of municipal solid waste: a critical review." *Renewable and Sustainable Energy Reviews* 104:439–469.

Chem READY. 2020. "How industrial wastewater pretreatment works." https://www.getchem-ready.com/water-facts/how-industrial-wastewater-pretreatment-works/

Chen, Le, Wei Fang, Jinsong Liang, Mohammad Nabi, Yajing Cai, Qingyan Wang, Panyue Zhang, and Guangming Zhang. 2023. "Biochar application in anaerobic digestion: performances, mechanisms, environmental assessment and circular economy." *Resources, Conservation and Recycling* 188:106720.

Chu, Chen-Yeon, Tan-Phat Vo, and Tsung-Hsien Chen. 2020. "A novel of biohythane gaseous fuel production from pineapple peel waste juice in two-stage of continuously stirred anaerobic bioreactors." *Fuel* 279:118526. doi: https://doi.org/10.1016/j.fuel.2020.118526

Clarke, S. 1999. "An introduction to anaerobic infections." *Biomedical Scientist* 43:192–193.

Cooper, C. David. 2014. *Introduction to environmental engineering*. Waveland Press.

Cremonez, Paulo André, Joel Gustavo Teleken, Thompson Ricardo Weiser Meier, and Helton José Alves. 2021. "Two-stage anaerobic digestion in agroindustrial waste treatment: a review." *Journal of Environmental Management* 281:111854.

Dai, Xiaohu, Chongliang Hu, Dong Zhang, and Yinguang Chen. 2017. "A new method for the simultaneous enhancement of methane yield and reduction of hydrogen sulfide production in the anaerobic digestion of waste activated sludge." *Bioresource Technology* 243:914–921.

Department of Energy and Environment. n.d. "Fertilizer impacts the environment." https://doee.dc.gov/service/fertilizer-impacts-environment

Dinopoulou, Georgia, Thomasine Rudd, and John N Lester. 1988. "Anaerobic acidogenesis of a complex wastewater: I. The influence of operational parameters on reactor performance." *Biotechnology and Bioengineering* 31(9):958–968.

Douard, Frédéric. 2014. "dinaMetan, logiciel d'optimisation des unités de méthanisation." https://www.bioenergie-promotion.fr/36030/dinametan-outil-doptimisation-des-unites-de-methanisation/

Duan, Nina, Bin Dong, Bing Wu, and Xiaohu Dai. 2012. "High-solid anaerobic digestion of sewage sludge under mesophilic conditions: feasibility study." *Bioresource Technology* 104:150–156.

Eaton, Andrew D., Mary Ann H. Franson, Lenore S. Clesceri, Eugene W. Rice, and Arnold E. Greenberg. 2005. *Standard Methods for the Examination of Water & Wastewater*, 21st ed. Washington, DC: American Public Health Association, American Water Works Association, and Water Environment Federation.

Ebner, Jacqueline H., Rodrigo A. Labatut, Jeffrey S. Lodge, Anahita A. Williamson, and Thomas A. Trabold. 2016. "Anaerobic co-digestion of commercial food waste and dairy manure: Characterizing biochemical parameters and synergistic effects." *Waste Management* 52: 286–294.

Elliott, Allan, and Talat Mahmood. 2012. "Comparison of mechanical pretreatment methods for the enhancement of anaerobic digestion of pulp and paper waste activated sludge." *Water Environment Research* 84(6):497–505.

Engelhart, M., M. Krüger, J. Kopp, and N. Dichtl. 2000. "Effects of disintegration on anaerobic degradation of sewage excess sludge in downflow stationary fixed film digesters." *Water Science and Technology* 41(3):171–179.

EPA. 2012. "Increasing anaerobic digester performance with codigestion." https://www.epa.gov/sites/default/files/2014-12/documents/codigestion.pdf

EPA. 2023. "How does anaerobic digestion work?", accessed 14/7. https://www.epa.gov/agstar/how-does-anaerobic-digestion-work

ESI. 2020. "Case study – threemile canyon farms." *Engineering & Energy Solutions*. Accessed June 10, 2025. https://www.esisolutions.com/free-anaerobic-digester-gas-to-rng-case-study

European Investment Bank. 2023. "Energy overview 2023." Luxembourg: EIB. https://www.eib.org/attachments/lucalli/20220286_energy_overview_2023_en.pdf.

Feng, Kai, Qiao Wang, Huan Li, Yangyang Zhang, Zhou Deng, Jianguo Liu, and Xinrui Du. 2020. "Effect of fermentation type regulation using alkaline addition on two-phase anaerobic digestion of food waste at different organic load rates." *Renewable Energy* 154:385–393. doi: https://doi.org/10.1016/j.renene.2020.03.051.

Ferry, James G. 2010. "The chemical biology of methanogenesis." *Planetary and Space Science* 58(14–15):1775–1783.

Franke-Whittle, Ingrid H., Andreas Walter, Christian Ebner, and Heribert Insam. 2014. "Investigation into the effect of high concentrations of volatile fatty acids in anaerobic digestion on methanogenic communities." *Waste Management* 34(11):2080–2089.

GENeco. 2024a. "Food waste." https://www.geneco.uk.com/our-services/food-waste/

GENeco. 2024b. "Gas to grid: facts and figures." https://www.geneco.uk.com/media/d14pu5h2/bio-bus-infographic.pdf

Gianico, A., C. M. Braguglia, A. Gallipoli, and G. Mininni. 2015. "Innovative two-stage mesophilic/thermophilic anaerobic degradation of sonicated sludge: performances and energy balance." *Environmental Science and Pollution Research* 22:7248–7256.

Gills Onions. 2024. "Gills onions advanced energy recovery system." https://www.energy.gov/sites/prod/files/2014/03/f10/waste_reardon.pdf

Gómez Camacho, Carlos Enrique, Bernardo Ruggeri, Lorenzo Mangialardi, Marco Persico, and Andrea Cristina Luongo Malavé. 2019. "Continuous two-step anaerobic digestion (TSAD) of organic market waste: rationalising process parameters." *International Journal of Energy and Environmental Engineering* 10(4):413–427. doi: 10.1007/s40095-019-0312-1

GPS Renewables. 2024. "The full stack clean fuels company." https://gpsrenewables.com

Greater Bay Area. n.d. *Environmental protection and sustainable development*. "Guangdong-hong kong-macao greater bay area - environmental protection and sustainable development." https://www.bayarea.gov.hk/en/opportunities/environmental.html

Green Create. 2024a. "Our green solution: biomethane." https://www.green-create.com/green-solutions/biomethane/

Green Create. 2024b. "Wijster, Netherlands: the largest anaerobic digestion plant in Europe to process 100% poultry manure." https://www.green-create.com/locations/wijster/

Group Op der Beeck. 2024a. "Energy production and nutrient recovery – Op de Beeck NV." http://www.groupopdebeeck.com/en/ennu_energieproductie/

Group Op der Beeck. 2024b. "Figures." https://www.groupopdebeeck.com/en/cijfers/

Group Op der Beeck. 2024c. "Organic waste processing – Biogas Solutions NV." https://www.groupopdebeeck.com/en/wat/

Hamzawi, N., K. J. Kennedy, and D. D. McLean. 1998. "Anaerobic digestion of co-mingled municipal solid waste and sewage sludge." *Water Science and Technology* 38(2):127–132.

Hartmann, Hinrich, and Birgitte Kiær Ahring. 2006. "Strategies for the anaerobic digestion of the organic fraction of municipal solid waste: an overview." *Water Science and Technology* 53(8):7–22.

Hassan, Gamal, Aly Al-Aneny, Ahmed Afify, Mohamed El-Liethy, Sherien Elagroudy, and Fatma El-Gohary. 2021. "Production of biofuels (H_2&CH_4) from food leftovers via dual-stage anaerobic digestion: enhancement of bioenergy production and determination of metabolic fingerprinting of microbial communities." *Egyptian Journal of Chemistry*. doi: 10.21608/ejchem.2021.67579.3493

Hassan, Gamal Kamel, Bahaa Ahmed Hemdan, and Fatma A. El-Gohary. 2020. "Utilization of food waste for bio-hydrogen and bio-methane production: influences of temperature, OLR, and in situ aeration." *Journal of Material Cycles and Waste Management* 22(4):1218–1226. doi: 10.1007/s10163-020-01014-5

Hitachi Zosen Inova. 2023. "New AI-based system for early detection of digester biology problems." https://www.hz-inova.com/new-ai-based-system-for-early-detection-of-digester-biology-problems/

Hitachi Zosen Inova. 2024. "DPM AI system: AI-based system for early detection of digester biology problems." https://www.hz-inova.com/wiki/dpm-ai-system/

Kan, Eunsung. 2013. "Effects of pretreatments of anaerobic sludge and culture conditions on hydrogen productivity in dark anaerobic fermentation." *Renewable Energy* 49:227–231.

Keenan Recycling. 2024a. "Biofuel – feedstock for anaerobic digestion." https://keenanrecycling.co.uk/services/biofuel/

Keenan Recycling. 2024b. "Nationwide commercial food waste collection service." https://keenanrecycling.co.uk

Khanthong, Kamonwan, Rahul Kadam, Taeyoung Kim, and Jungyu Park. 2023. "Synergetic effects of anaerobic co-digestion of food waste and algae on biogas production." *Bioresource Technology* 382:129208. doi: https://doi.org/10.1016/j.biortech.2023.129208

Kiener, Andreas, and Thomas Leisinger. 1983. "Oxygen sensitivity of methanogenic bacteria." *Systematic and Applied Microbiology* 4(3):305–312.

Kim, Woong, Seung Gu Shin, Juntaek Lim, and Seokhwan Hwang. 2013. "Effect of temperature and hydraulic retention time on volatile fatty acid production based on bacterial community structure in anaerobic acidogenesis using swine wastewater." *Bioprocess and Biosystems Engineering* 36:791–798.

Koyande, A. K., Chew, K. W., Rambabu, K., Tao, Y., Chu, D.-T., & Show, P.-L. (2019). "Microalgae: a potential alternative to health supplementation for humans." *Food Science and Human Wellness* 8(1):16–24. doi: https://doi.org/10.1016/j.fshw.2019.03.001

Li, Lei, Ying Xu, Xiaohu Dai, and Lingling Dai. 2021. "Principles and advancements in improving anaerobic digestion of organic waste via direct interspecies electron transfer." *Renewable and Sustainable Energy Reviews* 148:111367.

Li, Lu, Yu Qin, Zhe Kong, Jing Wu, Kengo Kubota, and Yu-You Li. 2019. "Characterization of microbial community and main functional groups of prokaryotes in thermophilic anaerobic co-digestion of food waste and paper waste." *Science of the Total Environment* 652:709–717. doi: https://doi.org/10.1016/j.scitotenv.2018.10.292

Li, Rongping, Shulin Chen, and Xiujiu Li. 2010. "Biogas production from anaerobic co-digestion of food waste with dairy manure in a two-phase digestion system." *Applied Biochemistry and Biotechnology* 160:643–654.

Li, Yebo, Stephen Y. Park, and Jiying Zhu. 2011. "Solid-state anaerobic digestion for methane production from organic waste." *Renewable and Sustainable Energy Reviews* 15(1):821–826.

Liang, Jialin, Liwen Luo, Jonathan W. C. Wong, and Di He. 2022. "Recent advances in conductive materials amended anaerobic co-digestion of food waste and municipal organic solid waste: roles, mechanisms, and potential application." *Bioresource Technology* 360:127613. doi: https://doi.org/10.1016/j.biortech.2022.127613

Lin, Lili, Rong Yan, Yongqiang Liu, and Wenju Jiang. 2010. "In-depth investigation of enzymatic hydrolysis of biomass wastes based on three major components: cellulose, hemicellulose and lignin." *Bioresource Technology* 101(21):8217–8223.

Liu, Chao, Wen Wang, Naveed Anwar, Zonghu Ma, Guangqing Liu, and Ruihong Zhang. 2017. "Effect of organic loading rate on anaerobic digestion of food waste under mesophilic and thermophilic conditions." *Energy & Fuels* 31(3):2976–2984.

Liu, Kaili, Longyi Lv, Weiguang Li, Zhijun Ren, Pengfei Wang, Xiaoyang Liu, Wenfang Gao, Li Sun, and Guangming Zhang. 2023. "A comprehensive review on food waste anaerobic co-digestion: research progress and tendencies." *Science of the Total Environment* 878:163155. doi: https://doi.org/10.1016/j.scitotenv.2023.163155

Luo, Liwen, Guneet Kaur, Jun Zhao, Jun Zhou, Suyun Xu, Sunita Varjani, and Jonathan W. C. Wong. 2021. "Optimization of water replacement during leachate recirculation for two-phase food waste anaerobic digestion system with off-gas diversion." *Bioresource Technology* 335:125234. doi: https://doi.org/10.1016/j.biortech.2021.125234

Mamimin, Chonticha, Maraike Probst, María Gómez-Brandón, Sabine Marie Podmirseg, Heribert Insam, Alissara Reungsang, and O. Sompong. 2019. "Trace metals supplementation enhanced microbiota and biohythane production by two-stage thermophilic fermentation." *International Journal of Hydrogen Energy* 44(6):3325–3338.

Mamimin, Chonticha, Apinya Singkhala, Prawit Kongjan, Benjaporn Suraraksa, Poonsuk Prasertsan, Tsuyoshi Imai, and O. Sompong. 2015. "Two-stage thermophilic fermentation and mesophilic methanogen process for biohythane production from palm oil mill effluent." *International Journal of Hydrogen Energy* 40(19):6319–6328.

Meegoda, Jay N., Brian Li, Kush Patel, and Lily B. Wang. 2018. "A review of the processes, parameters, and optimization of anaerobic digestion." *International Journal of Environmental Research and Public Health* 15(10). doi: https://doi.org/10.3390/ijerph15102224

Meng, Ying, Fubo Luan, Hairong Yuan, Xue Chen, and Xiujin Li. 2017. "Enhancing anaerobic digestion performance of crude lipid in food waste by enzymatic pretreatment." *Bioresource Technology* 224:48–55.

Moset, Veronica, Morten Poulsen, Radziah Wahid, Ole Højberg, and Henrik Bjarne Møller. 2015. "Mesophilic versus thermophilic anaerobic digestion of cattle manure: methane productivity and microbial ecology." *Microbial Biotechnology* 8(5):787–800.

Mussoline, Wendy, Giovanni Esposito, Andrea Giordano, and Piet Lens. 2013. "The anaerobic digestion of rice straw: a review." *Critical Reviews in Environmental Science and Technology* 43(9):895–915.

Nagao, Norio, Nobuyuki Tajima, Minako Kawai, Chiaki Niwa, Norio Kurosawa, Tatsushi Matsuyama, Fatimah Md Yusoff, and Tatsuki Toda. 2012. "Maximum organic loading rate for the single-stage wet anaerobic digestion of food waste." *Bioresource Technology* 118:210–218.

Nazir, Muhammad Shahzad, Ali Jafer Mahdi, Muhammad Bilal, Hafiz M. Sohail, Nisar Ali, and Hafiz M. N. Iqbal. 2019. "Environmental impact and pollution-related challenges of renewable wind energy paradigm – a review." *Science of the Total Environment* 683:436–444. doi: https://doi.org/10.1016/j.scitotenv.2019.05.274

Negri, Camilla, Marina Ricci, Massimo Zilio, Giuliana D'Imporzano, Wei Qiao, Renjie Dong, and Fabrizio Adani. 2020. "Anaerobic digestion of food waste for bio-energy production in China and Southeast Asia: a review." *Renewable and Sustainable Energy Reviews* 133:110138. doi: https://doi.org/10.1016/j.rser.2020.110138

Nguyen, Phuoc-Dan, Nguyen-Sang Truong Tran, Thuy-Truc Nguyen, Bao-Trong Dang, Minh-Tam Thi Le, Xuan-Thanh Bui, Fumitoshi Mukai, Hidemasa Kobayashi, and Huu Hao Ngo. 2021. "Long-term operation of the pilot scale two-stage anaerobic digestion of municipal biowaste in Ho Chi Minh City." *Science of the Total Environment* 766:142562. doi: https://doi.org/10.1016/j.scitotenv.2020.142562

Olleco. 2024a. "Anaerobic digestion." https://www.olleco.co.uk/sustainability/anaerobic-digestion
Olleco. 2024b. "Olleco acquires third anaerobic digestion facility." https://www.olleco.co.uk/media/2018/03/29/olleco-acquires-third-anaerobic-digestion-facility
Panigrahi, Sagarika, and Brajesh K. Dubey. 2019. "A critical review on operating parameters and strategies to improve the biogas yield from anaerobic digestion of organic fraction of municipal solid waste." *Renewable Energy* 143:779–797.
Paranjpe, Archana, Seema Saxena, and Pankaj Jain. 2023. "A review on performance improvement of anaerobic digestion using co-digestion of food waste and sewage sludge." *Journal of Environmental Management* 338:117733. doi: https://doi.org/10.1016/j.jenvman.2023.117733
Permastore. 2024. "Brocklesby limited chooses permastore glass-fused-to-steel tanks for its biogas facility in North Cave, East Yorkshire, UK." https://www.permastore.com/brocklesby-limited-chooses-permastore-glass-fused-to-steel-tanks-for-its-biogas-facility-in-north-cave-east-yorkshire-uk/
Petrogas. 2017. "Biogas conditioning." https://www.petrogas.nl/process-systems/biogas-conditioning/
Plan ET. 2024a. "Biomethane: a diverse and renewable energy source." https://planet-biogas.com/biomethane/
Plan ET. 2024b. "Groot Zevert." https://planet-biogas.com/reference/groot-zevert/
Plan ET. 2024c. "Home page." https://planet-biogas.com
Plan ET. 2024d. "Sustainable energy from industrial organic waste." https://planet-biogas.com/wp-content/uploads/2023/06/PlanET_Industriebroschuere_EN_March2023-1.pdf
Prodeval. 2023. "1st forum of the polish biomethane organization." https://bioch4.org/wp-content/uploads/2023/10/14.-Martin-VRTISKA-Prodeval.pdf
Prodeval. 2024a. "Biomethane / RNG production." https://www.prodeval.com/en/biomethane-production
Prodeval. 2024b. "Biomethane engineering for local energy production." https://www.prodeval.com/en
proweps envirotec. 2024. "Project portfolio." https://proweps-envirotec.com/project-portfolio/
Rajagopal, Rajinikanth, Daniel I. Massé, and Gursharan Singh. 2013. "A critical review on inhibition of anaerobic digestion process by excess ammonia." *Bioresource Technology* 143:632–641.
Rajput, Asad Ayub, and Zeshan Sheikh. 2019. "Effect of inoculum type and organic loading on biogas production of sunflower meal and wheat straw." *Sustainable Environment Research* 29(1):4. doi: https://doi.org/10.1186/s42834-019-0003-x
ReFood. 2024a. "Doncaster." https://refood.co.uk/facilities/doncaster/
ReFood. 2024b. "The process." https://refood.co.uk/the-process/
ReFood. 2024c. "Recycle your food waste and ReEnergise your business." https://refood.co.uk
Reuters. 2023. "The other greenhouse gas." https://www.reuters.com/graphics/FOOD-WASTE/METHANE/gdpzwqwqovw/
Rincon, Barbara, Lissette Travieso, Enrique Sanchez, María de los Ángeles Martín, Antonio Martín, Francisco Raposo, and Rafael Borja. 2007. "The effect of organic loading rate on the anaerobic digestion of two-phase olive mill solid residue derived from fruits with low ripening index." *Journal of Chemical Technology & Biotechnology: International Research in Process, Environmental & Clean Technology* 82(3):259–266.
Rowan, Max, Great C. Umenweke, Emmanuel I Epelle, Inioluwa Christianah Afolabi, Patrick U. Okoye, Burcu Gunes, and Jude A. Okolie. 2022. "Anaerobic co–digestion of food waste and agricultural residues. An overview of feedstock properties and the impact of biochar addition." *Digital Chemical Engineering* 4:100046.
SDGs.un.org. n.d. "The 17 goals." https://sdgs.un.org/goals

Severn Trent. 2024a. "Food waste recycling." https://www.stgreenpower.co.uk/food-waste-recycling

Severn Trent. 2024b. "Severn trent green power." https://www.stgreenpower.co.uk

Severn Trent. 2024c. "Severn trent green power cassington AD facility." https://www.stgreenpower.co.uk/where-we-operate/cassington-ad-facility

Shang, Zezhou, Rui Wang, Xiyi Zhang, Yongle Tu, Chenjing Sheng, Huan Yuan, Lei Wen, Yulu Li, Jing Zhang, and Xiaojiao Wang. 2023. "Differential effects of petroleum-based and bio-based microplastics on anaerobic digestion: a review." *Science of the Total Environment* 875:162674.

Sheriff, I., et al. 2024. "Extraction and analytical methods of microplastics in wastewater treatment plants: isolation patterns, quantification, and size characterization techniques." *Desalination and Water Treatment* 318(100399):100399. doi: https://doi.org/10.1016/j.dwt.2024.100399

Shi, Xiao-Shuang, Jian-Jun Dong, Jun-Hong Yu, Hua Yin, Shu-Min Hu, Shu-Xia Huang, and Xian-Zheng Yuan. 2017. "Effect of hydraulic retention time on anaerobic digestion of wheat straw in the semicontinuous continuous stirred-tank reactors." *BioMed Research International* 2017:6.

Soares, Juliana Ferreira, Tássia Carla Confortin, Izelmar Todero, Flávio Dias Mayer, and Marcio Antonio Mazutti. 2020. "Dark fermentative biohydrogen production from lignocellulosic biomass: technological challenges and future prospects." *Renewable and Sustainable Energy Reviews* 117:109484. doi: https://doi.org/10.1016/j.rser.2019.109484

Solé-Bundó, Maria, Fabiana Passos, Maycoll S Romero-Güiza, Ivet Ferrer, and Sergi Astals. 2019. "Co-digestion strategies to enhance microalgae anaerobic digestion: a review." *Renewable and Sustainable Energy Reviews* 112:471–482.

Sood, S., R. Singhal, S. Bhat, and A. Kumar. 2011. "2.19- inoculum preparation." In *Comprehensive biotechnology* (3rd ed.), edited by Murray Moo-Young, 230–243. Pergamon.

Stams, Alfons J. M., and Caroline M. Plugge. 2009. "Electron transfer in syntrophic communities of anaerobic bacteria and archaea." *Nature Reviews Microbiology* 7(8):568–577.

STRABAG. 2024a. "Dry digestion in the LARAN® plug-flow digester." https://www.strabag-umwelttechnik.com/databases/internet/_public/content.nsf/web/FA01F708E945624CC1257EE50033F358

STRABAG. 2024b. "References biogas plants." https://www.strabag-umwelttechnik.com/databases/internet/_public/files.nsf/SearchView/D09A19B020B46DA3C1258A5800278ACB/$File/97285001.pdf

Thakur, Hemant, Rashmi, Narendra Kumar Verma, Vikas Sharma, Shrawan Kumar, Atul Dhar, Tulika Prakash, and Satvasheel Powar. 2023. "Anaerobic co-digestion of food waste, bio-flocculated sewage sludge, and cow dung in CSTR using E(C2)Tx synthetic consortia." *Environmental Technology & Innovation* 32:103263. doi: https://doi.org/10.1016/j.eti.2023.103263

The World Bank Group. (2024). "Global Food Loss and Waste." https://datatopics.worldbank.org/what-a-waste/global_food_loss_and_waste.html

Tsigkou, Konstantina, Dimitris Zagklis, Panagiota Tsafrakidou, Paraskevi Zapanti, Georgios Manthos, Konstantina Karamitou, Constantina Zafiri, and Michael Kornaros. 2021. "Expired food products and used disposable adult nappies mesophilic anaerobic co-digestion: biochemical methane potential, feedstock pretreatment and two-stage system performance." *Renewable Energy* 168:309–318. doi: https://doi.org/10.1016/j.renene.2020.12.062

U.S. Environmental Protection Agency. 2001. "Total, fixed, and volatile solids in water, solids, and biosolids." https://www.epa.gov/sites/default/files/2015-10/documents/method_1684_draft_2001.pdf

U.S. Environmental Protection Agency. 2019. "How does anaerobic digestion work?" https://www.epa.gov/agstar/how-does-anaerobic-digestion-work

U.S. Environmental Protection Agency. 2023. "Quantifying methane emissions from landfilled food waste." https://www.epa.gov/land-research/quantifying-methane-emissions-landfilled-food-waste

U.S. Environmental Protection Agency. 2024. "Anaerobic system design and technology." https://www.epa.gov/agstar/anaerobic-system-design-and-technology

Van, Dinh Pham, Takeshi Fujiwara, Bach Leu Tho, Pham Phu Song Toan, and Giang Hoang Minh. 2020. "A review of anaerobic digestion systems for biodegradable waste: configurations, operating parameters, and current trends." *Environmental Engineering Research* 25(1):1–17.

Van Lier, Jules B., Nidal Mahmoud, and Grietje Zeeman. 2008. "Anaerobic wastewater treatment." In *Biological Wastewater Treatment: Principles, Modelling and Design*, 415–456. London: IWA Publishing.

Vanguard Renewables. 2024a. "Goodrich family farm." https://www.datocms-assets.com/97698/1689280890-goodrichfarm_casestudy.pdf

Vanguard Renewables. 2024b. "Harnessing the power of waste." https://www.vanguardrenewables.com

Vanguard Renewables. 2024c. "Our Story." https://www.vanguardrenewables.com/our-story

Veolia. 2024a. "Biothane® Advanced UASB." https://www.biothanesolutions.com/technologies/biothane-advanced-uasb

Veolia. 2024b. "Biothane® Anaerobic treatment." https://www.veoliawatertech.com/sites/g/files/dvc3601/files/document/2022/03/40233%2C150169_BiothaneAD_brochure.pdf

Vergara, Sintana E., and George Tchobanoglous. 2012. "Municipal solid waste and the environment: a global perspective." *Annual Review of Environment and Resources* 37:277–309.

Verma, Shefali. 2002. "Anaerobic digestion of biodegradable organics in municipal solid wastes." *Columbia University* 7(3):98–104.

Wainaina, Steven, Lukitawesa, Mukesh Kumar Awasthi, and Mohammad J. Taherzadeh. 2019. "Bioengineering of anaerobic digestion for volatile fatty acids, hydrogen or methane production: a critical review." *Bioengineered* 10(1):437–458.

Wang, Dou, Jing Ai, Fei Shen, Gang Yang, Yanzong Zhang, Shihuai Deng, Jing Zhang, Yongmei Zeng, and Chun Song. 2017. "Improving anaerobic digestion of easy-acidification substrates by promoting buffering capacity using biochar derived from vermicompost." *Bioresource Technology* 227:286–296.

Wang, Ling-Ling, Long-Fei Wang, Xue-Mei Ren, Xiao-Dong Ye, Wen-Wei Li, Shi-Jie Yuan, Min Sun, Guo-Ping Sheng, Han-Qing Yu, and Xiang-Ke Wang. 2012. "pH dependence of structure and surface properties of microbial EPS." *Environmental Science & Technology* 46(2):737–744.

Wang, Yaya, Bing Zang, Guoxue Li, and Yu Liu. 2016. "Evaluation the anaerobic hydrolysis acidification stage of kitchen waste by pH regulation." *Waste Management* 53:62–67.

Weltec Biopower. 2024a. "Arneburg biomethane refinery." https://www.weltec-biopower.com/plant-operation-service/own-plants/arneburg.html

Weltec Biopower. 2024b. "Biogas production and utilisation process." https://www.weltec-biopower.com/fileadmin/user_upload/weltec/08_Infocenter/Prospekte/englisch/Fotostory_Abfall_engl.pdf

Weltec Biopower. 2024c. "Company." https://www.weltec-biopower.com/company.html

Wu, Xiao, Wanying Yao, and Jun Zhu. 2010. "Effect of pH on continuous biohydrogen production from liquid swine manure with glucose supplement using an anaerobic sequencing batch reactor." *International Journal of Hydrogen Energy* 35(13):6592–6599.

Xu, Rong, Kai Zhang, Pu Liu, Aman Khan, Jian Xiong, Fake Tian, and Xiangkai Li. 2018. "A critical review on the interaction of substrate nutrient balance and microbial community structure and function in anaerobic co-digestion." *Bioresource Technology* 247:1119–1127.

Yan, Shoubao, Jun Li, Xiangsong Chen, Jingyong Wu, Pingchao Wang, Jianfeng Ye, and Jianming Yao. 2011. "Enzymatical hydrolysis of food waste and ethanol production from the hydrolysate." *Renewable Energy* 36(4):1259–1265.

Yi, Jing, Bin Dong, Jingwei Jin, and Xiaohu Dai. 2014. "Effect of increasing total solids contents on anaerobic digestion of food waste under mesophilic conditions: performance and microbial characteristics analysis." *PloS One* 9(7):e102548.

Zhao, Danyang, Binghua Yan, Chao Liu, Bin Yao, Lin Luo, Yang, Lichao Liu, Fan Wu, and Yaoyu Zhou. 2021. "Mitigation of acidogenic product inhibition and elevated mass transfer by biochar during anaerobic digestion of food waste." *Bioresource Technology* 338:125531.

Zhu, Xinyu, Songmei Zhu, Ze Zhao, Xiaoxi Kang, and Feng Ju. 2023. "Microbiome dynamics during anaerobic digestion of food waste and the genetic potential for poly (lactic acid) co-digestion." *Chemical Engineering Journal* 473:145194. doi: https://doi.org/10.1016/j.cej.2023.145194

Zorg Biogas. 2024a. "Continuous stirred-tank reactor (CSTR)." https://zorg-biogas.com/production-technology/classic-cstr-technology

Zorg Biogas. 2024b. "Portfolio." https://zorg-biogas.com/about-zorg/portfolio

9 Synergies in the Water–Energy–Food Nexus

Transforming challenges into opportunities for food waste valorisation

César Ramírez-Márquez, Alma Yunuen Raya-Tapia, Xate Geraldine Sánchez-Zarco, Brenda Cansino-Loeza, and José María Ponce-Ortega

9.1 INTRODUCTION

The trio of water, energy, and food, collectively referred to as the Water–Energy–Food (WEF) Nexus, is foundational to human prosperity, poverty alleviation, and sustainability (Raya-Tapia et al. 2023b). These indispensable elements form a tightly interwoven tapestry, encapsulating the intricate relationships and interdependencies among Water–Energy (WE), Energy–Food (EF), Food–Water (FW), and the broader synergies within the WEF spectrum (Scott et al. 2015).

Within this nexus, water assumes a multifunctional role. It not only supports agricultural efforts through irrigation but also significantly contributes to electrical energy generation (Tovar-Facio et al. 2023; Sánchez-Zarco et al 2023). Similarly, energy is vital for the cultivation, processing, and distribution of food and plays a pivotal role in water conservation and management (Bazilian et al. 2011). However, ensuring a consistent and sufficient supply of these WEF resources faces challenges, including climate variations, rising population, urban expansion, and economic growth (Rasul 2014). As the global population grows and urban areas transform, the demand for these essential resources surges accordingly.

In response to these growing demands, the WEF Nexus methodology has been thoughtfully designed to address potential resource scarcities and enhance societal well-being (Bazilian et al. 2011). This approach aligns with the Sustainable Development Goals (SDGs) and has gained prominence for effectively managing WEF resources (Endo et al. 2015).

DOI: 10.1201/9781032718507-9

The discourse delves into the interplay between food waste valorisation and the WEF Nexus (see Figure 9.1), underscoring how judicious management transforms waste into a linchpin connecting water, energy, and food systems (Gomez et al. 2008). By converting organic material into valuable resources, we optimize water use in a sustainable loop. Beyond merely addressing waste reduction, this synergy bolsters security across all three sectors, epitomizing a shift from challenges to opportunities.

Various tools and methodologies facilitate navigating the complexities of the WEF Nexus, incorporating food waste valorisation to cohesively address interconnected challenges. Among these tools are life cycle assessment (LCA) systems (Finnveden et al. 2009), sustainability assessment methodologies like the Leopold Matrix or Environmental Impact Assessment (EIA) (Banyal et al. 2019), and social network analysis (SNA) tools (Kurian et al. 2018). This integration is crucial for devising innovative solutions and shaping a resilient future.

Such an approach fosters collaboration and advocates for solutions that are technically viable, socially equitable, and environmentally sustainable (Scoones et al. 2020). Integrating the principles of the WEF Nexus with Food Waste Valorisation enables the exploration of innovative pathways for sustainable development.

Within the WEF Nexus, viewing food waste as a valuable resource is a paradigm-shifting perspective (Gomez et al. 2008). Redefining food waste unveils opportunities within the WEF Nexus. Discarded organic materials can be harnessed to minimise waste and present a sustainable alternative to conventional energy sources.

At the same time, food waste can contribute to water conservation strategies. Water used in processing food waste can be treated and returned to the agricultural sector for irrigation, creating a sustainable loop (Strunz and Schindler 2018). Additionally, nutrients from food waste can be reclaimed through processes like composting and producing biofertilisers that enhance soil fertility and contribute

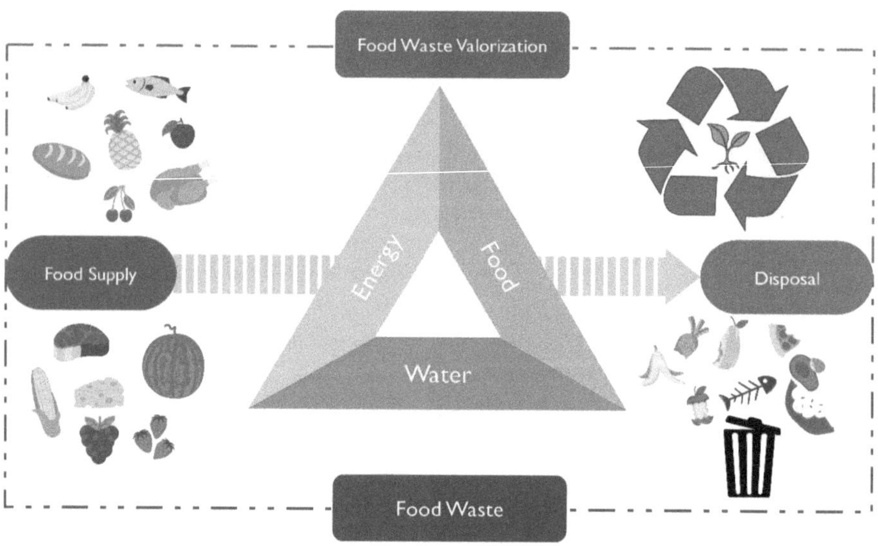

FIGURE 9.1 Food waste valorisation and the WEF Nexus.

to sustainable agriculture (Diacono et al. 2019). Food losses and food waste can be given added value as they are a source of potentially valuable bioactive compounds, such as oils, carotenoids, dietary fibers, enzymes, polyphenols, and vitamins. These compounds can be used in different industries, including the food industry, as natural food additives instead of synthetic additives.

Recognizing food waste as a valuable resource stimulates innovation and encourages the development of technologies that facilitate its integration within the WEF Nexus (Papargyropoulou et al. 2014). In essence, recognizing food waste as a resource within the WEF Nexus signifies a transformative shift that aligns economic, environmental, and social objectives.

The implications of food waste go beyond disposal, impacting both the environmental and economic realms (Parfitt et al. 2010). The global commitment to halve food waste by 2030 reflects a growing awareness and has stimulated research into the issue. In Europe, the amount of wasted food has been quantified in the millions of tons, with a hefty price tag, making up a significant portion of their total food production. This pattern is mirrored worldwide, with numerous studies from different countries shedding light on the magnitude and nuances of food waste, indicating a universal challenge that demands comprehensive strategies (Hoehn et al. 2023). Ecologically, wasting food equates to a direct depletion of natural resources. Economically, discarded food represents a tangible loss of financial assets. Addressing food waste can also give rise to new industries and job opportunities.

Navigating the domain of food waste valorisation offers opportunities but also presents technical challenges (Stenmarck et al. 2016). These challenges encompass various stages, from collection and segregation to processing and utilisation of the transformed resources. To overcome these technical challenges, a combination of research, innovation, and stakeholder collaboration is essential.

In the realm of food waste valorisation within the WEF Nexus, several regulatory and policy obstacles can impede the implementation of innovative solutions (Mourad 2016). To address these regulatory and policy obstacles, a unified and collaborative approach is needed.

Consumer perceptions and behaviors are pivotal in determining the direction of food waste valorisation initiatives within the WEF Nexus (Aschemann-Witzel et al. 2015). The success of products generated from food waste depends on consumer perceptions. Engaging and educating consumers can foster behaviors that reinforce and enhance the efficacy of food waste valorisation efforts within the interlinked domains of WEF.

Expanding on the fundamental understanding of the WEF Nexus and the transformative potential of food waste valorisation, this chapter aims to examine and clarify the intricate links between water, energy, and food systems, offering a thorough overview of existing tools, methodologies, and innovative solutions.

9.2 INNOVATIVE SOLUTIONS FOR FOOD WASTE VALORISATION WITHIN THE WEF NEXUS

Addressing food waste valorisation within the WEF Nexus requires a multifaceted approach that merges innovative solutions, sustainable practices, and collaborative efforts. By navigating through these complexities, the intertwined challenges in the nexus can be cohesively tackled. The subsequent sections delve into emerging technologies, sustainable practices, and public–private partnerships (PPP) that provide pioneering solutions for valorizing food waste in the context of the WEF Nexus.

9.2.1 Emerging technologies for food waste conversion

The transformative potential of emerging technologies to convert food waste into valuable assets is immense. Anaerobic digestion is at the forefront of these technologies, facilitating the conversion of organic waste into biogas. Smith et al. (2016) demonstrated the efficacy of this method in converting agricultural and food waste into clean energy. Another innovative technology is the use of microbial fuel cells, which convert chemical energy in food waste directly into electricity through the metabolic processes of microorganisms (Logan and Rabaey 2012).

In recent years, pyrolysis and gasification have gained prominence. These thermochemical processes enable the conversion of food waste into biochar, syngas, and bio-oil. Biochar produced from food waste can be used as a soil conditioner, while syngas and bio-oil can be utilised as renewable energy sources (Bridgwater 2012). Another promising option is the hydrothermal carbonisation of food waste, which produces a carbon-rich solid fuel (hydrochar) that can replace fossil fuels in the steel production process (Munir et al. 2023). Moreover, advances in enzymatic conversion processes facilitate the extraction of high-value products like bioactive peptides and antioxidants from food waste (Mirabella et al. 2014).

Technological innovations are also exploring the conversion of food waste into bioplastics and biopolymers. Through fermentation processes, food waste can be transformed into biodegradable plastics, offering a sustainable alternative to petroleum-based plastics (Kumar et al. 2009). Additionally, emerging research is focusing on the use of food waste as a substrate for cultivating edible fungi and insects, thereby contributing to the creation of alternative protein sources (Glover and Sexton 2015). This is the case of biomanufacturing systems where value-added products from different ranges such as metabolites, proteins, and enzymes from food waste are produced, and where fermentation is the core of the system (Kim et al. 2022). Fermentation is an important bioprocess that can replace petrochemical processes since novel bioproducts can be produced more sustainably and circularly.

The integration of these technologies in smart waste management systems can lead to optimized food waste valorisation. Leveraging artificial intelligence (AI) and machine learning algorithms can improve waste sorting and processing efficiency (Munir et al. 2023). Technologies such as blockchain can also contribute to traceability and transparency in food waste management supply chains.

Considering the evolving climate scenario, constant innovation and adaptation of these technologies become paramount. It is essential to stay abreast of the latest advancements in waste conversion technologies and continuously refine existing methodologies. Tailoring these technologies to suit different geographies, types of food waste, and energy requirements will be vital in ensuring their scalability and impact. Several technologies stand out for their efficiency, sustainability, and environmental friendliness in food waste valorisation. Anaerobic digestion efficiently converts organic waste into biogas, while microbial fuel cells directly transform food waste into electricity using microorganisms' metabolic processes. Pyrolysis and gasification generate renewable energy sources such as biochar, syngas, and bio-oil. Hydrothermal carbonisation produces hydrochar, a sustainable alternative to fossil fuels. Integrating these technologies into smart waste management systems, enhanced by AI, optimizes food waste valorisation and aligns with environmental sustainability goals.

9.2.2 Sustainable Practices in Food Waste Management

Sustainable management of food waste within the WEF Nexus necessitates practices that not only reduce waste generation but also facilitate its efficient utilisation. Composting, while a traditional method, has undergone several advancements to enhance its efficiency and nutrient recovery (Zhang et al. 2007). Vermicomposting, which involves the use of earthworms to expedite the composting process, is an example of a sustainable practice that has been refined over time.

The adoption of a circular economy approach to food waste management is essential. This perspective emphasizes the conversion of food waste into resources and seeks to close the loop in food production and consumption systems (Stenmarck et al. 2016). Incorporating practices such as eco-design, which considers the entire lifecycle of a product, can lead to packaging innovations that minimise waste generation.

Supply chain optimization is another sustainable practice that holds promise. By implementing technologies like predictive analytics and intelligent routing, perishable goods can be delivered more efficiently, thereby reducing the likelihood of spoilage and waste (Papargyropoulou et al. 2014). Additionally, adopting alternative food preservation methods, such as freeze-drying and vacuum packing, can significantly extend the shelf life of food products.

Consumer education and awareness are fundamental to driving sustainable practices in food waste management. Encouraging responsible consumption through campaigns, labeling, and public awareness programs can lead to a substantial reduction in food waste at the household level (Reynolds et al. 2019). Mobile applications that facilitate the sharing or donation of surplus food are also effective tools in promoting sustainable consumer behavior.

Legislation and policies can foster sustainable practices by incentivizing food waste reduction and valorisation. Implementing tax breaks or subsidies for businesses that invest in sustainable food waste management practices can drive industry-wide change. Similarly, stricter regulations regarding food waste disposal can motivate companies to seek alternative, sustainable solutions.

Incorporating sustainable practices requires an in-depth understanding of the social, economic, and environmental dimensions of food waste. By developing practices that are culturally relevant, economically viable, and environmentally beneficial, the WEF Nexus can harness the full potential of food waste valorisation.

9.2.3 Public–Private Partnerships for Sustainable Solutions

PPPs are instrumental in fostering innovative solutions for food waste valorisation within the WEF Nexus. By combining the resources, expertise, and reach of public and private entities, PPPs can create synergies that accelerate the adoption of sustainable practices (Mourad 2016). For instance, PPPs can enable the establishment of advanced biogas plants that effectively convert food waste into energy (Kaza et al. 2018).

The role of PPPs in funding and scaling innovative technologies cannot be overstated. By pooling resources, PPPs can facilitate the development of infrastructure and technologies necessary for food waste valorisation. Through strategic alliances, the risks, and rewards of investing in emerging technologies can be shared, thus enhancing the feasibility of such ventures. Furthermore, PPPs can foster research and development in food waste valorisation. By creating a collaborative environment, PPPs can encourage knowledge exchange, innovation, and the development of new technologies. Academic institutions, research organizations, and private companies can work together to advance the science and practice of food waste valorisation.

Government-backed initiatives can stimulate the participation of the private sector. By providing incentives, subsidies, or favorable regulatory frameworks, governments can create an enabling environment for the private sector to engage in food waste management (Mourad 2016). Examples include tax incentives for companies investing in food waste reduction technologies or infrastructure.

PPPs can also play a significant role in capacity building and workforce development. By fostering collaborations between educational institutions, industry, and government agencies, PPPs can ensure the development of a skilled workforce capable of implementing and managing food waste valorisation technologies and practices.

Engaging communities through PPPs is another approach that can lead to sustainable solutions. Community-driven initiatives, supported by PPPs, can encourage grassroots involvement in food waste management. Such partnerships can lead to localized solutions that are culturally relevant and sustainable.

PPPs offer a powerful avenue for harmonizing efforts across different sectors and stakeholders. By effectively leveraging the strengths of public and private entities, PPPs can drive forward the agenda of food waste valorisation within the WEF Nexus, ensuring a sustainable and resource-efficient future.

9.3 CASE STUDIES: EFFECTIVE FOOD WASTE VALORISATION INITIATIVES IN THE CONTEXT OF THE WEF NEXUS

9.3.1 Spain's Regional Innovations in Food Waste Valorisation and the WEF Nexus

The extensive research conducted by (Raya-Tapia et al. 2023a) provides a panoramic assessment of food waste valorisation efforts within the WEF Nexus across Spain's autonomous communities. This section condenses and explores the substantial data, highlighting unique initiatives and their quantifiable impacts as represented in a detailed heat map (see Figure 9.2).

The methodology for evaluating Spain's WEF nexus security revolves around a composite index based on 13 indicators from 2010 to 2020 (Table 9.1). This index captures the intricacies of resource availability and distribution, as well as inter-regional resource exchange, against the backdrop of Spain's significant water stress—aggravated by climate-induced droughts. The indicators, encompassing water availability, renewable resources, energy production and consumption,

TABLE 9.1
Comprehensive list of the 13 indicators used to compute the WEF-waste index

$$W_{1,c,t} = \frac{w_{c,t}^{production} - w_{c,t}^{consumption}}{w_{c,t}^{production}} \quad \forall c \in C, t \in T \tag{9.1}$$

$$W_{2,c,t} = \frac{w_{c,t}^{consumption} - w_{c,t}^{import}}{w_{c,t}^{consumption}} \quad \forall c \in C, t \in T \tag{9.2}$$

$$W_{3,c,t} = \frac{w_{c,t}^{wtreated}}{w_{c,t}^{consumption}} \quad \forall c \in C, t \in T \tag{9.3}$$

$$W_{4,c,t} = \frac{w_{c,t}^{wreused}}{w_{c,t}^{consumption}} \quad \forall c \in C, t \in T \tag{9.4}$$

$$E_{1,c,t} = \frac{e_{c,t}^{generation} - e_{c,t}^{consumption}}{e_{c,t}^{generation}} \quad \forall c \in C, t \in T \tag{9.5}$$

$$E_{2,c,t} = \frac{e_{c,t}^{renewable}}{e_{c,t}^{generation}} \quad \forall c \in C, t \in T \tag{9.6}$$

$$E_{3,c,t} = \frac{e_{c,t}^{consumption} - e_{c,t}^{import}}{e_{c,t}^{consumption}} \quad \forall c \in C, t \in T \tag{9.7}$$

$$F_{1,c,g,t} = \frac{f_{c,g,t}^{production} - f_{c,g,t}^{consumption}}{f_{c,g,t}^{production}} \quad \forall c \in C, g \in G, t \in T \tag{9.8}$$

$$F_{2,c,t} = \frac{f_{c,t}^{production} - f_{c,g,t}^{import}}{f_{c,t}^{production}} \quad \forall c \in C, t \in T \tag{9.9}$$

$$F_{3,c,t} = \frac{f_{c,t}^{surface}}{f_{t}^{totalsurface}} \quad \forall c \in C, t \in T \tag{9.10}$$

$$F_{4,c,t} = \frac{f_{c,t}^{waste}}{f_{c,t}^{production}} \quad \forall c \in C, t \in T \tag{9.11}$$

$$R_{1,c,t} = \frac{r_{c,t}^{separatedwaste}}{r_{c,t}^{totwaste}} \quad \forall c \in C, t \in T \tag{9.12}$$

$$R_{2,c,t} = \frac{r_{c,t}^{reused\ waste}}{r_{c,t}^{totwaste}} \quad \forall c \in C, t \in T \tag{9.13}$$

$$\text{WEF} - \text{Waste index}_{c,t} = \sum_{n=1}^{N} \varphi_n \bullet I_{n,c,t} \tag{9.14}$$

$$\sum_{n=1}^{N} \varphi_n = 1 \tag{9.15}$$

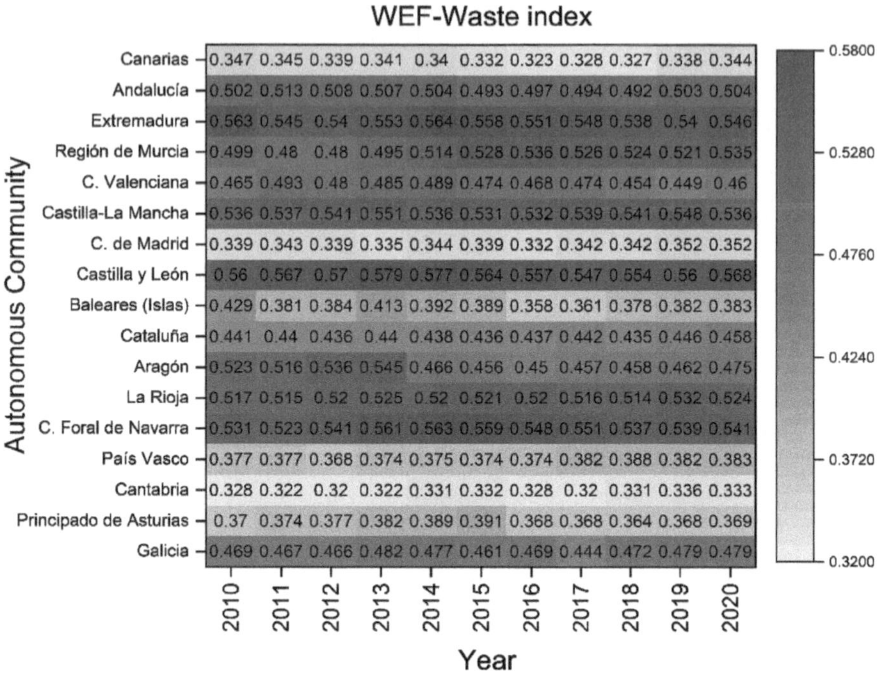

FIGURE 9.2 The impact of food waste valorisation in autonomous communities across Spain.

agricultural productivity, and waste management, are normalized and aggregated to reflect the performance of each autonomous community. This analysis highlights regional disparities and trends, providing a nuanced understanding that can guide the implementation of resource management policies in water-stressed industrialized countries.

In the Community of Madrid, food waste initiatives have significantly reduced waste from 509,624 tons in 2010 to 159,734 tons in 2020, alongside an increase in food production. In stark contrast, Cantabria, which initially had a high waste-to-production ratio, has seen considerable improvements through the implementation of waste reduction strategies.

Murcia's valorisation project stands as a beacon of sustainability, with its high reuse rate of treated wastewater in agriculture, achieving 91.4%. The region's focus on biogas production from food waste has not only improved waste management but also contributed to energy self-sufficiency.

In Andalusia, the approach to food waste valorisation has centered around the enhancement of the region's vast agricultural sector. By converting waste into biofertilisers, the community has seen an increase in soil fertility and a reduction in chemical fertiliser dependency, contributing to a more sustainable agricultural landscape.

The Valencian Community has also made strides in integrating food waste management with its robust agricultural sector, particularly in citrus production, which accounts for 56.7% of the EU's total. The region has leveraged food waste to supplement energy needs, particularly through the adoption of solar photovoltaic technology, which has seen an 80.4% increase in installation.

Catalonia's initiatives have been vital in addressing its water deficit, with significant investments in desalination and water reuse technologies. By integrating water resource management with food waste valorisation, Catalonia has effectively reduced its water stress while supporting its dense population and industrial activities.

The Balearic Islands present a unique case where the valorisation of food waste is essential for maintaining the delicate balance of its ecosystem. The islands have utilised treated wastewater for the irrigation of recreational areas and agriculture, ensuring the conservation of their limited freshwater resources.

Across these communities, the valorisation of food waste has been instrumental in addressing the interlinked challenges of water scarcity, energy demand, and food security. Key lessons include the critical role of regional adaptation of waste valorisation technologies and the value of cross-sector collaboration.

Best practices identified in the study involve the creation of closed-loop systems where waste from one process becomes the input for another, fostering circular economies within each community. Transparency in reporting and a commitment to public education have also been pivotal in achieving community buy-in and participation.

In closing, the multifaceted efforts detailed by (Raya-Tapia et al. 2023a) reflect a national commitment to the principles of the WEF Nexus. The diverse approaches across Spain's autonomous communities highlight the potential for tailored strategies to meet specific regional needs, setting a global example for sustainable resource management within the WEF Nexus framework.

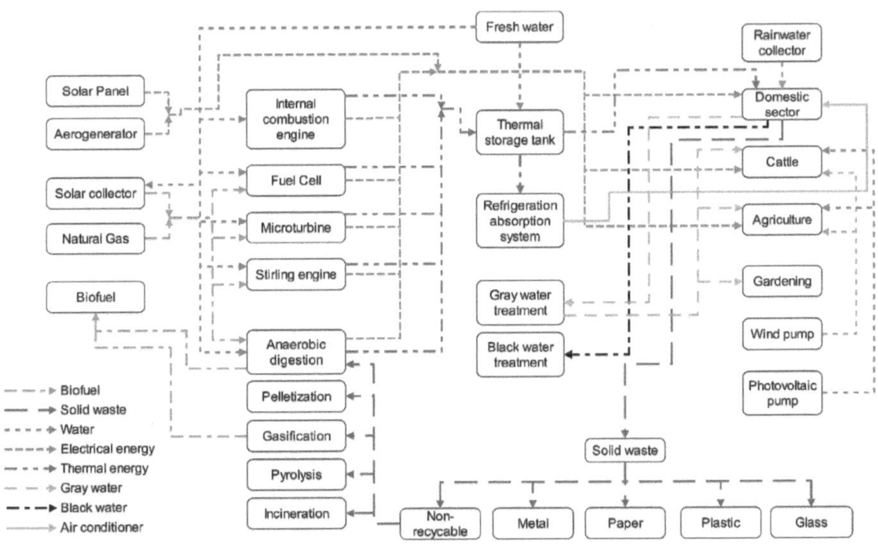

FIGURE 9.3 Superstructure for waste valorisation within the WEF Nexus in Cochoapa el Grande.

9.3.2 Cochoapa el Grande's Integrated Approach to Food Waste Valorisation within the WEF Nexus

The research by (Cansino-Loeza and Ponce-Ortega 2019) provides a clear demonstration of this strategy through a case study focused on "Cochoapa el Grande," Guerrero, Mexico, which is marked by significant resource constraints.

The core methodology of this research is the application of Mixed-Integer Linear Programming (MILP), which facilitates the design of a polygeneration system that simultaneously addresses the community's WEF requirements. The integration of waste valorisation into this system is critical, as it transforms food waste, typically an environmental burden, into a valuable asset within the WEF Nexus (see Figure 9.3).

Empirical data from the case study reveal the quantitative success of waste valorisation efforts. The optimization model leads to a strategic reduction in annual freshwater usage to about 3,060,000 m³. Furthermore, the model results indicate that even when prioritizing cost minimization, the increase in water consumption is limited to only 4% over the minimum water usage scenario, showcasing the efficiency of the waste valorisation mechanisms employed.

The cogeneration technologies, tailored to the energy component of the WEF Nexus, exhibit a reliance on the valorisation of waste materials, with food waste serving as a substrate for energy production (see Figure 9.4). The implementation of pelletization techniques converts a considerable amount of this waste into pellets. These substances not only fulfill a portion of the community's energy demand but also act as a source of economic return, reinforcing the financial viability of waste valorisation.

The work of (Cansino-Loeza and Ponce-Ortega 2019) thus emphasizes the substantial role that waste valorisation initiatives play within the WEF Nexus. It demonstrates that through the effective management and conversion of organic

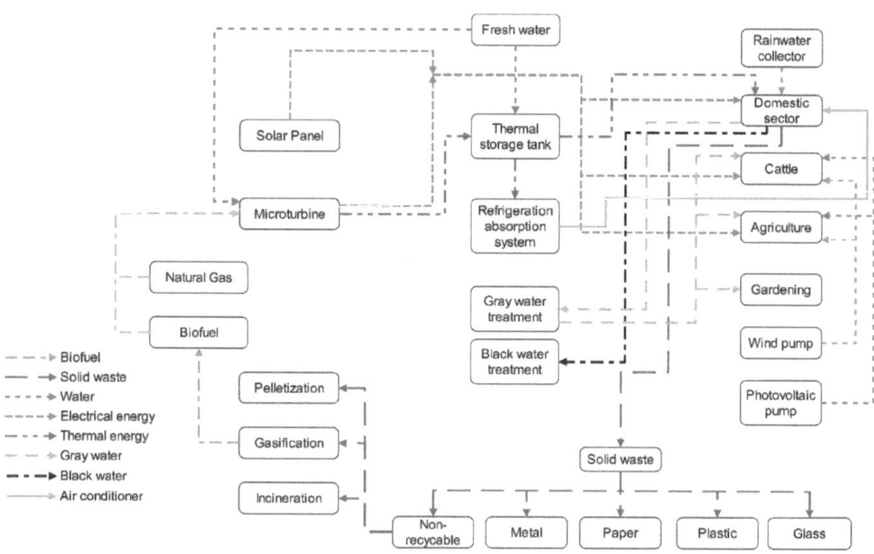

FIGURE 9.4 Optimized superstructure for waste valorisation within the WEF Nexus in Cochoapa el Grande.

waste, communities can enhance their sustainability and self-sufficiency, achieving significant gains in resource conservation and economic benefits. This case study offers a replicable model for integrating waste valorisation into the WEF Nexus, which can be adapted for use in other isolated or resource-limited communities globally.

9.4 POLICY AND GOVERNANCE FRAMEWORKS

The complexities and interdependencies within the WEF Nexus require strategic policy and governance frameworks to promote food waste valorisation. Effective policymaking can catalyse significant strides in managing food waste, aligning diverse initiatives to establish a cohesive approach to addressing interconnected challenges.

9.4.1 Policy recommendations for promoting food waste valorisation

Developing impactful policies for food waste valorisation entails recognizing the multifaceted aspects of the WEF Nexus. One recommendation is to introduce fiscal incentives, such as tax breaks or subsidies, to stimulate investments in valorisation technologies like anaerobic digestion and composting (Shahid and Hittinger 2021). By creating a favorable economic environment, governments can encourage businesses and municipalities to adopt sustainable practices.

Regulatory frameworks should advocate for Standardised food waste management procedures. Clear guidelines on food labeling, waste sorting, and recycling can ensure consistency and effectiveness in handling food waste. Standardisation can foster market conditions that support products derived from food waste, opening up new economic avenues (FAO 2019).

Strategic urban planning policies should seamlessly integrate food waste valorisation infrastructure. By incorporating systems for efficient collection, transportation, and processing of food waste, cities can enhance their sustainability profiles. Designing urban spaces with waste valorisation in mind can lead to systematic improvements in waste management (Paes et al. 2019).

Supporting research and development is paramount. Policymakers can channel funds into academic and private sector research aimed at uncovering innovative solutions for food waste valorisation. Research partnerships can foster collaborative approaches, resulting in breakthrough technologies (Papargyropoulou et al. 2014).

Consumer education and awareness campaigns are integral to policy frameworks. By encouraging responsible consumption and waste reduction, policymakers can influence behavior at the grassroots level, thus mitigating food waste generation (Quested et al. 2013).

Crafting cross-sectoral policies that holistically address energy, water, and food sectors is crucial for cohesive food waste valorisation strategies. By addressing the WEF Nexus in an integrated manner, policies can create synergies between sectors, enhancing overall effectiveness (Biggs et al. 2015).

International collaboration is vital for sharing knowledge and best practices. Harmonising policies across borders can lead to standardised and effective strategies for food waste valorisation on a global scale.

Adaptive policy frameworks that evolve with changing landscapes ensure long-term relevance. Periodic assessments and updates to policies can ensure that they continually drive innovation and sustainability in food waste valorisation efforts.

9.4.2 International and National Initiatives in Food Waste Reduction

International initiatives play a key role in steering global efforts toward food waste reduction. The United Nations' SDG 12 sets a clear agenda for halving food waste by 2030, emphasizing the importance of sustainable consumption patterns (Prokic et al. 2022).

The European Union's Circular Economy Action Plan is a testament to regional efforts in waste reduction. By outlining strategies to promote waste valorisation across member states, the plan has spurred concerted action toward sustainability (Esposito et al. 2020).

Nationally, countries have initiated policies that align with global directives. France's legislative measures against supermarket food waste have led to an increase in food donations to charities, demonstrating the impact of policy intervention (European Commission 2023).

The United States has leveraged alliances like the Food Waste Reduction Alliance (FWRA) to foster collaboration between food sector companies in efforts to reduce waste. FWRA has been instrumental in increasing food donations and recycling while reducing landfill disposal (Mourad 2016).

In South Korea, a volume-based waste fee system has successfully incentivized households to minimise food waste generation, showcasing policy effectiveness at the consumer level (Koh 2020).

Japan's Food Recycling Law and the United Kingdom's Waste and Resources Action Programme (WRAP) are examples of national strategies aimed at food waste reduction and recycling (Liu et al. 2016).

PPPs facilitated by national governments can scale up food waste reduction initiatives. By combining resources and expertise, these partnerships can amplify the impact of successful models.

Policy and governance frameworks are integral to fostering food waste valorisation within the WEF Nexus. Strategic alignment and collaboration between international and national initiatives can create a comprehensive and effective approach to managing food waste.

9.5 LEVERAGING ECONOMIC AND ENVIRONMENTAL BENEFITS OF FOOD WASTE VALORISATION IN THE WEF NEXUS

The process of food waste valorisation, wherein discarded organic materials are repurposed and transformed into valuable commodities, presents an opportunity to reap both economic and environmental benefits within the WEF Nexus. By understanding and strategically leveraging these benefits, societies can foster sustainable development, encourage investments, and ensure that food waste management aligns with broader societal goals. This section delves into the economic benefits and opportunities for investment in food waste valorisation projects, assesses the environmental merits, and explores the social implications of these endeavors.

9.5.1 Exploring the economic benefits of food waste valorisation

The economic potential of food waste valorisation within the WEF Nexus is substantial. By transforming food waste into valuable products such as bioenergy, biofertilisers, and animal feed, we can unlock significant economic gains (Mirabella et al. 2014). Innovative technologies such as anaerobic digestion and pyrolysis have shown promising results in the efficient extraction of energy from food waste (Bridgwater 2012). Strategically implemented food waste valorisation can also lead to cost savings for municipalities and businesses. By reducing the volume of waste directed to landfills, the costs associated with waste management and disposal can be significantly curtailed (Kaza et al. 2018). Simultaneously, the revenue generated from the sale of by-products such as compost and bioenergy can offer additional financial benefits. Furthermore, the agroindustry can capitalise on food waste by converting it into animal feed or agricultural supplements, thereby offsetting the costs of synthetic fertiliser and feeds (Tan et al. 2016). Small and medium enterprises (SMEs) can find new market niches in food waste valorisation, thus promoting entrepreneurship and stimulating local economies.

Market mechanisms such as carbon credits and subsidies can also enhance the economic viability of food waste valorisation. By aligning the valorisation processes with environmental goals, businesses can access new funding streams and enhance their sustainability credentials (Papargyropoulou et al. 2014). Moreover, the integration of food waste valorisation into existing supply chains can lead to increased resilience and risk mitigation. For instance, utilising food waste in bioenergy production can provide a stable energy supply and reduce dependency on fossil fuels (Smith et al. 2016).

The overall economic landscape of food waste valorisation underscores the need for holistic analyses and multi-stakeholder collaborations. By understanding the broader economic implications, policymakers and industry leaders can tailor strategies to maximise the economic benefits of food waste valorisation.

The economic impact of food waste valorisation is increasingly recognized in various sectors. Recent analyses estimate that globally, approximately one-third of food produced annually is wasted, which represents a significant economic loss and a substantial opportunity for valorisation. Specifically, the Food and Agriculture Organization (FAO) reported that around 14% of the world's food, valued at approximately $400 billion, is lost from post-harvest up to the retail level annually. Furthermore, the valorisation of food waste can mitigate environmental impacts significantly (Dongyu 2019). Ineffectual management of food waste contributes to around 8% of total anthropogenic greenhouse gas emissions (Bhatia et al. 2023). Therefore, transforming food waste into value-added products not only presents an opportunity to reduce environmental damage but also to enhance economic gains through the creation of new products and industries.

9.5.2 Opportunities for investment in food waste valorisation projects

Investing in food waste valorisation projects can yield significant returns and contribute to sustainable development. These investments can range from infrastructural developments, such as the construction of biogas plants, to the funding of research and development in food waste conversion technologies (Kaza et al. 2018).

PPPs can play a pivotal role in facilitating investments in food waste valorisation. Through these collaborations, risks and rewards can be shared, making it more feasible for private entities to invest in innovative solutions (Mourad 2016). Governments can stimulate private sector participation through incentives, subsidies, or favorable regulatory frameworks.

Impact investing, focusing on investments that generate social and environmental impact alongside financial returns, can be particularly relevant for food waste valorisation projects (Born and Brest 2013). Investors seeking to align their portfolios with the SDGs may find opportunities in ventures that tackle food waste while ensuring sustainable resource use. Moreover, green bonds and other financial instruments can be tailored to support food waste valorisation projects. By specifically targeting environmentally sustainable initiatives, these instruments can channel capital toward projects that might otherwise struggle to find funding (Leung et al. 2023).

International organizations and development banks can also play a role in fostering investments in food waste valorisation in developing countries. By providing financial support and technical expertise, these institutions can help bridge the gap between innovation and implementation (Shneiderman 2020).

At a grassroots level, crowdfunding platforms can enable smaller-scale food waste valorisation projects to access capital. This democratisation of funding can empower communities and foster localized solutions, allowing a diverse array of stakeholders to contribute to the valorisation effort (Mollick 2014).

Ultimately, a conducive investment climate, coupled with informed decision-making and strategic partnerships, can catalyse the scaling up of food waste valorisation projects, thereby maximising their impact across the WEF Nexus.

9.5.3 Assessing the Environmental Benefits and Social Implications of Food Waste Valorisation

The environmental benefits of food waste valorisation are closely aligned with the principles of circular economy and sustainable resource management. By converting food waste into energy, compost, or other products, the pressure on natural resources and ecosystems can be significantly reduced (Despoudi et al. 2021).

Carbon footprint reduction is one of the key environmental benefits of food waste valorisation. By utilizing food waste for bioenergy production, greenhouse gas emissions associated with traditional energy sources can be mitigated (Smith et al. 2016). Additionally, composting, and anaerobic digestion of food waste can sequester carbon and improve soil health (Brown et al. 2008). To delve deeper, it is crucial to discuss the specific mechanisms through which these valorisation processes impact the carbon cycle. For instance, transforming food waste into bioenergy not only prevents methane emissions from landfill decomposition but also replaces fossil fuels, thereby doubly reducing the carbon footprint. Furthermore, the enhancement of soil health through compost application leads to increased carbon sequestration, further mitigating climate change impacts. These discussions can elucidate the multifaceted benefits of food waste valorisation in reducing the carbon footprint and advancing sustainable environmental practices.

Water conservation is another critical aspect of food waste valorisation. Processes that repurpose food waste can reduce the demand for freshwater resources in agriculture and energy production (Hoekstra and Mekonnen 2012). By integrating food waste into water management strategies, a more sustainable and resilient water supply can be established.

Biodiversity can also be preserved through food waste valorisation. By reducing the need for additional agricultural land and minimising habitat destruction, food waste valorisation can indirectly contribute to the conservation of biodiversity (Tilman et al. 2011).

On the social front, food waste valorisation can lead to job creation and community development. By investing in local food waste processing facilities, new employment opportunities can emerge, contributing to local economies and fostering social cohesion (Montiel et al. 2021). Furthermore, the integration of food waste valorisation into educational and awareness campaigns can drive changes in consumer behavior and societal norms. By fostering an understanding of the value of food and the potential of waste, communities can be empowered to participate actively in creating a sustainable future (Shutaleva 2023).

Leveraging the economic and environmental benefits of food waste valorisation within the WEF Nexus requires a multifaceted approach. By acknowledging and strategically aligning the opportunities for investment, economic gains, environmental benefits, and social implications, we can foster a more sustainable, equitable, and prosperous future.

9.6 FUTURE DIRECTIONS AND CONCLUSION

Advancements in the WEF Nexus hinge upon anticipating the trajectory of food waste valorisation and mapping out the future with strategic insights. This section seeks to explore the horizons of research and innovation, anticipate upcoming trends in the field, and encapsulate the crux of the discourse by summarising key takeaways. Through this comprehensive approach, we aim to chart out a path that aligns with sustainability goals and ensures a resilient future.

9.6.1 Identifying Areas for Further Research and Innovation

The landscape of food waste valorisation within the WEF Nexus is expansive and demands rigorous inquiry into several potential avenues. First, there is a need for extensive research into optimizing the efficiency of existing food waste conversion technologies, such as anaerobic digestion and pyrolysis (Chen et al. 2008). Innovations in these technologies could significantly improve the yield and quality of the derived products. Furthermore, the development of novel materials from food waste, such as advanced bio-composites and nanomaterials, is a burgeoning field that warrants deeper investigation (Phiri et al. 2023). These materials can find applications in various industries, expanding the scope of food waste valorisation.

Integrating the Internet of Things (IoT) and blockchain technologies for real time monitoring and traceability of food waste is another area that requires further exploration (Saberi et al. 2019). These technologies can facilitate more efficient and transparent food waste management systems. In this context, (AI is emerging as a powerful tool for process optimisation and traceability by tracking a food product's journey through the supply chain in real time to ensure quality and food safety. These AI-based systems such as machine learning and big data analytics make it possible to identify and correct problems at early stages, minimising health and economic risks. For instance, the company Alibaba has developed a set of AI tools applied in waste collection in cities where the infrastructure is available; in monitoring animal farms, orchards, and crops, where the system allows farmers to optimize reproduction rates by raising a healthier herd and reducing mortality rates of newborns, in addition to crop and land management. Another example of this is LeanPath, where automated scales measure waste from the food preparation process in Google kitchens around the world. In a sustainability effort, Google set the goal of reducing food waste and costs in its more than 200 cafeterias and 1,000 kitchens, for example, by preparing eggs on the spot, making milk cartons smaller, and using leftovers to create new dishes, and inviting employees to take only the food they can eat. Additionally, research into region-specific strategies for food waste valorisation is crucial. Cultural, geographic, and economic factors can influence the feasibility and adoption of food waste valorisation practices, thus necessitating tailored approaches (Jurgilevich et al. 2016).

Economic modeling and lifecycle assessment studies that provide a comparative analysis of various food waste valorisation technologies can be invaluable. Such research can help decision-makers prioritize investments and policy interventions (Tonini et al. 2016).

Lastly, interdisciplinary research that bridges the gap between science, policy, and community engagement can lead to more holistic and sustainable solutions for food waste valorisation within the WEF Nexus (Galanakis 2020). Involving stakeholders from diverse fields can enhance the relevance and impact of research endeavors.

9.6.2 Predicting Future Trends in Food Waste Valorisation within the WEF Nexus

Anticipating future trends in food waste valorisation is crucial for steering the field toward sustainable and innovative pathways. One trend that is likely to gain momentum is the transition toward a circular economy model for food waste management (Geissdoerfer et al. 2017). This approach envisions a seamless integration of food production, consumption, and waste utilisation, minimising environmental footprints.

Decentralised food waste management systems could become increasingly prevalent, driven by the need for localized solutions and community engagement (Sharma et al. 2021). These systems empower communities to take ownership of their waste while promoting local innovation.

With an increasing focus on sustainability, investments in green technologies for food waste valorisation are likely to surge. These technologies, which prioritize environmental preservation, can attract support from governments, investors, and consumers (Zhang 2008).

Consumer-driven trends, such as the demand for transparency and sustainability in food production and waste management, could reshape industry practices (Pomponi and Moncaster 2017). This may lead to the rise of eco-labeling and certification programs for products derived from food waste.

Technological advancements in AI and machine learning can revolutionize food waste prediction, sorting, and processing, thus enhancing the efficiency and scalability of valorisation efforts (Dahiya et al. 2018).

Lastly, global policy alignments and international collaborations may bolster food waste valorisation initiatives. These collaborations can facilitate knowledge exchange, promote best practices, and drive collective action toward sustainability (Kurucz et al. 2017).

9.6.3 Summarizing the Key Takeaways and Contributions of the Chapter

This chapter aimed to delve deeply into the multifaceted dimensions of food waste valorisation within the WEF Nexus. The key takeaway is the recognition of food waste not as a mere byproduct but as a valuable resource with potential economic, environmental, and social benefits (Keegan and Breadsell 2021).

We underscored the transformative potential of integrating innovative technologies, such as anaerobic digestion and microbial fuel cells, into food waste management practices (Xin et al. 2018). These technologies not only contribute to resource recovery but also align closely with the principles of sustainability.

The exploration of the economic and environmental benefits of food waste valorisation revealed opportunities for investment, job creation, and environmental preservation (Elimelech and Phillip 2011). A circular approach towards food waste can lead to substantial economic gains while mitigating environmental impacts. Additionally, the importance of consumer perceptions, behavior, and PPPs was highlighted. Engaging stakeholders at all levels is crucial for the successful implementation of food waste valorisation initiatives (Despoudi et al. 2021).

In the quest for sustainability, navigating the regulatory and policy landscape emerged as a vital aspect. The development and harmonisation of policies can foster an enabling environment for food waste valorisation (Salemdeeb et al. 2017).

In conclusion, the chapter posits that a forward-looking and integrated approach toward food waste valorisation within the WEF Nexus can catalyse transformative changes. Through continuous research, innovation, and collaboration, we can unlock new potentials and ensure a sustainable and resilient future.

After reading this chapter, Do you think it is important to reduce food waste? Do you have any idea how much food is wasted in your home and what changes will you implement in your daily life to reduce food waste? What actions should the government promote to reduce food waste in the different productive sectors? How could the food waste generated be used? Should technological tools be applied to accelerate the reduction of food waste?

REFERENCES

Aschemann-Witzel, Jessica, Ilona De Hooge, Pegah Amani, Tino Bech-Larsen, and Marije Oostindjer. 2015. "Consumer-related food waste: causes and potential for action." *Sustainability* 7 (6): 6457–6477. doi:10.3390/su7066457.

Banyal, Shelly, Rk Aggarwal, and Sk Bhardwaj. 2019. "A review on methodologies adopted during environmental impact assessment of development projects." *Journal of Pharmacognosy and Phytochemistry* 8 (4): 2108–19. doi:10.22271/phyto.2019.v8.i4aj.9273.

Bazilian, Morgan, Holger Rogner, Mark Howells, Sebastian Hermann, Douglas Arent, Dolf Gielen, Pasquale Steduto, et al. 2011. "Considering the energy, water and food nexus: towards an integrated modelling approach." *Energy Policy* 39 (12): 7896–7906. doi: 10.1016/j.enpol.2011.09.039.

Bhatia, L., H. Jha, T. Sarkar, and P. K. Sarangi. 2023. "Food waste utilization for reducing carbon footprints towards sustainable and cleaner environment: a review." *International Journal of Environmental Research and Public Health*, 20 (3): 2318. doi:10.3390/ijerph20032318.

Biggs, Eloise M., Eleanor Bruce, Bryan Boruff, John M. A. Duncan, Julia Horsley, Natasha Pauli, Kellie McNeill, et al. 2015. "Sustainable development and the water–energy–food nexus: a perspective on livelihoods." *Environmental Science & Policy* 54 (December): 389–397. doi:10.1016/j.envsci.2015.08.002.

Born, Kelly, and Paul Brest. 2013. "When can impact investing create real impact?" *Stanford Social Innovation Review* 11. [object Object]: 2231. doi:10.48558/JFWQ-GZ70.

Bridgwater, A. V. 2012. "Review of fast pyrolysis of biomass and product upgrading." *Biomass and Bioenergy* 38 (March): 68–94. doi:10.1016/j.biombioe.2011.01.048.

Brown, Sally, Chad Kruger, and Scott Subler. 2008. "Greenhouse gas balance for composting operations." *Journal of Environmental Quality* 37 (4): 1396–1410. doi:10.2134/jeq 2007.0453.

Cansino-Loeza, Brenda, and José María Ponce-Ortega. 2019. "Involving the water–energy–food nexus in integrating low-income and isolated communities." *ACS Sustainable Chemistry & Engineering* 7 (1): 1399–1418. doi:10.1021/acssuschemeng.8b05134.

Chen, Ye, Jay J. Cheng, and Kurt S. Creamer. 2008. "Inhibition of anaerobic digestion process: a review." *Bioresource Technology* 99 (10): 4044–4064. doi:10.1016/j.biortech.2007.01.057.

Dahiya, Shikha, A. Naresh Kumar, J. Shanthi Sravan, Sulogna Chatterjee, Omprakash Sarkar, and S. Venkata Mohan. 2018. "Food waste biorefinery: sustainable strategy for circular bioeconomy." *Bioresource Technology* 248 (January): 2–12. doi:10.1016/j.biortech.2017.07.176.

Despoudi, Stella, Camelia Bucatariu, Semih Otles, Canan Kartal, Semih Otles, Stella Despoudi, Camelia Bucatariu, and Canan Kartal. 2021. "Food waste management, valorisation, and sustainability in the food industry." In *Food waste recovery*, 3–19. Elsevier. doi:10.1016/B978-0-12-820563-1.00008-1.

Diacono, Mariangela, Alessandro Persiani, Elena Testani, Francesco Montemurro, and Corrado Ciaccia. 2019. "Recycling agricultural wastes and by-products in organic farming: biofertilizer production, yield performance and carbon footprint analysis." *Sustainability* 11 (14): 3824. doi:10.3390/su11143824.

Dongyu, Q. 2019. "Moving forward on reducing food loss and waste." In Joachim von Braun, Marcelo Sánchez Sorondo, and Roy Steiner (eds), *Reduction of food loss and waste*, 25. Libreria Editrice Vaticana.

Elimelech, Menachem, and William A. Phillip. 2011. "The future of seawater desalination: energy, technology, and the environment." *Science* 333 (6043): 712–717. doi:10.1126/science.1200488.

Endo, Aiko, Kimberly Burnett, Pedcris Orencio, Terukazu Kumazawa, Christopher Wada, Akira Ishii, Izumi Tsurita, and Makoto Taniguchi. 2015. "Methods of the water-energy-food nexus." *Water* 7 (10): 5806–5830. doi:10.3390/w7105806.

Esposito, Benedetta, Maria Rosaria Sessa, Daniela Sica, and Ornella Malandrino. 2020. "Towards circular economy in the agri-food sector. A systematic literature review." *Sustainability* 12 (18): 7401. doi:10.3390/su12187401.

European Commission. Joint Research Centre. 2023. *Setting the scene for an EU initiative on food waste reduction targets: outcomes of consultation activities and analysis of efforts on food waste reduction.* LU: Publications Office. https://data.europa.eu/doi/10.2760/13859.

FAO, ed. 2019. "The state of food and agriculture 2019. Moving forward on food loss and waste reduction." In *Moving forward on food loss and waste reduction*, 2–13. The State of Food and Agriculture 2019. Rome: Food and Agriculture Organization of the United Nations. https://www.fao.org/3/ca6030en/ca6030en.pdf.

Finnveden, Göran, Michael Z. Hauschild, Tomas Ekvall, Jeroen Guinée, Reinout Heijungs, Stefanie Hellweg, Annette Koehler, David Pennington, and Sangwon Suh. 2009. "Recent developments in life cycle assessment." *Journal of Environmental Management* 91 (1): 1–21. doi:10.1016/j.jenvman.2009.06.018.

Galanakis, Charis M. 2020. "The food systems in the era of the coronavirus (COVID-19) pandemic crisis." *Foods* 9 (4): 523. doi:10.3390/foods9040523.

Geissdoerfer, Martin, Paulo Savaget, Nancy M. P. Bocken, and Erik Jan Hultink. 2017. "The circular economy – a new sustainability paradigm?" *Journal of Cleaner Production* 143 (February): 757–768. doi:10.1016/j.jclepro.2016.12.048.

Glover, Dominic, and Alexandra E. Sexton. 2015. "Edible insects and the future of food: a foresight scenario exercise on entomophagy and global food security." doi:10.1163/2210-7975_HRD-0148-2015066.

Gomez, Leonardo D., Clare G. Steele-King, and Simon J. McQueen-Mason. 2008. "Sustainable liquid biofuels from biomass: the writing's on the walls." *New Phytologist* 178 (3): 473–485. doi:10.1111/j.1469-8137.2008.02422.x.

Hoehn, Daniel, Ian Vázquez-Rowe, Ramzy Kahhat, María Margallo, Jara Laso, Ana Fernández-Ríos, Israel Ruiz-Salmón, and Rubén Aldaco. 2023. "A critical review on food loss and waste quantification approaches: is there a need to develop alternatives beyond the currently widespread pathways?" *Resources, Conservation and Recycling* 188 (January): 106671. doi:10.1016/j.resconrec.2022.106671.

Hoekstra, Arjen Y., and Mesfin M. Mekonnen. 2012. "The water footprint of humanity." *Proceedings of the National Academy of Sciences* 109 (9): 3232–3237. doi:10.1073/pnas.1109936109.

Jurgilevich, Alexandra, Traci Birge, Johanna Kentala-Lehtonen, Kaisa Korhonen-Kurki, Janna Pietikäinen, Laura Saikku, and Hanna Schösler. 2016. "Transition towards circular economy in the food system." *Sustainability* 8 (1): 69. doi:10.3390/su8010069.

Kaza, Silpa, Lisa C. Yao, Perinaz Bhada-Tata, and Frank Van Woerden. 2018. *What a waste 2.0: a global snapshot of solid waste management to 2050.* World Bank. doi:10.1596/978-1-4648-1329-0.

Keegan, Emily, and Jessica K. Breadsell. 2021. "Food waste and social practices in Australian households." *Sustainability* 13 (6): 3377. doi:10.3390/su13063377.

Kim, Nayeong, Jemin Jeon, Raylin Chen, and Xiao Su. 2022. "Electrochemical separation of organic acids and proteins for food and biomanufacturing." *Chemical Engineering Research and Design* 178 (February): 267–288. doi:10.1016/j.cherd.2021.12.009.

Koh, Tommy. 2020. *Governing cities: Asia's urban transformation.* In Kris Hartley, Glen Kuecker, Michael Waschak, Jun Jie Woo, and Charles Chao Rong Phua (eds). 1st ed. Routledge. doi:10.4324/9780429439940.

Kumar, Parveen, Diane M. Barrett, Michael J. Delwiche, and Pieter Stroeve. 2009. "Methods for pretreatment of lignocellulosic biomass for efficient hydrolysis and biofuel production." *Industrial & Engineering Chemistry Research* 48 (8): 3713–3729. doi:10.1021/ie801542g.

Kurian, Mathew, Kent E. Portney, Gerhard Rappold, Bryce Hannibal, and Solomon H. Gebrechorkos. 2018. "Governance of water-energy-food nexus: a social network analysis approach to understanding agency behaviour." In Stephan Hülsmann and Reza Ardakanian (eds), *Managing water, soil and waste resources to achieve sustainable development goals*, 125–147. Springer International Publishing. doi:10.1007/978-3-319-75163-4_6.

Kurucz, Elizabeth C., Barry A. Colbert, Florian Lüdeke-Freund, Antony Upward, and Bob Willard. 2017. "Relational leadership for strategic sustainability: practices and capabilities to advance the design and assessment of sustainable business models." *Journal of Cleaner Production* 140 (January): 189–204. doi:10.1016/j.jclepro.2016.03.087.

Leung, Ho Hon, Sally Mingle Yorke, and Artie Ng. 2023. "Social bonds as an innovative financial instrument for ESG initiatives." In Artie Ng and Jatin Nathwani (eds), *Financial and technological innovation for sustainability*, 1st ed., 54–74. Routledge. doi:10.4324/9781003288343-5.

Liu, Chen, Yasuhiko Hotta, Atsushi Santo, Matthew Hengesbaugh, Atsushi Watabe, Yoshiaki Totoki, Desmond Allen, and Magnus Bengtsson. 2016. "Food waste in Japan: trends, current practices and key challenges." *Journal of Cleaner Production* 133 (October): 557–564. doi:10.1016/j.jclepro.2016.06.026.

Logan, Bruce E., and Korneel Rabaey. 2012. "Conversion of wastes into bioelectricity and chemicals by using microbial electrochemical technologies." *Science* 337 (6095): 686–690. doi:10.1126/science.1217412.

Mirabella, Nadia, Valentina Castellani, and Serenella Sala. 2014. "Current options for the valorisation of food manufacturing waste: a review." *Journal of Cleaner Production* 65 (February): 28–41. doi:10.1016/j.jclepro.2013.10.051.

Mollick, Ethan. 2014. "The dynamics of crowdfunding: an exploratory study." *Journal of Business Venturing* 29 (1): 1–16. doi:10.1016/j.jbusvent.2013.06.005.

Montiel, Ivan, Alvaro Cuervo-Cazurra, Junghoon Park, Raquel Antolín-López, and Bryan W. Husted. 2021. "Implementing the United Nations' sustainable development goals in international business." *Journal of International Business Studies* 52 (5): 999–1030. doi:10.1057/s41267-021-00445-y.

Mourad, Marie. 2016. "Recycling, recovering and preventing 'food waste': competing solutions for food systems sustainability in the United States and France." *Journal of Cleaner Production* 126 (July): 461–477. doi:10.1016/j.jclepro.2016.03.084.

Munir, Muhammad Tajammal, Najam Ul Saqib, Bing Li, and Muhammad Naqvi. 2023. "Food waste hydrochar: an alternate clean fuel for steel industry." *Fuel* 346 (August): 128395. doi:10.1016/j.fuel.2023.128395.

Paes, Luis Alberto Bertolucci, Barbara Stolte Bezerra, Rafael Mattos Deus, Daniel Jugend, and Rosane Aparecida Gomes Battistelle. 2019. "Organic solid waste management in a circular economy perspective – a systematic review and SWOT analysis." *Journal of Cleaner Production* 239 (December): 118086. doi:10.1016/j.jclepro.2019.118086.

Papargyropoulou, Effie, Rodrigo Lozano, Julia K. Steinberger, Nigel Wright, and Zaini Bin Ujang. 2014. "The food waste hierarchy as a framework for the management of food surplus and food waste." *Journal of Cleaner Production* 76 (August): 106–115. doi:10.1016/j.jclepro.2014.04.020.

Parfitt, Julian, Mark Barthel, and Sarah Macnaughton. 2010. "Food waste within food supply chains: quantification and potential for change to 2050." *Philosophical Transactions of the Royal Society B: Biological Sciences* 365 (1554): 3065–3081. doi:10.1098/rstb.2010.0126.

Phiri, Resego, Sanjay Mavinkere Rangappa, Suchart Siengchin, Oluseyi Philip Oladijo, and Hom Nath Dhakal. 2023. "Development of sustainable biopolymer-based composites for lightweight applications from agricultural waste biomass: a review." *Advanced Industrial and Engineering Polymer Research* 6 (4): 436–450. doi:10.1016/j.aiepr.2023.04.004.

Pomponi, Francesco, and Alice Moncaster. 2017. "Circular economy for the built environment: a research framework." *Journal of Cleaner Production* 143 (February): 710–718. doi:10.1016/j.jclepro.2016.12.055.

Prokic, D., J. Stepanov, Lj. Curcic, N. Stojic, and M. Pucarevic. 2022. "The role of circular economy in food waste management in fulfilling the united nations' sustainable development goals." *Acta Universitatis Sapientiae, Alimentaria* 15 (1): 51–66. doi:10.2478/ausal-2022-0005.

Quested, T.E., E. Marsh, D. Stunell, and A.D. Parry. 2013. "Spaghetti soup: the complex world of food waste behaviours." *Resources, Conservation and Recycling* 79 (October): 43–51. doi:10.1016/j.resconrec.2013.04.011.

Rasul, Golam. 2014. "Food, water, and energy security in south asia: a nexus perspective from the Hindu Kush Himalayan region☆." *Environmental Science & Policy* 39 (May): 35–48. doi:10.1016/j.envsci.2014.01.010.

Raya-Tapia, Alma Yunuen, Brenda Cansino-Loeza, Xate Geraldine Sánchez-Zarco, César Ramírez-Márquez, Mariano Martín, and José María Ponce-Ortega. 2023a. "A spatial and temporal assessment of resource security in the water, energy, food and waste nexus in Spain." *Sustainable Production and Consumption* 39 (July): 109–122. doi:10.1016/j.spc.2023.05.008.

Raya-Tapia, Alma Yunuen, Brenda Cansino-Loeza, Xate Geraldine Sánchez-Zarco, César Ramírez-Márquez, and José María Ponce-Ortega. 2023b. "A systematic approach for assessing water, energy and food security during the COVID-19 pandemic: case study of Mexico." *Environment, Development and Sustainability* 25 (12): 14381–14409. doi:10.1007/s10668-022-02671-2.

Reynolds, Christian, Liam Goucher, Tom Quested, Sarah Bromley, Sam Gillick, Victoria K. Wells, David Evans, et al. 2019. "Review: consumption-stage food waste reduction interventions – what works and how to design better interventions." *Food Policy* 83 (February): 7–27. doi:10.1016/j.foodpol.2019.01.009.

Saberi, Sara, Mahtab Kouhizadeh, Joseph Sarkis, and Lejia Shen. 2019. "Blockchain technology and its relationships to sustainable supply chain management." *International Journal of Production Research* 57 (7): 2117–2135. doi:10.1080/00207543.2018.1533261.

Salemdeeb, Ramy, Erasmus K. H. J. Zu Ermgassen, Mi Hyung Kim, Andrew Balmford, and Abir Al-Tabbaa. 2017. "Environmental and health impacts of using food waste as animal feed: a comparative analysis of food waste management options." *Journal of Cleaner Production* 140 (January): 871–880. doi:10.1016/j.jclepro.2016.05.049.

Sánchez-Zarco, Xate Geraldine, and José María Ponce-Ortega. 2023. "Water-energy-food-ecosystem nexus: an optimization approach incorporating life cycle, security and sustainability assessment." *Journal of Cleaner Production* 414 (August): 137534. doi:10.1016/j.jclepro.2023.137534.

Scoones, Ian, Andrew Stirling, Dinesh Abrol, Joanes Atela, Lakshmi Charli-Joseph, Hallie Eakin, Adrian Ely, et al. 2020. "Transformations to sustainability: combining structural, systemic and enabling approaches." *Current Opinion in Environmental Sustainability* 42 (February): 65–75. doi:10.1016/j.cosust.2019.12.004.

Scott, Christopher A., Mathew Kurian, and James L. Wescoat. 2015. "The water-energy-food nexus: enhancing adaptive capacity to complex global challenges." In Mathew Kurian and Reza Ardakanian (eds), *Governing the nexus*, 15–38. Springer International Publishing. doi:10.1007/978-3-319-05747-7_2.

Shahid, Kanwal, and Eric Hittinger. 2021. "Techno-economic optimization of food waste diversion to treatment facilities to determine cost effectiveness of policy incentives." *Journal of Cleaner Production* 279 (January): 122634. doi:10.1016/j.jclepro.2020.122634.

Sharma Hari Bhakta, Kumar Raja Vanapalli, Biswajit Samal, V.R. Sankar Cheela, Brajesh K. Dubey, and Jayanta Bhattacharya. 2021. "Circular economy approach in solid waste management system to achieve UN-SDGs: solutions for post-COVID recovery." *Science of The Total Environment* 800 (December): 149605. doi:10.1016/j.scitotenv.2021.149605.

Shneiderman Ben. 2020. "Bridging the gap between ethics and practice: guidelines for reliable, safe, and trustworthy human-centered AI systems." *ACM Transactions on Interactive Intelligent Systems* 10 (4): 1–31. doi:10.1145/3419764.

Shutaleva Anna. 2023. "Ecological culture and critical thinking: building of a sustainable future." *Sustainability* 15 (18): 13492. doi:10.3390/su151813492.

Smith Pete, Steven J. Davis, Felix Creutzig, Sabine Fuss, Jan Minx, Benoit Gabrielle, Etsushi Kato, et al. 2016. "Biophysical and economic limits to negative CO_2 emissions." *Nature Climate Change* 6 (1): 42–50. doi:10.1038/nclimate2870.

Stenmarck Åsa, Carl Jensen, Tom Quested, and Graham Moates. 2016. "Estimates of European food waste levels." [object Object]. doi:10.13140/RG.2.1.4658.4721.

Strunz Sebastian, and Harry Schindler. 2018. "Identifying barriers toward a post-growth economy – a political economy view." *Ecological Economics* 153 (November): 68–77. doi:10.1016/j.ecolecon.2018.06.017.

Tan Hwei-Ting, Kendall R. Corbin, and Geoffrey B. Fincher. 2016. "Emerging technologies for the production of renewable liquid transport fuels from biomass sources enriched in plant cell walls." *Frontiers in Plant Science* 7 (December). doi:10.3389/fpls.2016.01854.

Tilman David, Christian Balzer, Jason Hill, and Belinda L. Befort. 2011. "Global food demand and the sustainable intensification of agriculture." *Proceedings of the National Academy of Sciences* 108 (50): 20260–20264. doi:10.1073/pnas.1116437108.

Tonini Davide, Lorie Hamelin, and Thomas F. Astrup. 2016. "Environmental implications of the use of agro-industrial residues for biorefineries: application of a deterministic model for indirect land-use changes." *GCB Bioenergy* 8 (4): 690–706. doi:10.1111/gcbb.12290.

Tovar-Facio Javier, Brenda Cansino-Loeza, and José María Ponce-Ortega. 2022. "Management of renewable energy sources." In *Sustainable design for renewable processes*, 3–31. Elsevier. doi:10.1016/B978-0-12-824324-4.00004-4.

Xin Xiaodong, Yingqun Ma, and Yu Liu. 2018. "Electric energy production from food waste: microbial fuel cells versus anaerobic digestion." *Bioresource Technology* 255 (May): 281–287. doi:10.1016/j.biortech.2018.01.099.

Zhang, R., H. Elmashad, K. Hartman, F. Wang, G. Liu, C. Choate, and P. Gamble. 2007. "Characterization of food waste as feedstock for anaerobic digestion." *Bioresource Technology* 98 (4): 929–935. doi:10.1016/j.biortech.2006.02.039.

Zhang ZhongXiang. 2008. "Asian energy and environmental policy: promoting growth while preserving the environment." *Energy Policy* 36 (10): 3905–3924. doi:10.1016/j.enpol.2008.07.015.

Index

Note: **Bold** page numbers refer to tables and *italic* page numbers refer to figures.

J

K

L

M

FSC
www.fsc.org
MIX
Papier | Fördert gute Waldnutzung
FSC® C083411

Kolibri 360 GmbH
Ferdinand-Jühlke-Straße 7
99095 Erfurt, Deutschland
produktsicherheit@kolibri360.de